Hydrological Extremes in a Warming Climate: Nonstationarity, Uncertainties and Impacts

Hydrological Extremes in a Warming Climate: Nonstationarity, Uncertainties and Impacts

Editors

Rajesh R. Shrestha
Mohammad Reza Najafi

MDPI • Basel • Beijing • Wuhan • Barcelona • Belgrade • Manchester • Tokyo • Cluj • Tianjin

Editors
Rajesh R. Shrestha
Environment and Climate Change Canada
University of Victoria
Victoria
Canada

Mohammad Reza Najafi
Department of Civil and Environmental Engineering
Western University
London
Canada

Editorial Office
MDPI
St. Alban-Anlage 66
4052 Basel, Switzerland

This is a reprint of articles from the Special Issue published online in the open access journal *Water* (ISSN 2073-4441) (available at: www.mdpi.com/journal/water/special_issues/hydrological_extremes_warming_climate_nonstationary_uncertainties).

For citation purposes, cite each article independently as indicated on the article page online and as indicated below:

LastName, A.A.; LastName, B.B.; LastName, C.C. Article Title. *Journal Name* **Year**, *Volume Number*, Page Range.

ISBN 978-3-0365-4644-5 (Hbk)
ISBN 978-3-0365-4643-8 (PDF)

Contents

About the Editors

Rajesh R. Shrestha

Dr. Rajesh R Shrestha is a Research Scientist at Environment and Climate Change Canada and an Adjunct Professor at the Department of Geography, University of Victoria. Prior to his current position, he was a Research Hydrologist at the Pacific Climate Impacts Consortium, University of Victoria. His research focuses on assessing the cascading impacts of climate change on hydrologic systems that include snowpack, streamflow and water quality constituents, by integrating state-of-the-science climate model downscaling with process-based hydrologic and water quality models, and complementary machine learning and statistical methods. He is interested in developing methods or tools for identifying the key drivers and controls of hydro-climatic variability and change, including extreme events (floods and drought). His recent research includes the assessment of climate change impacts on snow drought, extreme streamflow and its predictability, and implications on the river temperature regime.

Mohammad Reza Najafi

Dr. Mohammad Reza Najafi is an Associate Professor in the Department of Civil and Environmental Engineering at Western University. Prior to joining Western, he was a Research Scientist and a Postdoctoral Researcher at the Pacific Climate Impacts Consortium (University of Victoria), and the Byrd Polar Research Centre (Ohio State University). Dr. Najafi's research team develops and integrates state-of-the-art statistical and process-based approaches to understand and predict the spatial and temporal variability of natural hazard risks under climate change. We quantify the contribution of anthropogenic factors and internal climate variability to changes in the historical and projected hydroclimatic extremes. Our group characterizes the interactions between hazards and infrastructure systems in space and time to assess the corresponding compounding and cascading risks. This will lead to the development of effective mitigation and adaptation strategies. Our interdisciplinary research contributes to the assessment of climate change impacts on extreme events, floods and droughts, climate model downscaling, multi-modelling, uncertainty quantification and communication, Bayesian inference, remote sensing, distributed hydrologic modelling and multivariate extremes.

water

MDPI

Editorial

Special Issue: Hydrological Extremes in a Warming Climate: Nonstationarity, Uncertainties and Impacts

Rajesh R. Shrestha [1,*] and Mohammad Reza Najafi [2]

1 Watershed Hydrology and Ecology Research Division, Environment and Climate Change Canada, University of Victoria, Victoria, BC V8N 1V8, Canada

2 Department of Civil and Environmental Engineering, Western University, London, ON N6A 5B9, Canada; mnajafi7@uwo.ca

* Correspondence: rajesh.shrestha@ec.gc.ca

Citation: Shrestha, R.R.; Najafi, M.R. Special Issue: Hydrological Extremes in a Warming Climate: Nonstationarity, Uncertainties and Impacts. *Water* **2022**, *14*, 1658. https://doi.org/10.3390/w14101658

Received: 16 May 2022
Accepted: 19 May 2022
Published: 23 May 2022

1. Introduction

The intensification of global water cycle, associated with anthropogenic climate change, is affecting the characteristics of hydrologic extreme events throughout the world. With the increases in the intensity of extreme precipitation, persistent low precipitation and evaporative water demand at different spatial and temporal scales, hydrologic extremes (floods and droughts) have become more likely and more severe in many regions [1,2]. The changes in precipitation and evapotranspiration rates are projected to continue and intensify in a warmer future, and further exacerbate the risks associated with floods and droughts. In snow-dominated regions of the world, hydrologic extremes are further influenced by the transitions from snow towards rainfall-dominated regimes [3], along with exceptionally low snow conditions or snow drought [4] and changes in the frequency and severity of rain-on-snow conditions [5]. Additionally, the risks associated with the climate-induced changes in extremes could be exacerbated by the direct human impacts, such as floodplain development and land use change in some river basins. Thus, understanding the historical and future trajectories of hydrologic extremes is crucial for water resources and disaster risk management, such as reservoir storage management and flood and drought preparedness, as well as planning for adaptation measures.

In this context, the nonstationarity of hydrologic extremes is highly relevant, as it can significantly alter the magnitude and frequency of extreme events [6,7]. Furthermore, hydrologic extremes often result from a combination of interacting physical processes, referred to as compound events, and risk assessment methods that consider a single driver and/or hazard in isolation can potentially lead to an underestimation of the associated risks [8]. However, addressing nonstationarity and compound events pose a number of challenges, such as selecting an appropriate modelling strategy, handling uncertainties, and understanding and communicating the associated concepts and risks.

This Special Issue comprises a collection of 11 papers that provide advances in various aspects of climate change impacts on hydrologic extremes, including both drivers (temperature, precipitation and snow) and effects (peak flow, low flow, water temperature). The studies cover a broad range of topics on hydrologic extremes, including hydro-climatic controls, trends, homogeneity, nonstationarity, compound events and associated uncertainties, over both historical and future climates.

2. Summary of This Special Issue

Precipitation is a main driver of hydrologic extremes, and future changes in precipitation indices can be expected to have implications on both floods and droughts. In this respect, Khoi et al. [9] analyzed spatio-temporal changes in the intensity, duration and frequency of maximum and minimum precipitation over Ho Chi Minh City, Vietnam. The projections from statistically downscaled Global Climate Models (GCMs) from the Coupled

Model Intercomparison Project 5 (CMIP5) representative concentration pathways (RCP) 8.5 ensemble indicated generally increasing future trends in most extreme indices, with more statistically significant trends and higher rate of increases for the intermediate future period (2051–2080) compared to the near future period (2021–2050). They also found higher trends and more statistically significant increases in the extreme precipitation intensity and frequency indices than the duration indices.

Arctic and subarctic regions of the world have been experiencing enhanced hydrologic changes in response to the amplified warming and moisture transport to the region. In this respect, Shrestha et al. [10] analyzed historical trends in annual mean flow, minimum flow, maximum flow and its timing for stations across the permafrost region of Canada. The results revealed significant warming for the majority of stations over both cold and warm seasons, and precipitation increases for some of the stations. In response, nearly half of stations exhibited significant minimum flow increases, while the number stations with significant trends in mean flow, maximum flow and its timing were relatively smaller. Further, by using a multiple linear regression (MLR) framework, they showed the dominant controls of precipitation on mean and maximum flow, and temperature on minimum flow.

In snow-dominated regions, the change in volume, extent and duration of snowpack can be expected to have considerable effects on the streamflow extreme response. To this end, Wagner et al. [11] analyzed trends of temperature, snow water equivalent (SWE) and streamflow extremes for selected rivers in the Yakima River Basin in the Pacific Northwest US. They found increasing trends in winter air temperature, accompanied by decreasing trends in SWE accumulation and a shift to an earlier peak SWE. The implications of these changes were reflected in streamflow extremes in terms of increase in winter maximum streamflow and decrease in summer maximum and minimum streamflow. Future projections indicated a continuation of the historical patterns that lead to above freezing winter temperatures at most stations by 2060, and a transition of the basin to rain-dominant hydrologic regime. Furthermore, Dibike et al. [12] investigated the spatial variations and relative importance of precipitation, temperature and SWE drivers on annual maximum flow and mean spring flow across snow-dominated river basins of western Canada. By using a MLR framework, they found that the annual maximum SWE is the most important predictor of both flow variables. They also analyzed the ability of the MLR model to project future streamflow changes by comparing with the previous studies in the region that used process-based hydrological models. The results were both consistent and inconsistent, and they urged caution in using regression models for future hydrologic projections.

Studying a rainfall dominated basin in Malaysia, Tan et al. [13] quantified the projected impacts of climate change on hydrological extreme flows and environmental flow components using a large set of indicators. They showed increases in future projections of precipitation, streamflow, maximum and minimum temperature across the basin based on a hydrological model driven by bias-adjusted CMIP6 GCM simulations. Overall, extreme high flows showed more sensitivity to changes in climatic factors compared to the normal and low flows. Further, they highlighted the different behavior of simulated future hydroclimatic extremes based on high- and low-resolution model outputs.

In the context of compound hydrologic extremes, Bennett et al. [14] investigated changes in concurrent extreme events (heat wave, drought, low flow and flood) in the Colorado River basin under historical-to-future (1970–1999, 2070–2099) RCP8.5 scenario. They projected increases in the future intensity and magnitude of concurrent events within critical regions of the basin, with temperature-driven extremes (heatwaves and drought) strongest and spatially coherent, and precipitation-driven extremes (flooding and low flows) less strong and more spatially variable across the basin. They also found an increase in the magnitude of all concurrent events from synoptic (5 days) to annual time scales, ranging from large increases for heatwaves and drought, to a smaller increase for low flows. Heatwave also affects glacier runoff and river water temperature, which was analyzed by Pelto et al. [15] using glacier runoff, discharge and water temperature records from the recent late summer heatwave events in the Nooksack river basin, located at the northwestern US–Canada border. The results indicated variable increases in discharge and water temperature across different areas of the basin in response to heatwave driven glacier runoff that account for about a third of total discharge. For the heavily glaciated northern sub-basin, discharge increase was relatively larger and water temperature increase was relatively smaller compared to the unglaciated southern sub-basin. With the ongoing glacier area loss and declining glacier runoff, the study suggested increased frequency of low flow extremes and high water temperatures that could exceed the tolerance levels of aquatic species.

Wang et al. [16] assessed the compounding effects of riverine and coastal flooding, the impacts of climate change on the corresponding drivers and the associated uncertainties, at Stephenville Crossing, a coastal-estuarine region in eastern Canada. They setup and calibrated a two-dimensional hydraulic model that combined with a hydrological model was applied to determine historical and projected flood characteristics (such as depths and extent) under various scenarios. The results suggested possible underestimations of future flood risks associated with projected intensity–duration–frequency curves generated based on statistically downscaled GCMs compared with the ones derived from convection-permitting regional climate model simulations. Temporal patterns of storm events had a major impact on flood characteristics and therefore design storm method can be considered a main source of uncertainty. Future increases in both drivers of flooding can further exacerbate the impacts of their concurrent occurrences. Besides, through a bivariate statistical analysis they showed the underestimations of compound flood risks when the interdependencies between driving mechanisms were not considered.

Nonstationarity of the hydroclimatic factors can lead to projected increases in the frequency and severity of floods and droughts, and subsequently challenge water resources management. In this respect, Xie et al. [17] developed a framework to consider different driving factors for nonstationary design flood volume estimation and represent the nonstationary spatial correlation of the flood events. Studying the cascade reservoirs in the Han River basin in China, they showed the long-term impacts of climate change and population growth on the regional hydrological characteristics, and subsequently the flood risks that can be misrepresented by the traditional design flood estimation methods base on stationarity assumption. They also found that the cascade reservoir regulation can reduce flow peaks and decrease flood volumes. Pasek and Marton [18] assessed the functional water volumes of a reservoir in Czech Republic during extreme hydrological conditions. They evaluated the uncertainties associated with the input variables including water inflows, hydrographs, bathymetric curves, and water losses due to evaporation and dam seepage. To design the functional volumes of multi-purpose reservoir and characterize the uncertainties, they linked a simulation-optimization model of the reservoir, to determine the optimal storage volume, with a simulation model that transforms the flood discharge and determines the retention volume of the reservoir. The study highlighted the significant effects of uncertainties in the storage volume and retention volume estimations, and the importance of considering climate change uncertainties and nonstationary flow conditions for reservoir management.

Regional flood frequency analysis (RFA) is a widely recognized approach to tackle the limitations associated with data availability at specific locations for flood quantile estimations for structure/infrastructure design. Identification of homogenous flood regions is a common RFA step prior to pooling flood information between similar catchments. Zhang and Stadnyk [19] evaluated multiple attributes, including geographic proximity, flood seasonality, physiographic variables, monthly precipitation and temperature patterns, to identify homogenous regions for RFA at 186 sites across Canada. They showed that the identification of homogenous regions relies on local hydrological complexities, representation of the primary flood mechanisms and geographic clustering of the sites. Catchments across eastern Canada form small geographic regions while areas in northern Canada, that are snowmelt dominated, are sensitive to temperature variations signifying the importance of monthly temperature pattern. They also found that the identification of homogenous regions can be a challenge across the Prairies and western Canada due the complex physiographic characteristics.

3. Conclusions

The intensification of global water cycle is affecting climate and hydrologic extreme in different regions of the world and this Special Issue provides critical information towards understanding the historical and projected future changes. The studies covered regions in Asia, Europe and North America, and included a range of precipitation, temperature, snow and streamflow extreme variables. The papers also demonstrated an intensification of the precipitation, temperature and streamflow extremes in the future climate, and as well as their relative controls and interactions. For snow-dominated regions, the studies highlighted the role of decreasing snowpack volume on both winter and spring maximum flow. The papers also emphasized the compounding effects of climate and hydrologic extremes, for example, temperature-driven (heatwaves, drought and elevated water temperature) and precipitation-driven (flooding) concurrent extremes, and their implications on water resources management. Additionally, the studies highlighted the importance of considering hydro-climatic nonstationarity and associated uncertainties in water resources risk assessment. Overall, the studies contributed to a growing body of knowledge on the changing hydro-climatic and hydrologic extremes, as well as methods to characterize and quantify the extremes and associated uncertainties. The advances in understanding and quantifying extremes is critical towards an effective water resources management, and planning adaptation strategies in a warming climate.

Funding: This research received no external funding.

Institutional Review Board Statement: Not applicable.

Informed Consent Statement: Not applicable.

Data Availability Statement: Not applicable.

Acknowledgments: The co-editors would like to thank all of the authors, reviewers, editors for their contributions. We are also grateful to *Water* journal staffs for their help and guidance toward this successful Special Issue on "Hydrological extremes in a warming climate: nonstationarity, uncertainties and impacts".

Conflicts of Interest: The authors declare no conflict of interest.

References

1. Douville, H.; Raghavan, K.; Renwick, J.; Allan, R.P.; Arias, P.A.; Barlow, M.; Cerezo-Mota, R.; Cherchi, A.; Gan, T.Y.; Gergis, J.; et al. Chapter 8: Water Cycle Changes. In *Climate Change 2021: The Physical Science Basis. Contribution of Working Group I to the Sixth Assessment Report of the Intergovernmental Panel on Climate Change*; Cambridge University Press: Cambridge, UK, 2021.
2. Caretta, M.A.; Mukherji, A.; Renwick, J.; Betts, R.A.; Gelfan, A.; Hirabayashi, Y.; Lissner, T.K.; Cherchi, A.; Gunn, E.L.; Liu, J.; et al. Chapter 4: Water. In *Climate Change 2022: Impacts, Adaptation and Vulnerability. Contribution of Working Group II to the Sixth Assessment Report of the Intergovernmental Panel on Climate Change*; Cambridge University Press: Cambridge, UK, 2022.
3. Berghuijs, W.A.; Woods, R.A.; Hrachowitz, M. A Precipitation Shift from Snow towards Rain Leads to a Decrease in Streamflow. *Nat. Clim. Change* **2014**, *4*, 583. [CrossRef]
4. Shrestha, R.R.; Bonsal, B.R.; Bonnyman, J.M.; Cannon, A.J.; Najafi, M.R. Heterogeneous Snowpack Response and Snow Drought Occurrence across River Basins of Northwestern North America under 1.0 °C to 4.0 °C Global Warming. *Clim. Change* **2021**, *164*, 40. [CrossRef]
5. Musselman, K.N.; Lehner, F.; Ikeda, K.; Clark, M.P.; Prein, A.F.; Liu, C.; Barlage, M.; Rasmussen, R. Projected Increases and Shifts in Rain-on-Snow Flood Risk over Western North America. *Nat. Clim. Change* **2018**, *8*, 808–812. [CrossRef]
6. Milly, P.; Betancourt, J.; Falkenmark, M.; Hirsch, R.; Kundzewicz, Z.; Lettenmaier, D.; Stouffer, R. Stationarity Is Dead: Whither Water Management? *Science* **2008**, *319*, 573–574. [CrossRef] [PubMed]
7. Shrestha, R.R.; Cannon, A.J.; Schnorbus, M.A.; Zwiers, F.W. Projecting Future Nonstationary Extreme Streamflow for the Fraser River, Canada. *Clim. Change* **2017**, *145*, 289–303. [CrossRef]
8. Zscheischler, J.; Westra, S.; van den Hurk, B.J.J.M.; Seneviratne, S.I.; Ward, P.J.; Pitman, A.; AghaKouchak, A.; Bresch, D.N.; Leonard, M.; Wahl, T.; et al. Future Climate Risk from Compound Events. *Nat. Clim. Change* **2018**, *8*, 469–477. [CrossRef]

9. Khoi, D.N.; Trong Quan, N.; Thi Thao Nhi, P.; Nguyen, V.T. Impact of Climate Change on Precipitation Extremes over Ho Chi Minh City, Vietnam. *Water* **2021**, *13*, 120. [CrossRef]
10. Shrestha, R.R.; Pesklevits, J.; Yang, D.; Peters, D.L.; Dibike, Y.B. Climatic Controls on Mean and Extreme Streamflow Changes Across the Permafrost Region of Canada. *Water* **2021**, *13*, 626. [CrossRef]
11. Wagner, A.M.; Bennett, K.E.; Liston, G.E.; Hiemstra, C.A.; Cooley, D. Multiple Indicators of Extreme Changes in Snow-Dominated Streamflow Regimes, Yakima River Basin Region, USA. *Water* **2021**, *13*, 2608. [CrossRef]
12. Dibike, Y.B.; Shrestha, R.R.; Johnson, C.; Bonsal, B.; Coulibaly, P. Assessing Climatic Drivers of Spring Mean and Annual Maximum Flows in Western Canadian River Basins. *Water* **2021**, *13*, 1617. [CrossRef]
13. Tan, M.L.; Liang, J.; Samat, N.; Chan, N.W.; Haywood, J.M.; Hodges, K. Hydrological Extremes and Responses to Climate Change in the Kelantan River Basin, Malaysia, Based on the CMIP6 HighResMIP Experiments. *Water* **2021**, *13*, 1472. [CrossRef]
14. Bennett, K.E.; Talsma, C.; Boero, R. Concurrent Changes in Extreme Hydroclimate Events in the Colorado River Basin. *Water* **2021**, *13*, 978. [CrossRef]
15. Pelto, M.S.; Dryak, M.; Pelto, J.; Matthews, T.; Perry, L.B. Contribution of Glacier Runoff during Heat Waves in the Nooksack River Basin USA. *Water* **2022**, *14*, 1145. [CrossRef]
16. Wang, S.; Najafi, M.R.; Cannon, A.J.; Khan, A.A. Uncertainties in Riverine and Coastal Flood Impacts under Climate Change. *Water* **2021**, *13*, 1774. [CrossRef]
17. Xie, Y.; Guo, S.; Xiong, L.; Tian, J.; Xiong, F. Nonstationary Design Flood Estimation in Response to Climate Change, Population Growth and Cascade Reservoir Regulation. *Water* **2021**, *13*, 2687. [CrossRef]
18. Paseka, S.; Marton, D. The Impact of the Uncertain Input Data of Multi-Purpose Reservoir Volumes under Hydrological Extremes. *Water* **2021**, *13*, 1389. [CrossRef]
19. Zhang, Z.; Stadnyk, T.A. Investigation of Attributes for Identifying Homogeneous Flood Regions for Regional Flood Frequency Analysis in Canada. *Water* **2020**, *12*, 2570. [CrossRef]

water

MDPI

Article

Impact of Climate Change on Precipitation Extremes over Ho Chi Minh City, Vietnam

Dao Nguyen Khoi [1,2], Nguyen Trong Quan [2], Pham Thi Thao Nhi [2] and Van Thinh Nguyen [3,*]

1 Faculty of Environment, University of Science, Vietnam National University Ho Chi Minh City, Ho Chi Minh City 700000, Vietnam; dnkhoi@hcmus.edu.vn
2 Institute for Computational Science and Technology, Ho Chi Minh City 700000, Vietnam; quannguyen201294@gmail.com (N.T.Q.); nhi.ptt@icst.org.vn (P.T.T.N.)
3 Department of Civil and Environmental Engineering, Seoul National University, 1 Gwanak-ro, Gwanak-gu, Seoul 151-744, Korea
* Correspondence: vnguyen@snu.ac.kr; Tel.: +82-2-880-7355

Abstract: In the context of climate change, the impact of hydro-meteorological extremes, such as floods and droughts, has become one of the most severe issues for the governors of mega-cities. The main purpose of this study is to assess the spatiotemporal changes in extreme precipitation indices over Ho Chi Minh City, Vietnam, between the near (2021–2050) and intermediate (2051–2080) future periods with respect to the baseline period (1980–2009). The historical extreme indices were calculated through observed daily rainfall data at 11 selected meteorological stations across the study area. The future extreme indices were projected based on a stochastic weather generator, the Long Ashton Research Station Weather Generator (LARS-WG), which incorporates climate projections from the Coupled Model Intercomparison Project 5 (CMIP5) ensemble. Eight extreme precipitation indices, such as the consecutive dry days (CDDs), consecutive wet days (CWDs), number of very heavy precipitation days (R20mm), number of extremely heavy precipitation days (R25mm), maximum 1 d precipitation amount (RX1day), maximum 5 d precipitation amount (RX5day), very wet days (R95p), and simple daily intensity index (SDII) were selected to evaluate the multi-model ensemble mean changes of extreme indices in terms of intensity, duration, and frequency. The statistical significance, stability, and averaged magnitude of trends in these changes, thereby, were computed by the Mann-Kendall statistical techniques and Sen's estimator, and applied to each extreme index. The results indicated a general increasing trend in most extreme indices for the future periods. In comparison with the near future period (2021–2050), the extreme intensity and frequency indices in the intermediate future period (2051–2080) present more statistically significant trends and higher growing rates. Furthermore, an increase in most extreme indices mainly occurs in some parts of the central and southern regions, while a decrease in those indices is often projected in the north of the study area.

Keywords: extreme precipitation; LARS-WG; CMIP5; spatiotemporal changes; climate change

Citation: Khoi, D.N.; Trong Quan, N.; Thi Thao Nhi, P.; Nguyen, V.T. Impact of Climate Change on Precipitation Extremes over Ho Chi Minh City, Vietnam. *Water* **2021**, *13*, 120. https://doi.org/10.3390/w13020120

Received: 24 November 2020
Accepted: 4 January 2021
Published: 7 January 2021

1. Introduction

According to the Global Risks Report 2019 published by the World Economic Forum (WEF), extreme weather events (e.g., heavy rainfall or heat waves) and the failure of climate-change mitigation and adaptation have remained as the top three risks in terms of likelihood and impact on the environment and human being [1]. In addition, the Intergovernmental Panel on Climate Change (IPCC) states in their Fifth Assessment Report (AR5) that the global average surface temperature is expected to rise, and precipitation is likely to occur more intensely and frequently in the 21st century [2,3]. In other words, as a result of greenhouse gas emissions (GHGs), the evidence has shown that a growing shift of temperature and humidity at global near-surface and troposphere layer could be potential for precipitation changes in the frequency, intensity, and duration of extreme events around the world [4–6]. It should be noticed that extreme climate events have always posed ever-intensifying threats to most aspects of natural resources and daily life, such as natural ecosystems and biodiversity [7–9]; human life, community health and social care systems [10,11]; energy supply and demand, transportation, and urbanization [12–14]; economics and agricultural productions [15–17], etc. Therefore, the study on extreme climate events has been of great concern to the scientific community and governments around the globe for planning and implementing efficient mitigation and adaptation strategies.

To identify and examine the extreme events and their variation in the context of climate change, quantitative approaches for evaluating the extreme climatic indices have been developed and widely used by various researchers [18–20]. For consolidating the definition of climate extreme indices on a global scale, as well as facilitating research to compare differences in climate extremes across different study areas, a set of 27 extreme climatic indices computed from daily temperature and precipitation data series has been proposed and highly recommended by the Expert Team on Climate Change Detection and Indices [21]. Afterward, there is a large volume of publications using these proposed indices to address the concerned issues on the analysis of climate extremes in the context of climate change [22,23]. Particularly, in the study of Tian et al. [24], trend analysis of temporal and spatial variability of extreme precipitation indices has attracted more attention. Therefore, this study is focused on assessment of the spatiotemporal variation of extreme precipitation over Ho Chi Minh City caused by climate change. The future climate change scenarios are usually produced using two downscaling techniques based on the outputs of general circulation models (GCMs), namely, statistical downscaling technique and dynamical downscaling technique. In comparison to the dynamical downscaling technique, the statistical downscaling technique is simple to apply to different regions at the station scale and requires fewer computing resources [25]. In this study, the well-known stochastic weather generator, LARS-WG, which incorporates climate projections from the Coupled Model Intercomparison Project 5 [26] ensemble, was used. Instead of analyzing by individual model, a methodology of multi-model ensemble mean is applied to compute future precipitation data for the assessment of the impact of climate change.

Ho Chi Minh City (HCMC) is the largest economic and financial center of Vietnam. Although the City occupies just 0.6% of the country's area and contains a total population of around 8.8 million inhabitants in 2018 (approximately 8.34% of the population of Vietnam), it occupies about 23% GDP of Vietnam (2019). Besides its rapid economic growth, HCMC has been encountering many challenges, varying from natural extreme weather to anthropogenic impacts. Additionally, HCMC has been ranked in the top 20 coastal cities that is considered to be vulnerable and severely affected by climate change, and the estimated cost for mitigation and adaptation of climate change impacts is expected to be the highest in the East Asia region [27,28]. Furthermore, according to the projected scenarios of Asian Development Bank (ADB), over 61% of the city area will be covered by regular flooding, and HCMC will be entirely submerged under water due to sea level rise by 2050 [29]. Therefore, given the high extra cost linked to climate change, it is profoundly essential to consider the potential climate-related risks as an integral part of urban management and planning [30]; particularly, a localized case study is highly recommended for the analysis of changes in extreme precipitation events [31–33].

Eventually, the aim of this study is (1) to compute and evaluate the spatial and temporal trends in extreme precipitation indices over HCMC in the future period (2021–2080); (2) to assess the spatiotemporal changes of trends in those extreme indices with the effect of climate change. An expectation of this study is to obtain informative assessment of the spatial and temporal variability of extreme precipitation events in HCMC during the current and future periods, in order to contribute scientific references to the governors for planning strategies to integrate climate change mitigation and adaptation frameworks for the city development.

2. Study Area and Data Preparation

2.1. Study Area

Located in the south of Vietnam, HCMC (Figure 1) has been recognized as an extraordinarily economic and financial center in Vietnam, and also a vibrant metropolis with the rapid industrialization, urbanization, and agricultural intensification. The climate in HCMC is identified as the tropical monsoon regime with specificity of two distinguished seasons, rainy and dry. In the dry season from November to April, the city obtains a low rainfall (approximately of 10–15% of the total annual rainfall), high evaporation, and high temperature (around 29–30 °C). On the other hand, the average rainfall during the rainy season from May to October generally accounts for approximately 85–90% of the total annual rainfall, which has approximately varied from 1000 to 1600 mm in recent years as shown in Figure 2.

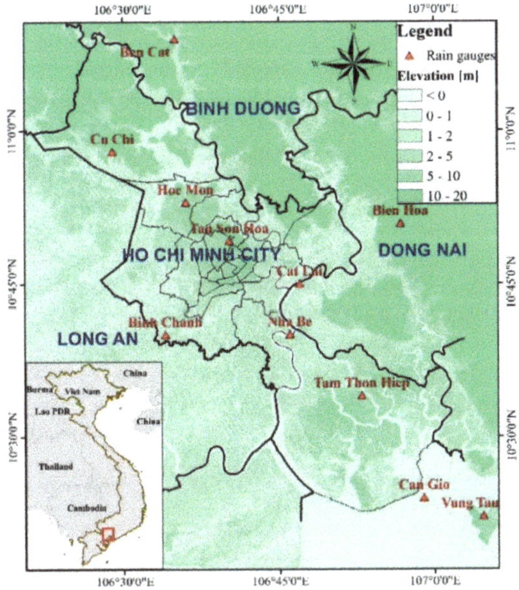

Figure 1. Map of Ho Chi Minh City and the location of 11 meteorological stations.

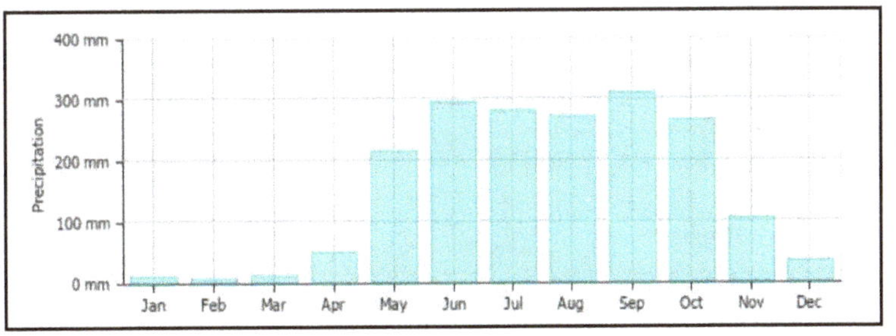

Figure 2. Average monthly precipitation in Ho Chi Minh City, Vietnam (source: weather-and-climate.com).

With an area of 2095 km^2, HCMC is located on the delta formed by the Saigon and Dong Nai Rivers. Its topography is mostly plain, including small hills; approximately 50% of its elevation is lower than 1.5 m above mean sea level [34]; around 16% of its total area is covered by a dense network of rivers and canals about 8000 km in length. In addition, the water levels in these widespread waterways are primarily affected by a combination of semi-diurnal tides, extreme monsoon rainfalls and storms across the city. Consequently, the low-lying terrain accompanying the open and unrestrained water systems in HCMC poses serious threats of flooding vulnerability to the city, especially in the context of climate change [35].

2.2. Data Preparation

The historical daily precipitation data during the period of 1980–2017 provided by the Hydro-Meteorological Data Center of Vietnam were collected at 11 meteorological stations in HCMC and adjacent regions, as shown in Figure 1.

In this study, the basic quality controls and homogeneity assessment were applied to both the observed and projected datasets in order to secure the completeness, reliability, and consistency of the data before computing selected extreme precipitation indices and performing trend analysis. To assess the quality and homogeneity of meteorological data, as well as to compute the climatic extreme indices, the RClimdex statistical toolkit and RHtests toolkit [36] are used. The RClimadex and RHtests are the libraries for computing 27 core extreme climate indices [21] and testing the data homogenization. These toolkits are accepted and recommended by the World Meteorology Organization [37].

3. Methodology

3.1. Extreme Precipitation Indices

To assess the spatial and temporal variability of trends in extreme rainfall events across HCMC, the well-known extreme precipitation indices suggested by the Expert Team on Climate Change Detection and Indices (ETCCDI) were used in this study. Since the HCMC is located in tropical climate region, a set of 8 appropriate extreme precipitation indices (Rx1day, Rx5day, CDD, CWD, R20mm, R25mm, R95p, and SDII), as shown in Table 1, is chosen in order to evaluate the elemental characteristics of extreme precipitation events, such as intensity, duration, and frequency. Particularly, R20mm and R25mm indicate the frequency of extreme rainfall; CDD and CWD describe the duration of extreme rainfall; RX1day, RX5day, R95p, and SDII present the intensity of extreme rainfall. These extreme indices were selected based on the previous studies [38].

Table 1. List of precipitation extreme indices used in this study.

Types	Indices	Name	Definitions	Unit
Intensity indices	RX1day	Max 1-day precipitation amount	Monthly maximum 1-day precipitation	mm
	RX5day	Max 5-day precipitation amount	Monthly maximum 5-day precipitation	mm
	R95p	Very wet days	Annual total precipitation when precipitation > 95th percentile	mm
	SDII	Simple daily intensity index	Annual total precipitation divided by the number of wet days	mm/day
Frequency indices	R20mm	Number of heavy precipitation days	Annual count of days when precipitation > 20 mm	days
	R25mm	Number of very heavy precipitation days	Annual count of days when precipitation > 25 mm	days
Duration indices	CDD	Consecutive dry days	Maximum number of consecutive days with precipitation < 1 mm	days
	CWD	Consecutive wet days	Maximum number of consecutive days with precipitation > 1 mm	days

These selected indices were computed by RClimdex tool for each station of the 11 weather stations in HCMC based on the corresponding datasets of historical observation and the future projections. Thereby, the trend analysis approach was used to provide valuable information about the spatial and temporal change of trends in extreme precipitation events due to the impact of climate change in the study region.

3.2. *Trend Analysis*

3.2.1. Mann-Kendall Test

For analyzing the underlying tendency in climate and hydrological data series, the Mann-Kendall (M-K) nonparametric test [39,40] and Sen's estimator [41] are highly recommended by the World Meteorological Organization (WMO) and widely applied in most studies on the trend analysis of hydro-meteorological data [32,42]. Additionally, the non-parametric M-K test was selected for the present study instead of parametric tests, because this method is suitable for the hydro-meteorological data with a non-normal distribution, and it has low sensitivity to abrupt breaks due to the inhomogeneity of data [42]. The M-K hypothesis test is based on the assumptions that the data need to be independent and identically distributed, i.e., there is no serial autocorrelation remains in the datasets [43]. The statistic parameter (Z-score) computed from the M-K test was used to detect the movement of trends (increasing or decreasing tendency); the corresponding p-value indicates the level of statistical significance, whereas Sen's slope was calculated to represent the magnitude of trends.

Prior to applying the M-K test and Sen's estimator, the study datasets were examined for the existence of serial autocorrelation to reduce the adverse impact on the detection of deterministic trends. In this study, the trend-free pre-whitening approach (TF-PW) was conducted in order to remove the serial autocorrelation effects; the detailed definition and procedure of this method can be found in [44,45].

3.2.2. Trend Strength and Stability

An advantage of using the M-K test for trend analysis is the test results being rarely affected by outliers; it is not restricted to a specific sample distribution. However, the determined sign and magnitude of a trend are strongly depended on a selected time period, i.e., the trend's test results will be changed as the time span of the dataset is varied. Hence, it is challenging to evaluate the tendency of a time series due to the instability of the test results [31,32]. Therefore, the traditional M-K test should be modified for assessing a long-term persistence of the statistical significance and stability of the trend. The choice of studying trends in 25-year moving periods was intended as a cross-validation technique for time series data. The rainfall data of a station were recognized to have a predominantly upward (or downward) trend if the number of 25-year moving time series with significantly upward (or downward) trends was higher than the number of significantly downward (or upward) trends. In this study, the M-K test and Sen's estimator were sequentially applied at each station for each interval of 25-year moving time series of extreme precipitation indices within the historical (1980–2017), the baseline (1980–2009), near future (2021–2050) and intermediate future (2051–2080) periods. In detail, there are 14 series of 25-year intervals regarding the historical observed data from the period of (1980–2004) to (1993–2017), 6 series of 25-year intervals regarding simulated baseline data from the period of (1980–2004) to (1985–2009), and 12 series of 25-year intervals regarding simulated future data from the period of (2021–2045) to (2026–2050) and from the period of (2051–2075) to (2056–2080), respectively.

The statistical significance of a trend was classified by the p-value, which is computed by the M-K test following statistical classification as a strongly significant trend if $p < 0.1$, a weakly significant trend if $0.1 \leq p \leq 0.2$, and an insignificant trend if $p > 0.2$ [32]. In addition, a particular trend, e.g., significant increasing/decreasing or insignificant trend, with the greatest number of 25-year moving periods, was recognized as the dominant trend at a specific station for spatial analysis. Thereby, the stability (S) of a significant increasing/decreasing trend in each extreme precipitation indices at a station was expressed as the percentage of the 25-year moving time series, in which their trends were significantly increasing/decreasing during a particular period. The determined stability of trends was categorized as an unstable trend if $0\% \leq S \leq 15\%$; a poor trend if $15\% \leq S \leq 25\%$; a stable trend if $25\% \leq S \leq 50\%$; strongly stable if $50\% \leq S \leq 75\%$; very strongly stable if $S \geq 75\%$. In this study, the arithmetic average of the slope values, which were computed by Sen's estimator for each interval of 25-year moving periods, was taken to express an average magnitude of a trend in each extreme precipitation indices at each station during the particular periods. The detailed description of this approach could be found in [31,32].

3.3. LARS-WG Downscaling Tool

There are various statistical downscaling techniques, such as the Statistics Downscaling Model (SDSM), Automated Statistical Downscaling (ASD), delta change methods, etc.; among them, LARS-WG developed by Semenov and Stratonovitch [46] was selected for this study because it can provide a better performance on reproducing monthly meteorological variables than other statistical downscaling techniques [47]. Basically, LARS-WG uses observed meteorological data (daily rainfall, temperature) from a specified site to estimate a set of parameters for fitting probability distributions, which is then used to generate synthetic weather time series of a tributary length by randomly selecting values from appropriate distributions [48]. Based on the coordinates of the meteorological station, the LARS-WG tool automatically determines the match between the station and GCM grid cell to generate a site-specific scenario of climate change using the GCM outputs, even though the size of the GCM grid cells is larger than that of the study area. To produce the climate change scenarios, the distribution parameters for a specific site were perturbed with the monthly delta factors of each future period derived from the GCM output. The delta factors are estimated as relative changes in precipitation. The LARS-WG has been widely applied as a downscaling tool to generate future climate data from general circulation models (GCMs) for climate change assessments in numerous studies, and it has confirmed a good efficiency in simulating the extreme precipitation [49–51]. More discussion on the applications of the LARS-WG tool for climatic extreme event analysis could be found in various studies, such as [46,52,53].

By using the historical observed rainfall data, LARS-WG was sequentially applied to estimate a set of parameters for the semi-empirical probability distributions of precipitation data at each station in the study area. These probability distributions were used for fitting the sequences of wet and dry days and precipitation amount. Thereby, the site-calibrated parameters were used to simulate daily precipitation at each rain gauge for the baseline period. To evaluate the performance of LARS-WG in simulating the rainfall, statistical analysis of observed and simulated rainfall data was conducted. In order to generate the future climate scenarios, LARS-WG modifies the historical rainfall data by adding monthly change factors (the ratio of rainfall in the future and baseline periods) [52].

In this study, five GCMs from the Coupled Model Intercomparison Project Phase 5 (CMIP5) multi-model ensemble, which were incorporated in LARS-WG, were used to generate the climate change scenario for the study area. The RCP8.5 scenario was chosen for the present study, as it corresponds to the worst-case climate change scenario caused by the highest greenhouse emission and radiative forcing level (8.5 Wm^{-2}) in 2100. Since the purpose of this study is mainly focused on the analysis of extreme events, the RCP8.5 was considered as a plausible selection [54]. The list of five GCMs used in this study is presented in Table 2. The downscaled daily precipitation data from GCM outputs under RCP8.5 scenario was generated by LARS-WG for the near (2021–2050) and intermediate (2051–2080) future periods of the 21st century at each station of 11 meteorological stations in HCMC.

Table 2. The descriptions of the general circulation models (GCMs) from Coupled Model Intercomparison Project 5 (CMIP5) used in the study.

GCMs	Research Center	Country	Resolution
EC-EARTH	EC—Earth consortium	Europe	1.125° × 1.125°
HadGEM2-ES	UK Meteorological Office	UK	1.25° × 1.88°
GFDL-ESM2M	Geophysical Fluid Dynamics Laboratory (NOAA GFDL)	USA	2.5° × 2.0°
MIROC5	The University of Tokyo, National Institute for Environmental Studies, Japan Agency for Marine-Earth Science and Technology	Japan	1.39° × 1.41°
MPI-ESM-MR	Max Planck Institute for Meteorology	Germany	1.85° × 1.88°

3.4. Model Evaluation Metrics

In this study, the performance of LARS-WG was evaluated according to the relative bias (RB), the mean differences (MDs), and Willmott Score (WS) [55] between the observation-based and simulation-based series of corresponding extreme precipitation indices. The mathematical formulas of those evaluation metrics are as follows:

$$RB = \frac{\sum_{n=1}^{N}(Mod_n - Obs_n)}{\sum_{n=1}^{N}|Obs_n|} \quad (1)$$

$$MD = \frac{1}{N}\sum_{n=1}^{N}(Mod_n - Obs_n) \quad (2)$$

$$WS = 1 - \frac{\frac{1}{N}\sum_{n=1}^{N}(Mod_n - Obs_n)^2}{\frac{1}{N}\sum_{n=1}^{N}\left(\left|Mod_n - \overline{Obs}\right| + \left|Obs_n - \overline{Obs}\right|\right)^2} \quad (3)$$

where Mod_n and Obs_n are the values of the nth simulation-based and observation-based extreme precipitation indices in a sample of size N, respectively.

The mean difference (MD) is used for a direct comparison between the observed and generated extreme indices, while the two-sample Student's t-test is applied for testing the statistical significance of those differences at the 95% confidence level. The relative bias (RB) is used to measure the relative differences between the observed and generated extreme indices, as well as to determine the overestimated or underestimated property of the simulation. According to the study of Moriasi et al. [56], the values of RB $\leq \pm 25\%$ indicate an acceptable performance of the model and the MD values of 0 indicate a perfect performance of the model. The Willmott score (WS) or the Index of Agreement, whose value varies between 0 and 1, is used as a standardized measurement of the degree of model prediction error [57]. Particularly, the levels of model performance are classified in a poor level if $0 \leq WS < 0.3$, a good level if $0.3 \leq WS < 0.8$, and an excellent level if $0.8 \leq WS \leq 1.0$.

3.5. Spatial Interpolation

The maps of extreme rainfall indices were built using interpolated techniques of inverse distance weighting (IDW). This method had the assumption that the interpolated points are the most influenced by the nearest points and the least influenced by the farthest points. The IDW method was chosen for the present study, because it is widely applied to spatial interpolation of rainfall data [58]; particularly it is applied to HCMC, whose topography is mostly plain, and the orographic effect is negligible.

4. Results and Discussion

4.1. Model Validation

In this section, the ability of the LARS-WG tool for reproducing extreme precipitation indices was evaluated at each station for the baseline period of 1980–2009. Generally, the results indicated the ability of the LARS-WG tool for reproducing the extreme indices as its Willmott scores belonged to the good performance level (WS > 0.3) in most regions as shown in Figure 3; the results of the *t*-test indicated that only few stations in the study area (less than 36% of stations) obtained statistically significant differences between the observation and simulation values, as shown in Table 3. Particularly, regarding the indices CDD and CWD, the results of the Willmott scores indicated good agreement between the simulation and observation as the WS values exceeded 0.3 for the entire region (ranged from 0.31 to 0.57), whereas the index CWD shows the worst performance among other indices as only good WS values (ranged 0.33 to 0.62) obtained at 6 out of 11 stations in the research area. Furthermore, the relative bias indicated the underestimation of the index CDD for most regions, especially in the central parts, whereas the overestimation of the index CWD was observed in the same regions. Concerning the indices, RX1day and RX5day, the high WS values (ranged from 0.33 to 0.67) and small negative biases were apparent at most parts in the southern and central regions, whereas the lower WS values (ranged from 0.17 to 0.28), as well as large positive biases, occurred in small parts of the northern regions. About the indices, R20mm, R25mm, R95p, and SDII, the simulation generally showed good performance as the WS values exceeded 0.3 for most regions (ranged from 0.31 to 0.64), except in some parts in the central region the WS values were observed slightly weaker scores (ranged from 0.14 to 0.28). Additionally, Figure 3 and Table 3 show that the high values of WS are not always related to the lesser difference between simulation-based and observation-based extreme indices (low values of RB and MD) as shown in Equations (1)–(3).

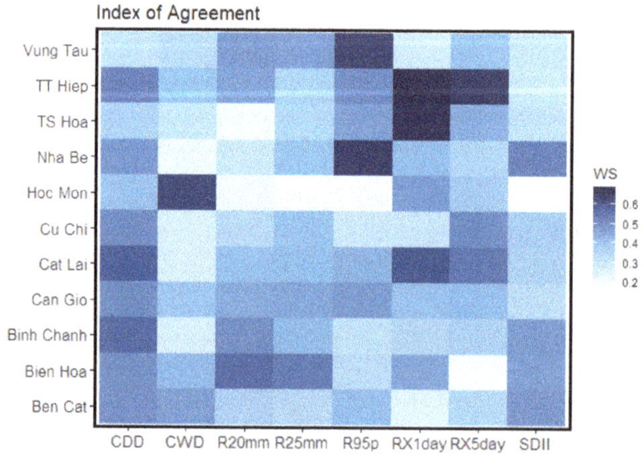

Figure 3. Heatmap of the Willmott scores between observed and simulated extreme indices at each station for the baseline period (1980–2009).

Table 3. Relative bias (%) and mean differences (unit based on each index) between simulation-based and observation-based extreme indices during the baseline period (1980–2009).

Station	CDD		CWD		R20mm		R25mm		R95p		RX1day		RX5day		SDII	
Paras	**RB**	**MD**	**RB**	**MD**	**RB**	**MD**	**RB**	**MD**	**RB**	**MD**	**RB**	**MD**	**RB**	**MD**	**RB**	**MD**
Ben Cat	6.8	6.0	−20.4	**−2.1 ***	−7.5	−2.3	−6.3	−1.4	−11	1.4	3.6	3.1	−0.1	−0.2	−0.5	−0.1
Bien Hoa	−32.2	**−33.2 ***	65.5	**3.7 ***	16.3	**3.9 ***	9.9	1.9	25.7	**101.6 ***	6.2	6.1	7.3	11.6	−15	**−2.8 ***
Binh Chanh	−7.9	−6.9	5	0.4	9.5	2.3	16.5	**3.0 ***	−7.3	−27.9	16.9	15.0	7.4	11.3	8.8	**1.2 ***
Can Gio	1.8	2.1	−8.2	−0.8	−52.7	**−11.7 ***	−35.2	**−3.8 ***	−38	−16.6	20.3	13.0	−17.5	**−23.6 ***	−35	**−6.0 ***
Cat Lai	−16.7	−15.8	2.4	0.2	−43.2	**−23.7 ***	−49	**−22.5 ***	−27	−98.2	1.4	1.7	−18.4	**−44.0 ***	−37	**−11.9 ***
Cu Chi	−19.3	**−18.5 ***	0.4	0.0	−1.4	−0.4	−1.4	−0.3	7.9	44.4	25.9	**22.0 ***	0.9	1.5	−4.2	−0.7
Hoc Mon	1.5	1.4	6.5	0.5	3.7	0.8	8.4	1.4	−3.2	−0.5	−1.7	−1.6	−4.5	−6.9	1.6	0.2
Nha Be	−13.7	−12.1	12.1	1.0	3.2	0.9	−0.4	−0.1	−8.9	−14.1	−5.9	−5.8	−5.1	−8.6	3.3	0.5
TS Hoa	−20.5	−15.9	15.8	**1.6 ***	−4	−1.3	−3.1	−0.8	5.6	24.4	−0.2	−0.3	−7.4	−13.4	−1.8	−0.3
TT Hiep	−11.2	−11.2	−11.6	−1.0	−0.6	−0.2	−4.5	−0.9	−14	−48.6	−5.5	−5.3	−10.1	−17.4	0.1	0.0
Vung Tau	−24.5	**−27.6 ***	−15.2	−1.5	5.4	1.3	7.1	1.3	8.5	31.6	10.2	11.3	9.5	16.7	4.7	0.7

* Statistically significant difference at the 95% confidence level.

In addition, Figure 6(6.1–6.8) show the spatial distribution of the observation (Figure 6(6.1a–6.8a)) and simulation (Figure 6(6.1b–6.8b)) extreme precipitation indices during the baseline period (1980–2009), and relative differences between the simulation and observation (Figure 6(6.1c–6.8c)). In general, the spatial distribution of the differences between simulation and observation under those indices had similar trends—here, large negative biases were apparent in small parts of the central and southern regions, while positive biases mainly occurred in other regions. General speaking, the results show that the output from the LARS-WG tool is acceptable for simulating the spatial pattern of the extreme indices.

4.2. *The Historical Observation of the Extreme Precipitation Indices*

In this section, the mean annual values of each extreme indices and the corresponding averaged slope of temporal trends during the historical period (1980–2017) at each station in the study area were listed in Table 4. Generally, most extreme indices exhibited downward trends at most stations of the study area during this period, which accounted for approximately 56.8% of the total number of the trends. However, only about one-third proportion (approximately 33%) of all trends were statistically significant; in which the significant downward and upward trends accounted for 19.3% and 13.6% proportion, respectively.

About the spatial distribution patterns, a majority of the significant trends frequently occurred in the northern (Cu Chi Dist.) and southern (Can Gio Dist.) parts of the study area, whereas most of the central regions demonstrated insignificant trends (Table 4).

4.3. *Projected Future Extreme Precipitation Indices*

In this study, the future extreme precipitation indices were calculated from the statistically downscaled outputs of five GCMs (EC-EARTH, HadGEM2-ES, GFDL-ESM2M, MIROC5, and MPI-ESM-MR) as mentioned in Section 3.3 above. The daily rainfall at each station in the study area was generated for the future period of 2021–2080 under the RCP8.5 emission scenario. The temporal variation of each spatially averaged extreme index for the entire study area during the future period was illustrated in Figure 4. According to the results, while two duration-based indices, CDD and CWD, show a high degree of agreement among the five GCMs, the other extreme precipitation indices are generally projected with high uncertainty levels, especially in the intermediate future period (2051–2080). Particularly, three GCMs such as MIROC5, MPI-ESM-MR, and EC-Earth projected higher values of extreme indices than two other GCMs, HadGEM2-ES and GFDL-ESM2M.

However, the multi-model ensemble mean generally presents the obviously increasing trends in value at the majority of extreme precipitation indices during both future periods lasting from 2021 to 2080 for the entire study area.

4.4. *Climate Change Impact on the Precipitation Extremes by the Averaged Multi-Model Ensemble*

4.4.1. Variability in Inter-Annual Changes

To produce more reliable results from simulation of extreme precipitation indices for assessing the impact of climate change, a multi-model ensemble (MME) at each station is prepared by taking the arithmetic mean of the extreme indices computed from the output from the five different GCMs mentioned in the previous section.

Table 4. Mean annual values with a standard variation of extreme precipitation indices and the corresponding averaged slope of trend at each station during the historical period of 1980–2017 in Ho Chi Minh City (HCMC).

Indices	Stats	CDD	CWD	R20mm	R25mm	R95p	RX1day	RX5day	SDII
Unit	Mean ± SD	days	days	days	days	mm	mm	mm	mm/day
	Avg. Trend	days/year	days/year	days/year	days/year	mm/year	mm/year	mm/year	mm/day/year
North Region									
Cu Chi	Mean ± SD	97.1 ± 37.8	8.4 ± 3.6	28.4 ± 6	22.1 ± 5.1	343.9 ± 192	86.7 ± 27.6	161.7 ± 34.9	16.9 ± 3.1
	Avg. Trend	−0.44	−0.11	**−0.24 ***	**−0.18 ***	**−12.63 ***	0.37	**−1.78 ***	**−0.09 ***
Hoc Mon	Mean ± SD	86.5 ± 43.6	7.9 ± 3.5	23.6 ± 8.8	17.2 ± 7.3	327.6 ± 211.5	90.1 ± 25.9	153.4 ± 42.7	15 ± 4
	Avg. Trend	−1.12	0.06	0.26	0.22	3.07	0.22	0.15	−0.09
Central Region									
Binh Chanh	Mean ± SD	82 ± 41.6	8.9 ± 2.7	24.8 ± 6.5	17.9 ± 5.9	370.9 ± 226.8	88 ± 33.5	155 ± 39.8	13.7 ± 2.1
	Avg. Trend	−1.28	0.01	0.2	**0.22 ***	3.05	−0.35	0.63	0.01
Cat Lai	Mean ± SD	89.8 ± 38.6	6.8 ± 3.6	53.5 ± 14.5	44.2 ± 13.1	405.7 ± 322.5	114.2 ± 32.2	232 ± 67.4	31.3 ± 5.2
	Avg. Trend	−0.63	**−0.11 ***	−0.21	−0.4	**−14.91 ***	−0.56	−2.83	**−0.37 ***
Nha Be	Mean ± SD	84.1 ± 41.7	8.5 ± 2.3	28.2 ± 7.2	21.6 ± 5.9	411.1 ± 251.6	98.8 ± 33.1	172.1 ± 41.1	15.7 ± 2.4
	Avg. Trend	−0.67	−0.06	−0.2	−0.16	−0.89	0.04	0.03	−0.05
Tan Son Hoa	Mean ± SD	73.8 ± 38	10.2 ± 2.5	32.4 ± 5.7	25.1 ± 4.6	464.2 ± 205.1	104 ± 28.4	185.6 ± 39.8	15.5 ± 1.7
	Avg. Trend	−0.32	**−0.07 ***	−0.05	−0.02	3.87	0.39	−0.44	0.01
South Region									
TT Hiep	Mean ± SD	93.1 ± 45.1	9.2 ± 3.6	26.2 ± 5	19.7 ± 4.7	364.8 ± 160.4	97.4 ± 39.8	168.8 ± 48.4	15.8 ± 2.4
	Avg. Trend	**−1.67 ***	**0.13 ***	−0.01	−0.03	1.51	−0.39	−0.89	**−0.15 ***
Can Gio	Mean ± SD	120.3 ± 32.9	8.9 ± 3.4	22.3 ± 8.2	11.9 ± 7.1	288.7 ± 255.7	69.7 ± 39	138.5 ± 34.3	16.4 ± 2.1
	Avg. Trend	−0.08	−0.03	−0.04	**0.4 ***	**16.62 ***	**2.01 ***	**1.97 ***	**−0.14 ***

Table 4. *Cont.*

Indices	Stats	CDD	CWD	R20mm	R25mm	R95p	RX1day	RX5day	SDII
Neighborhood Region									
Ben Cat	Mean ± SD	89.3 ± 35.9	10.1 ± 3.3	30.7 ± 7.5	22.1 ± 6.4	389.2 ± 233.1	90.4 ± 33.5	160.6 ± 38.4	15.8 ± 2
	Avg. Trend	−0.13	−0.05	0.22	**0.22 ***	4.9	0.16	0.58	0.02
Bien Hoa	Mean ± SD	95.2 ± 42.2	6.6 ± 2.9	25.6 ± 8.5	20.3 ± 6.7	331.2 ± 189.5	100.6 ± 29.4	162 ± 50	17.5 ± 2.8
	Avg. Trend	**−2.35 ***	**0.22 ***	**0.62 ***	**0.45 ***	**8.11 ***	0.55	**2.68 ***	**−0.11 ***
Vung Tau	Mean ± SD	108.9 ± 38.4	9.7 ± 3.4	23.1 ± 5.2	17.5 ± 4.2	358.5 ± 127.1	110.1 ± 44.1	173.5 ± 49.6	15.1 ± 2
	Avg. Trend	−0.49	0.03	**−0.29 ***	**−0.19 ***	−3.84	−1.52	−1.38	**−0.11 ***

* Statistically significant trend at the 95% confidence level.

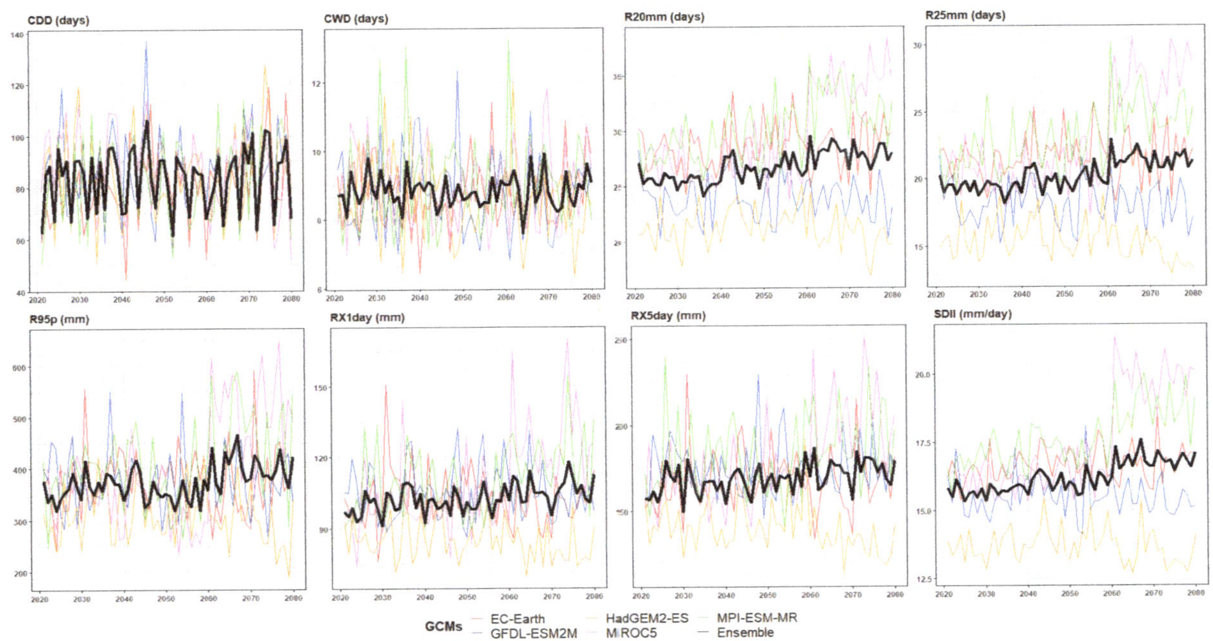

Figure 4. Temporal variation of the spatially averaged extreme precipitation indices for entire study area during the future period (2021–2080). Each color line represents a distinct model while the bold and black line indicates the multi-model ensemble mean.

In the following section, the temporal variation of percentage changes in each extreme precipitation indices between the future (2021–2080) and baseline (1980–2009) periods over the entire study area is analyzed and discussed. The temporal variation of changes in spatially averaged MME extreme indices and its corresponding magnitude of linear trend in each future period were illustrated in Figure 4 and Table 5. As shown in Table 5, the decreasing trend is projected in most extreme precipitation indices during the near future period, while the opposite tendency mainly occurs during the intermediate future period. However, except for the index CDD, the temporal variation of relative changes in other extreme indices is expected to be increased during both near (2021–2050) and intermediate (2051–2080) future periods, in which only four extreme indices, R20mm, R25mm, RX5day, and SDII, show the statistically significant trends.

Table 5. Trend in relative change between future and baseline periods (unit: % per decade).

Period	CDD	CWD	R20mm	R25mm	R95p	RX1day	RX5day	SDII
2021–2050	2.02	−0.25	−0.77	−0.34	−5.54	−0.69	2.21	−0.55
2051–2080	0.83	1.04	−0.26	1.82	0.28	0.79	2.69	0.65

Figure 5 illustrates the temporal variation of the relative changes in the spatially averaged extreme indices between the future and baseline periods over the entire study area. Figure 5 shows the multi-model ensemble mean with 5–95% interquartile range of five different GCMs under RCP8.5 scenario. Considering the duration extreme indices, the tendency of the index CDD presents an obviously increasing trend in both future periods; while CWD gradually tends to decrease in the near future period, then slightly increase in the intermediate future period. Particularly, the annual mean value of the percentage changes in CDD indicates an increase by 0.15% (from −16.05% to 16.80%) in the near future period (2021–2050), and by 0.7% (from −12.46% to 14.73%) in the intermediate future period (2051–2080). In contrast, the future index CWD is expected to increase by 4.07% (from −10.21% to 20.52%) and by 3.76% (from −8.63% to 17.34%) at the end of the near and intermediate future periods with respect to those of the reference period.

In regard to the indices of heavy and extreme heavy precipitation days, R20mm and R25mm, the annual changes in the future are comparatively similar with respect to the baseline period. The length of heavy and extremely heavy rainfall days is expected to be shortened during the near future period (2021–2050) with the projected decrease in R20mm and R25mm by −2.67% (from −18.46% to 10.52%) and −1.08% (from −18.79% to 13.66%), respectively. However, the MME predicts the indices R20mm and R25mm to be increased by 3.75% (from −18.74% to 25.59%) and 5.65% (from −21.29% to 31.26%) by 2080, respectively.

Concerning the indices RX1day and RX5day, it shows that the annual percentages of change in the near future period are slightly lower than those in the intermediate future period. Moreover, the index RX5day clearly shows the strongest increasing trend among other indices during both future periods with the mean value of percentage increase by 5.11% (from −8.85% to 20.78%) and 9.44% (from −10.33% to 28.37%) at the end of two future periods (in 2050 and 2080), respectively. Although the RX1day exhibits a downward trend during the early future period, an increase of 5.56% (from −9.67% to 21.84%) and 10.5% (from −10.04% to 10.50%) is predicted by the mean models at the end of the 2050 and 2080 periods, respectively.

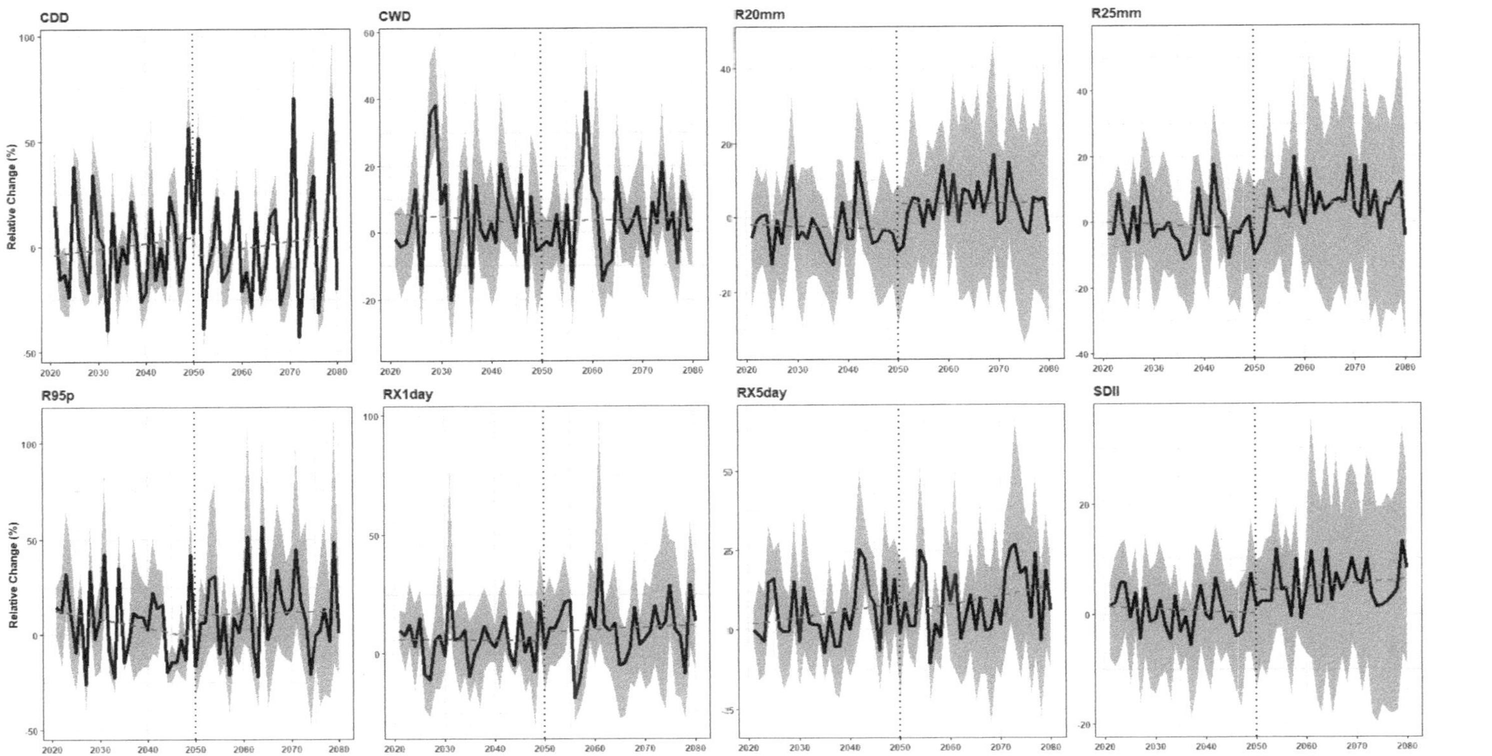

Figure 5. Temporal variation of the relative changes in the spatially averaged extreme precipitation indices over the entire study area. All indices are compared with respect to the baseline period (1980–2009). Solid black line represents the ensemble mean and the shading bounds indicate the range of the 5th and 95th quantiles of 5 different GCMs. The color dashed lines show the Mann-Kendall trends in each future period.

About the extreme intensity indices, the predicted percentage of changes in the indices R95p and SDII approximately tend to be comparable behavior in the future, with a decreasing trend in the near future period, and an increasing tendency in the later future period. Both indices are projected to be increased in the future periods; however, the increasing percentage of the index R95p is much larger than the increasing percentage of the index SDII. Meanwhile, the multi-model ensemble predicts an increase of 5.5% (from −13.23% to 23.72%) and 11.35% (from −17.94% to 43.67%) for the index R95p by 2050 and 2080, respectively. The percentage of change in SDII is projected to be 0.77% (from −10.34% to 8.94%) and 5.37% (from −11.10% to 20.78%) at the end of two future periods, respectively. Generally, the projected percentage changes, as well as the increasing rates in intensity and frequency indices for future extreme rainfall, are predicted to be larger than those in the duration indices, especially in the intermediate future period (2051–2080). In addition, the extreme precipitation indices show larger variation in the intermediate future period (2051–2080) in comparison to those in the near future period (2021–2050).

4.4.2. Distribution of Spatial Changes

In this section, the spatial distribution of the MME-projected mean relative changes in extreme precipitation indices between the near (2021–2050) and intermediate (2051–2080) future periods and the baseline period (1980–2009) over the study area is shown in Figure 6 including Figure 6(6.1–6.8). Furthermore, the statistical significance, stability, and magnitude of the trend in the distribution of spatial changes were determined and computed using the M-K statistical test and the Sen's estimator for particular 25-year moving time spans within the future periods. Afterwards, the dominated trend during these future periods at each station is also illustrated in these figures.

Generally, in comparison with the baseline period, the absolute changes of extreme indices in the intermediate future period (2051–2080) show notably higher than those in the near future period (2021–2050). In which, the largest absolute changes mainly occur in the southern regions, while some parts of the northern and central areas often exhibit little changes (i.e., lowest absolute changes). In terms of relative changes in extreme rainfall indices in the same projection period, the multi-model ensemble predicts an increase in most indicators in the south and the central regions, and a decline of those indices in the north of the study area.

Particularly, regarding the index CDD, the relative changes between the future and historical periods, along with the statistical significance, stability, and magnitude of trends, are shown in Figure 6(6.1). During the near future period (2021–2050), the averaged multi-model ensemble predicts an increase in the index CDD for most regions in the central and southern parts of the study area, whereby a variation of percentage changes in indices is from 0.4 to 20.6%; the averaged growth rate is from 0.4 to 13.1 [days.decade^{-1}]. However, only three stations located in the southern region present significant increasing trends during the near future period, in which the strongest and most stable upward trend in CDD is predicted to occur at the Tam Thon Hiep Station. In contrast, a decrease ranging from −20.1 to −8.7% is predicted in the northern regions, while Cu Chi Station shows a slight increase by 3.7%; even though none of the downward trends is statistically significant. In the intermediate future period (2051–2080), the spatial distribution of the CDD is similar with that of the near future period in the study area, in which the positive percentage changes concentrate in the central and southern parts, while negative values are distributed mostly in the northern regions of the city. During this intermediate future period, the mean percentage of changes in CDD is predicted to decrease from −20.1 to −1.7% at Cu Chi, Hoc Mon, Can Gio, Bien Hoa, and Ben Cat, and to increase from 1.82 to 21.8% at the other stations, whereby the averaged magnitude of increasing trend is ranging between 0.1 and 10.4 [days.decade^{-1}] while the decreasing rate is from −2.9 to −0.8 [days.decade^{-1}].

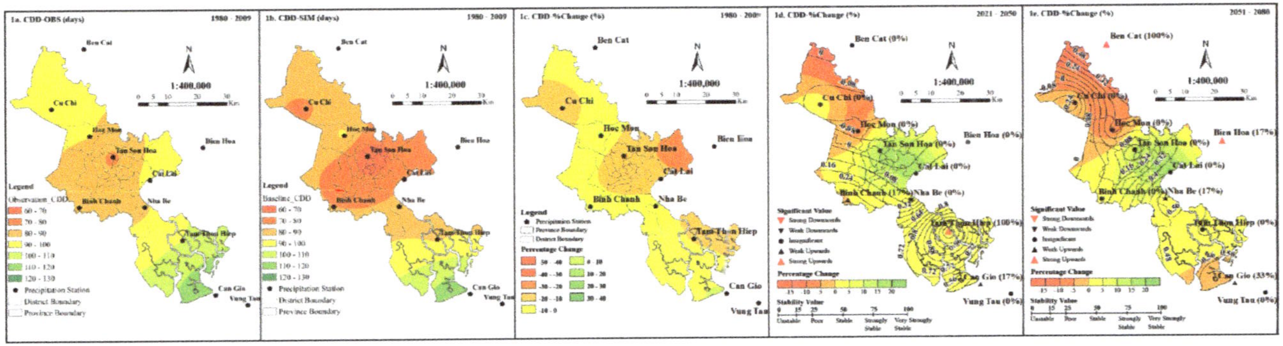

6.1 Spatial distribution of the observation (1a) and simulation (1b) extreme precipitation index (CDD) during the baseline period (1980–2009); relative differences between the simulation and observation (1c); relative changes between the near (2021–2050) and intermediate (2051–2080) future periods (1d,1e) with respect to the baseline period.

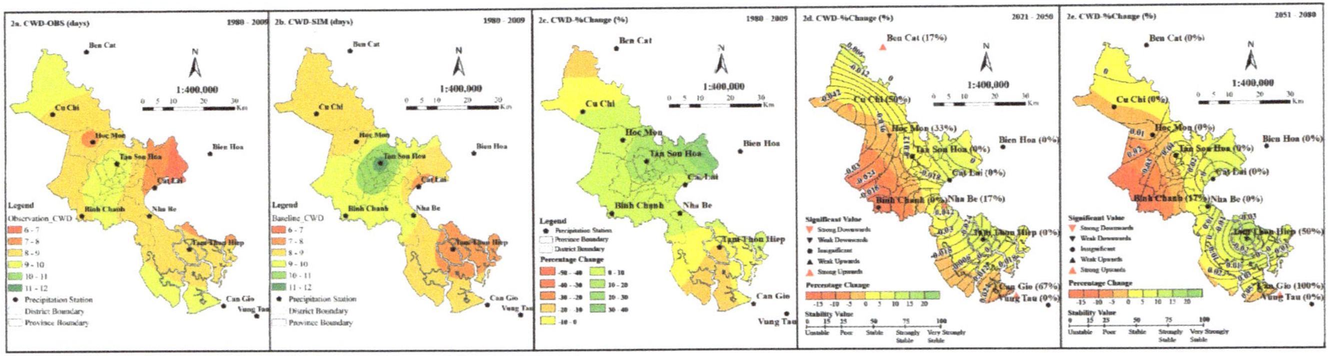

6.2 Spatial distribution of the observation (2a) and simulation (2b) extreme precipitation index (CWD) during the baseline period (1980–2009); relative differences between the simulation and observation (2c); relative changes between the near (2021–2050) and intermediate (2051–2080) future periods (2d,2e) with respect to the baseline.

Figure 6. *Cont.*

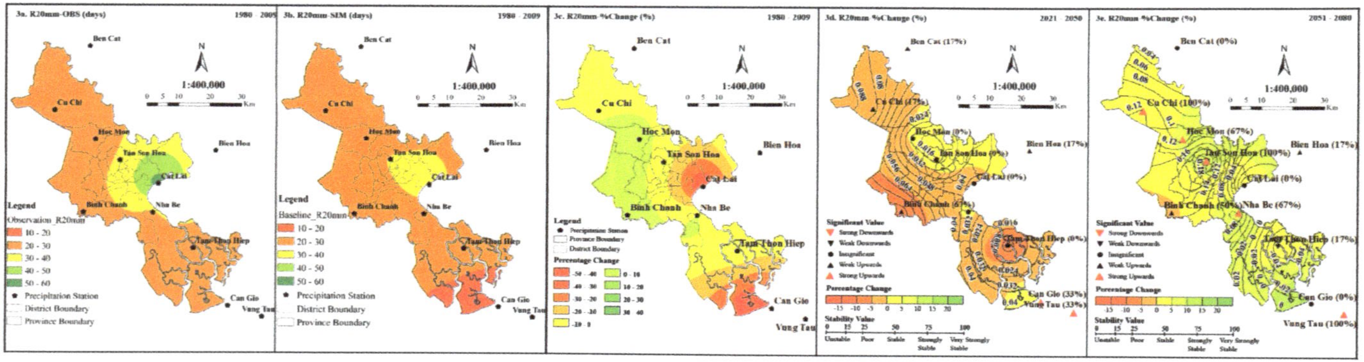

6.3. Spatial distribution of the observation (3a) and simulation (3b) extreme precipitation index (R20mm) during the baseline period (1980–2009); relative differences between the simulation and observation (3c); relative changes between the near (2021–2050) and intermediate (2051–2080) future periods (3d,3e) with respect to the baseline.

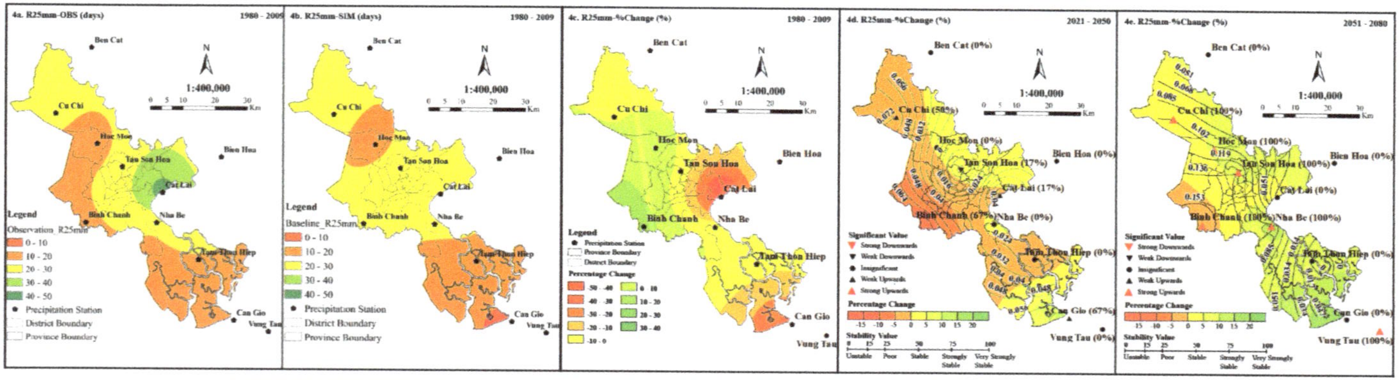

6.4. Spatial distribution of the observation (4a) and simulation (4b) extreme precipitation index (R25mm) during the baseline period (1980–2009); relative differences between the simulation and observation (4c); relative changes between the near (2021–2050) and intermediate (2051–2080) future periods (4d,4e) with respect to the baseline.

Figure 6. *Cont.*

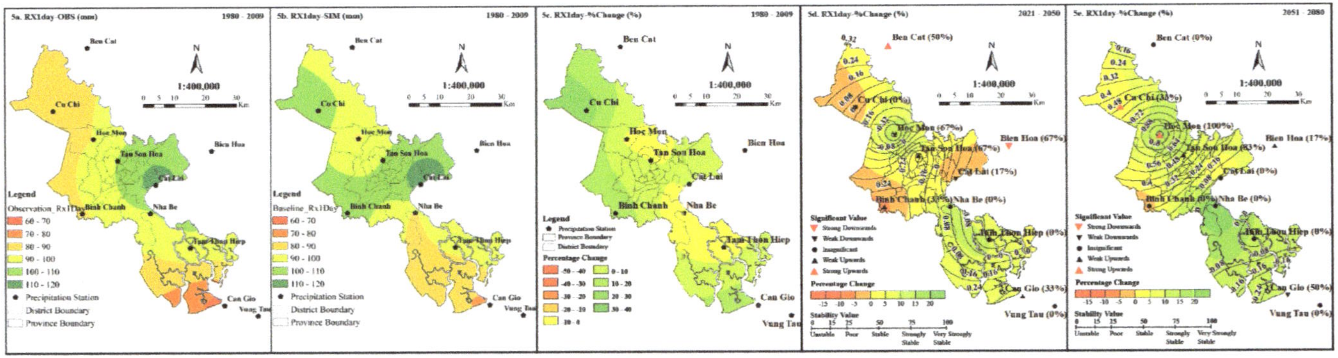

6.5. Spatial distribution of the observation (5a) and simulation (5b) extreme precipitation index (RX1day) during the baseline period (1980–2009); relative differences between the simulation and observation (5c); relative changes between the near (2021–2050) and intermediate (2051–2080) future periods (5d,5e) with respect to the baseline.

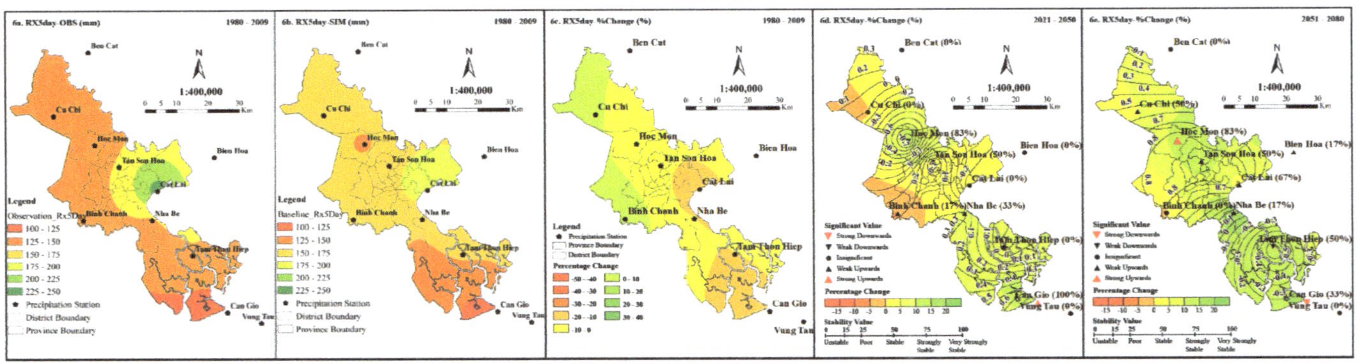

6.6. Spatial distribution of the observation (6a) and simulation (6b) extreme precipitation index (RX5day) during the baseline period (1980–2009); relative differences between the simulation and observation (6c); relative changes between the near (2021–2050) and intermediate (2051–2080) future periods (6d,6e) with respect to the baseline.

Figure 6. *Cont.*

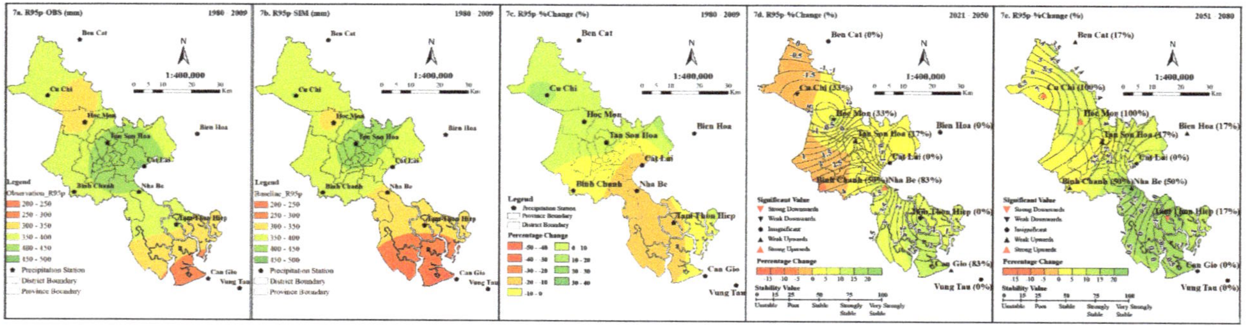

6.7. Spatial distribution of the observation (7a) and simulation (7b) extreme precipitation index (R95p) during the baseline period (1980–2009); relative differences between the simulation and observation (7c); relative changes between the near (2021–2050) and intermediate (2051–2080) future periods (7d,7e) with respect to the baseline period.

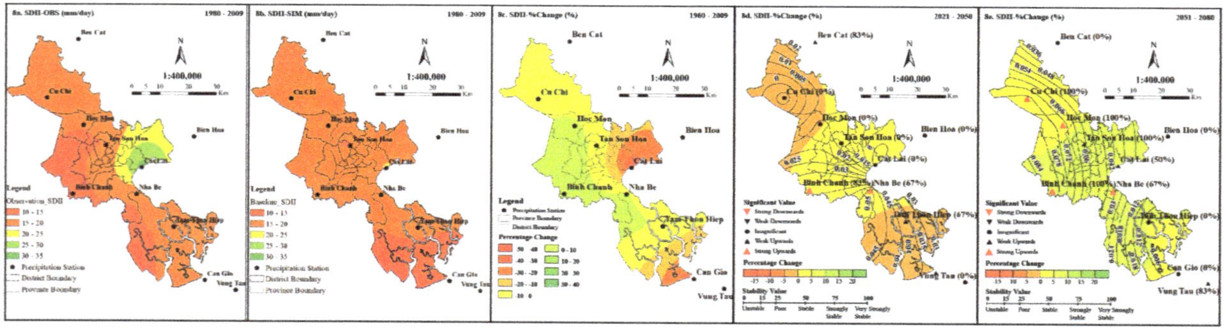

6.8. Spatial distribution of the observation (8a) and simulation (8b) extreme precipitation indices (SDII) during the baseline period (1980–2009); relative differences between the simulation and observation (8c); relative changes between the near (2021–2050) and intermediate (2051–2080) future periods (8d,8e) with respect to the baseline.

Figure 6. Spatial distribution of the observation and simulation extreme precipitation indices (CDD, CWD, R20mm, R25mm, RX1day, RX5day, R95p and SDII) during the baseline period (1980–2009); relative differences between the simulation and observation; relative changes between the near (2021–2050) and intermediate (2051–2080) future periods with respect to the baseline period. In the sub-figures d and e, the contour lines indicate the averaged magnitude of trend in temporal variation of future relative changes (unit: day.year^{-1}); the up/down triangle symbols represent the significant levels of the dominating trend during the future period at corresponding stations; the percentage numbers inside the parenthesis indicate the stability of trend at corresponding stations.

In regard to the index CWD, the projected results shown in Figure 6(6.2) present the west-to-east increasing trend for the percentage of changes in extreme indices with respect to both future periods. Generally, the MME predicts few differences between the spatial distribution as well as the volume of relative change of the index CWD in both future periods. Particularly, in comparison with the historical period, the index CWD is predicted to be decreased from −18 to −0.2%, mainly in the central regions and small parts of other areas, while it is predicted to be increased from 0.9 to 11.4% mostly in the southern regions in both future periods. Additionally, the trend in changes of the projected index CWD shows statistically significant and stable tendency occurring at more stations in the near future period (2021–2050) than that in the later future period (2051–2080).

The spatial distribution of annual mean relative change in the indices R20mm and R25mm between the future and historical periods is shown in Figure 6(6.3,6.4). Generally, in comparison to the historical period, the MME predicts a decrease in both extreme indices in the near future period, while an increase in both indices is expected in the intermediate future period in most regions of the study area. Additionally, the projection results indicate that the future spatial distribution of percentage changes in the index R20mm is comparable with that in the index R25mm. Particularly, in the near future period (2021–2050), the MME predicts a decrease in the index R20mm varying from 2.1% to 10.2%, and in the index R25mm varying from 1.1% to 13% over most regions, especially in the central (Binh Chanh) and southern (Tam Thon Hiep) area. In comparison, an increase from 0.5 to 6% and 0.26 to 13.2% in the rest of this location is corresponding to the indices R20mm and R25mm, respectively (Figure 6(6.3d,6.4d)). For the same projection period, an increasing trend in the temporal changes of the indices' values is projected to be dominated at most stations with the average rates ranging from 0.4 to 0.9 [days.decade^{-1}] for the index R20mm, and from 0.1 to 0.7 [days.decade^{-1}] for the index R25mm. In the near future period, the stable and strongly upward trends are projected to mainly occur in the central and southern regions, while the downward trends are weak and unstable. In the intermediate future period (2051–2080), the MME predicts increasing from 0.5 to 17.3% for the index R20mm in most regions (with high values mostly distributed in the central and southern parts of the study area), while it presents a nuance with an increase ranging from 1.5 to 21.9% for the index R25mm in the similar spatial patterns. Additionally, during the intermediate future period, a slight decrease of −1.6% and −4.2% is projected to occur in a small part of the central area for R20mm and R25mm, respectively (Figure 6(6.3e,6.4e)). In this period, both extreme indices, R20mm and R25mm, show strongly significant and stable increasing trends at most stations and equally distributed across the study area, in which the averaged magnitude of dominated trends is ranging from 0.5 to 2.1, and from 1 to 1.6 [days.decade^{-1}] for the indices, R20mm and R25mm, respectively.

The relative changes in the maximum 1-day and 5-day total rainfall amount (RX1day and RX5day) between the future and the historical periods are shown in Figure 6(6.5d,e–6.6d,e). The simulation results show that the spatial distribution of the indices RX1day and RX5day in each future period is relatively similar.

Generally, in the near future period (2021–2050), the multi-model ensemble predicts an average increase in RX1day from 3.5 to 14.4%, and of RX5day from 0.2 to 12.5% in most regions, especially in the southern parts of the study area, while a decrease occurring in small parts of the central and northern regions is expected from −6.9 to −1.1% and −5.4 to −0.4% for the indices RX1day and RX5day, respectively. In this period, the statistically significant and stable increasing trends are dominated, and present high likelihood of occurrence in the central and southern parts of the study area, along with the averaged growing rates from (3.2–4.2) [mm/decade^{-1}] and (1.8–8.5) [mm.decade^{-1}] for the indices RX1day and RX5day, respectively.

In the intermediate future period (2051–2080), the multi-model ensemble predicts an increase in both extreme indices RX1day and RX5day to be fluctuated across entire study area from 1.3 to 28.2% and 2.8 to 18.5%, respectively; particularly, the highest values frequently concentrate in the southern regions. Besides, a slight decrease (less than 2.8%) on average for both indices is also presented in the small parts of central regions. Additionally, the indices, RX1day and RX5day, are projected to demonstrate the statistically significant and stable upward trends in the northern and central regions, wherein an increasing rate of (0.8–10.1) [mm.decade^{-1}] for the index RX1day, and of (2.1–9.1) [mm.decade^{-1}] for the index RX5day; the stable and downward trends are predicted in the southern parts with a decreasing rate of −3.5 [mm.decade]$^{-1}$ for the index RX1day and of (−1.9 to −7.0) [mm.decade^{-1}] for the index RX5day.

Considering changes in the future intensity indices, the relative differences between the indices R95p and SDII in the future and historical period are shown in Figure 6(6.7d,e–6.8d,e). In the period 2021–2050, the MME predicts a decrease in the index R95p from −4.3 to −5.8% on average in some locations of the central and northern regions, and an increase from 0.01 to 12.5% in the rest of study area, especially in the southern regions (Figure 6(6.7d)). During the same period, the index R95p exhibits statistically significant and stable downward trend in the north of the study area with the averaged rates between −21.3 and −31.7 [days.decade^{-1}], while the stable and significant upward trends are projected for the central and southern regions with the increasing rates varying from 21.1 to 33.5 [days.decade^{-1}]. In the period 2051–2080, an average increase from 0.02 to 22.3% is forecasted by the MME over the entire study area, along with the northwestern-to-southeastern tendency of growth in the extreme indices (Figure 6(6.7e)). In this period, the statistically significant and stable trends are projected to be occurred in most regions. Moreover, in contrast to the near future period, the statistically strong and stable upwards (11.7 to 71.2) [days.decade^{-1}] are projected with high likelihood of presence in the northern regions, while the statistically strong but poor-stability downward trends (−7.6 to −15.8) [days.decade^{-1}] are predicted to appear in the southern regions of the city during this intermediate future period.

The mean models predict a decrease in the index SDII from −0.65% to −3.77% on average in the north and south of the study area, while an increase from 0.1 to 4.8% is projected over entire central regions in the period 2021–2050 (Figure 6(6.8d)). During this period, SDII presents the domination of the statistically significant and stable increasing trends at most stations in the central (strong upwards) and southern regions (weak upwards), along with the averaged magnitude ranging from 0.1 to 0.4 [mm.day^{-1}.decade^{-1}]. In the period 2051–2080, the MME forecasts an increase from 2.8 to 9.1% on average over the entire study area, with high values being mainly observed in the central regions. During the same projection period, the statistically significant and stable upward trends are dominated at most stations over the study area (mainly distributed in the north and central regions) with the averaged magnitude of trends varying from 0.3 to 0.8 [mm.day^{-1}.decade^{-1}] (Figure 6(6.8e)).

5. Discussion

The results of the current study on the spatiotemporal trend in changes of extreme precipitation indices between the future (2021–2080) and historical (1980–2009) periods indicate a general increase in most extreme indices over most regions in Ho Chi Minh City, especially in some parts of the southern and central areas. Currently, the studies on the impacts of climate change on extreme weather events in Vietnam are still limited in terms of quantity and quality of the research, especially studies using the extreme indices proposed by ETCCDI. Most of these studies were conducted for the entire Vietnam territory instead of focusing only on a specific location such as HCMC. To the best of our knowledge, our study seems to be the first reports assessing and mapping the changes of these extreme indices over HCMC for the future period in the context of climate change. The findings in our paper are consistent with previous remarkable related studies [58–60]. Particularly, our main results are in agreement with the report on Climate Change and Sea Level Rise Scenarios for Vietnam, which was a well-known and widely cited research work on this topic and published by the Ministry of Natural Resources and Environment of Vietnam [59], in which the authors only focused on the changes in RX1day and RX5day by the mid (2046–

2065) and end (2080–2099) of 21st century with respect to the baseline period (1986–2005) under RCP4.5 and RCP8.5 scenarios. According to the report by MONRE, the RX1day and RX5day are expected to increase from 10 to 70% by the mid-century under RCP8.5 scenario; in which, the indices for HCMC area are projected to increase approximately between 10 and 30% with respect to the baseline period. In terms of the intensity extreme indices, our projection results agree with previous notable studies conducted by Ngo-Duc et al. [60] on applying regional climate models (RegCM4) with quantile mapping bias correction for evaluating extreme weather events over Southeast Asia regions for the mid-century period [60,61]. Here, the authors suggested an overall increase in CDD, CWD, SDII, R20mm, and R95p in the southern parts of Vietnam, especially in the rainy season. These results are reaffirmed by Tangang et al. [62] in their recent study on future changes in extreme precipitation indices under global warming of 2 °C; in which, the significant and robust increasing changes in frequency and intensity indices are expected to occur in the mid-future period [59]. From these results, the central and southern regions of the city are warned to be threaten by flooding, while the plan of water management for rural areas in the northern sides should be prepared in order to address and mitigate the impact on agricultural sectors by potentially extreme droughts.

In this study, although the performance of the statistical downscaling tool and its incorporated GCMs was evaluated and confirmed to be able to reproduce and generate extreme precipitation indices, the model biases and uncertainties still remain in the projection results, especially in the projected spatial changes. Therefore, it is essential to apply proper bias correction methods, as well as more precise downscaling methods for maximizing the model performance. Moreover, for the approach of multi-model ensemble, the unweighted averaging multi-models used in this study are the simplest way but not effective when a better skillful model could be underestimated (or vice versa). Thereby, further research on individual model evaluation to create a weighted multi-model ensemble based on the ranked models could be a promising approach in order to optimize the simulation performance [63–65]. In addition, the relevant climate teleconnections (such as El Nino Southern Oscillation or Sea Surface Temperature) have a significant contribution to the behavior of climate extremes in the study area; therefore, their impacts on precipitation extremes should be taken into account in further research.

6. Conclusions

In this paper, the projected temporal and spatial changes in selected ETCCDI extreme precipitation indices between the future (2021–2080) and baseline period (1980–2009) in Ho Chi Minh City (Vietnam) were evaluated in the context of climate change. The projected extreme indices were computed based on daily rainfall data downscaled from five different GCMs under an RCP8.5 scenario using the LARS-WG tool. The performance of the LARS-WG tool for the simulation of extreme precipitation indices has been examined using the model evaluation metrics, the Mann–Kendall statistical test and the Sen's estimator, which were also applied to analyze the significance, stability, and magnitude of trends in future changes. A summary of the major findings of this study is as follows:

- The LARS-WG tool has presented adequate performance for reproducing the extreme precipitation indices in most stations during the historical period (1980–2017), especially for simulating the spatial distribution of those indices. Notwithstanding, the uncertainty in simulation results was probably inevitable. Notably, the simulation tended to underestimate extreme indices in some parts of the central and southern regions, while overestimating these indices in the northern regions. In addition, the performance for simulating intensity indices is better than that for duration and frequency indices, especially for CWD and R20mm.
- In the historical period (1980–2017), the high values of temporally mean extreme indices were frequently observed in the central regions, while the low values of these indices were mainly distributed in the northern and southern regions. During this period, the decreasing trend in extreme indices were regularly observed at most

stations in the study area. However, only 33% of this trend was significant and mainly distributed in the north and south parts of the study area.

- In the future periods, the projected extreme precipitation indices were computed from the downscaling from five different GCMs (EC-EARTH, HadGEM2-ES, GFDL-ESM2M, MIROC5, and MPI-ESM-MR) under the RCP8.5 emission scenario. Afterwards, the multi-model ensemble mean was calculated for evaluating the spatiotemporal changes in the near (2021–2050) and intermediate (2051–2080) future extreme indices with respect to the simulated indices in the baseline period (1980–2009). Generally, in comparison with the historical period, the temporally relative changes in most extreme precipitation indices are predicted to be increased during both future periods (2021–2080), but the index CDD. In which, the extreme intensity and frequency indices present a stronger magnitude and statistically significant increasing trends than those of extreme duration indices during the future periods. The spatial distribution of changes in future projected extreme indices across the study area is relatively systematic. In which, the highest values of mean absolute changes are frequently observed in the southern regions, while the lowest values of mean absolute changes are regularly observed in the northern and central areas of the study area in the future periods.

Moreover, the percentage changes in extreme indices in the intermediate future period (2051–2080) present noticeably higher values than those in the near future period (2021–2050) over most regions of the study area. In terms of relative changes in the same projection period, the multi-model ensemble reasonably predicts an increase in most extreme indices, and this mainly occurs in the southern regions. Meanwhile, a decrease in these indices is often projected in the northern and some parts in the central region of the study area. Furthermore, most stations obtain statistically insignificant and unstable trends in extreme precipitation indices during the near future period (2021–2050), but present statistically strong significance and high stability during the intermediate future period (2051–2080), especially in the central and southern parts of the study area.

Consequently, the prediction of increasing trends in extreme precipitation indices, especially for intensity and frequency of heavy and extreme heavy rainfall events, has posed the potential risk of more frequent floods in the central and southern regions of HCMC. Meanwhile, the projection shows a decrease in consecutive dry days in the northern areas, which implies the potential risk of severe droughts occurring in the northern regions of HCMC. Therefore, the simulation results in this paper could contribute a reasonable assessment and detailed potential-risk maps based on the changes in extreme precipitation indices across Ho Chi Minh City for the near and intermediate future periods in the 21st century. Although it is essential to reaffirm that these findings should be taken into consideration with caution due to the sources of uncertainty, such as the GCMs and statistical downscaling tool applied in this study, the achieved results could be recognized as the informative scientific references for the city governors to develop and implement comprehensive assessment, adaptation, and mitigation responding to the impact of extreme weather events in the context of climate change in Ho Chi Minh City.

Author Contributions: Conceptualization, D.N.K. and V.T.N.; data curation, N.T.Q. and P.T.T.N.; formal analysis, N.T.Q. and D.N.K.; funding acquisition, V.T.N.; investigation, N.T.Q. and P.T.T.N.; methodology, D.N.K. and N.T.Q.; software, N.T.Q. and P.T.T.N.; supervision, V.T.N. All authors have read and agreed to the published version of the manuscript.

Funding: The study was supported by the Department of Science and Technology of Ho Chi Minh City, managed by Institute for Computational Science and Technology under the funding number 05/2019/HĐ-KHCNTT.

Institutional Review Board Statement: Not applicable.

Informed Consent Statement: Not applicable.

Data Availability Statement: Not applicable.

Acknowledgments: The study was supported by the Department of Science and Technology of Ho Chi Minh City, managed by Institute for Computational Science and Technology under the contract number 05/2019/HĐ-KHCNTT. The authors also would like to thank the anonymous reviewers for their valuable and constructive comments to improve our manuscript.

Conflicts of Interest: The authors declare no conflict of interest.

References

1. World Economic Forum. *The Global Risks Report 2019*, 14th ed.; The World Economic Forum: Geneva, Switzerland, 2019; ISBN 978-1-944835-15-6.
2. IPCC. Summary for Policymakers. In *Intergovernmental Panel on Climate Change, editor. Clim Chang 2013—Phys Sci Basis [Internet]*; Cambridge University Press: Cambridge, UK, 2013; pp. 1–30. Available online: https://www.cambridge.org/core/product/identifier/CBO9781107415324A009/type/book_part (accessed on 5 October 2020).
3. IPCC. Summary for Policymakers. 2014. Available online: https://www.ipcc.ch/report/ar5/wg2/summary-for-policymakers (accessed on 5 October 2020).
4. Zhai, P.; Zhang, X.; Wan, H.; Pan, X. Trends in total precipitation and frequency of daily precipitation extremes over China. *J. Clim.* **2005**, *18*, 1096–1108. [CrossRef]
5. Bengtsson, L.; Rana, A. Long-term change of daily and multi-daily precipitation in southern Sweden. *Hydrol. Process.* **2014**, *28*, 2897–2911. [CrossRef]
6. Du, H.; Xia, J.; Zeng, S.; She, D.; Liu, J. Variations and statistical probability characteristic analysis of extreme precipitation events under climate change in Haihe River Basin, China. *Hydrol. Process.* **2014**, *28*, 913–925. [CrossRef]
7. Peñuelas, J.; Ciais, P.; Canadell, J.G.; Janssens, I.A.; Fernández-Martínez, M.; Carnicer, J.; Obersteiner, M.; Piao, S.; Vautard, R.; Sardans, J. Shifting from a fertilization-dominated to a warming-dominated period. *Nat. Ecol. Evol.* **2017**, *1*, 1438–1445. [CrossRef] [PubMed]
8. Hutchison, C.; Gravel, D.; Guichard, F.; Potvin, C. Effect of diversity on growth, mortality, and loss of resilience to extreme climate events in a tropical planted forest experiment. *Sci. Rep.* **2018**, *8*, 15443. [CrossRef] [PubMed]
9. Harrison, P.A.; Dunford, R.W.; Holman, I.P.; Cojocaru, G.; Madsen, M.S.; Chen, P.-Y.; Pedde, S.; Sandars, D. Differences between low-end and high-end climate change impacts in Europe across multiple sectors. *Reg. Environ. Chang.* **2019**, *19*, 695–709. [CrossRef]
10. Mishra, S.R.; Bhandari, P.M.; Issa, R.; Neupane, D.; Gurung, S.; Khanal, V. Climate change and adverse health events: Community perceptions from the Tanahu district of Nepal. *Environ. Res. Lett.* **2015**, *10*, 034007. [CrossRef]
11. Curtis, S.; Fair, A.; Wistow, J.; Val, D.V.; Oven, K. Impact of extreme weather events and climate change for health and social care systems. *Environ. Health* **2017**, *16*, 128. [CrossRef]
12. Rogelj, J.; Luderer, G.; Pietzcker, R.C.; Kriegler, E.; Schaeffer, M.; Krey, V.; Riahi, K. Energy system transformations for limiting end-of-century warming to below 1.5 °C. *Nat. Clim. Chang.* **2015**, *5*, 519–527. [CrossRef]
13. Bauer, N.; Mouratiadou, I.; Luderer, G.; Baumstark, L.; Brecha, R.J.; Edenhofer, O.; Kriegler, E. Global fossil energy markets and climate change mitigation—An analysis with REMIND. *Clim. Chang.* **2016**, *136*, 69–82. [CrossRef]
14. Carvajal, P.E.; Anandarajah, G.; Mulugetta, Y.; Dessens, O. Assessing uncertainty of climate change impacts on long-term hydropower generation using the CMIP5 ensemble—The case of Ecuador. *Clim. Chang.* **2017**, *144*, 611–624. [CrossRef]
15. Hassapoyannes, K. Mitigation Challenges and Options in Agriculture. In *Global Challenges for Future Food and Agricultural Policies*; Blandford, D., Hassapoyannes, K., Eds.; World Scientific: Singapore, 2019; pp. 249–266.
16. Hassapoyannes, K.; Blandford, D. Agriculture and Climate Change: National and International Policy Response. In *Global Challenges for Future Food and Agricultural Policies*; Blandford, D., Hassapoyannes, K., Eds.; World Scientific: Singapore, 2019; pp. 217–248.
17. Mendelsohn, R. Adapting to Climate Change in Agriculture: Challenges and Options. In *Global Challenges for Future Food and Agricultural Policies*; Blandford, D., Hassapoyannes, K., Eds.; World Scientific: Singapore, 2019; pp. 267–280.
18. Halmstad, A.; Najafi, M.R.; Moradkhani, H. Analysis of precipitation extremes with the assessment of regional climate models over the Willamette River Basin, USA. *Hydrol. Process.* **2013**, *27*, 2579–2590. [CrossRef]
19. Rana, A.; Moradkhani, H. Spatial, temporal and frequency based climate change assessment in Columbia River Basin using multi downscaled-scenarios. *Clim. Dyn.* **2016**, *47*, 579–600. [CrossRef]
20. Rana, A.; Uvo, C.B.; Bengtsson, L.; Sarthi, P.P. Trend analysis for rainfall in Delhi and Mumbai, India. *Clim. Dyn.* **2012**, *38*, 45–56. [CrossRef]
21. ETCCDI Climate Change Indices. Available online: http://etccdi.pacificclimate.org/list_27_indices.shtml (accessed on 2 September 2018).
22. Karl, T.R.; Nicholls, N.; Ghazi, A. CLIVAR/GCOS/WMO Workshop on Indices and Indicators for Climate Extremes Workshop Summary. In *Weather Clim Extrem*; Springer: Dordrecht, The Netherlands, 1999; pp. 3–7. Available online: http://link.springer.com/10.1007/978-94-015-9265-9_2 (accessed on 15 September 2020).
23. Peterson, T.C.; Folland, C.C.; Gruza, G.; Hogg, W.; Mokssit, A.; Plummer, N. *Report on the activities of the Working Group on Climate Change Detection and Related Rapporteurs 1998–2001*; Springer: Southampton, UK, 2001; p. 144.

24. Tian, Y.E.; Xu, Y.; Booij, M.J.; Zhang, Q.; Lin, S. Trends in precipitation extremes and long-term memory of runoff records in Zhejiang, east China. In *IAHS Publication No. 344 (Hdro-climatology: Variability and Change)*; Franks, S.W., Boegh, E., Blyth, E., Hannah, D.M., Yilmaz, K.K., Eds.; IAHS Press: Wallingford, UK, 2011; pp. 227–232.
25. Wilby, R.L.; Dawson, C.W. *SDSM 4.2-A Decision Support Tool for the Assessment of Regional Climate Change Impacts, Version 4.2 User Manual*; Lancaster University, Lancaster/Environment Agency of England and Wales: Lancaster, UK, 2007; pp. 1–94.
26. Coupled Model Intercomparison Project Phase 5. Available online: https://pcmdi.llnl.gov/mips/cmip5/ (accessed on 20 August 2018).
27. ADB. Ho Chi Minh City Adaptation to Climate Change. 2010. Available online: https://www.adb.org/publications/ho-chi-minh-city-adaptation-climate-change-summary-report (accessed on 25 September 2020).
28. Pillai, P.; Philips, B.R.; Shyamsundar, P.; Ahmed, K.; Wang, L. *Climate Risks and Adaptation in Asian Coastal Megacities: A Synthesis report (English)*; World Bank Group: Washington, DC, USA, 2010.
29. KULP, S.A.; Strauss, B.H. New elevation data triple estimates of global vulnerability to sea-level rise and coastal flooding. *Nat. Commun.* **2019**, *10*, 4844. Available online: http://www.nature.com/articles/s41467-019-12808-z (accessed on 25 September 2020).
30. Storch, H.; Downes, N.; Thinh, N.-X.; Thamm, H.-P.; Phi, H.-L.; Thuc, T.; Thị Hiền Thuận, N.; Emberger, G.; Goedecke, M.; Welsch, J.; et al. Adaptation planning framework to climate change for the urban area of Ho Chi Minh City, Vietnam. In Proceedings of the Fifth Urban Research Symposium, 'Cities and Climate Change: Responding to an Urgent Agenda', Marseille, France, 28–30 June 2009.
31. Lupikasza, E. Spatial and temporal variability of extreme precipitation in Poland in the period 1951–2006. *Int. J. Climatol.* **2010**, *30*, 991–1007. [CrossRef]
32. Wu, C.; Huang, G.; Yu, H.; Chen, Z.; Ma, J. Spatial and temporal distributions of trends in climate extremes of the Feilaixia catchment in the upstream area of the Beijiang River Basin, South China. *Int. J. Climatol.* **2014**, *34*, 3161–3178. [CrossRef]
33. Shi, J.; Cui, L.; Wen, K.; Tian, Z.; Wei, P.; Zhang, B. Trends in the consecutive days of temperature and precipitation extremes in China during 1961–2015. *Environ. Res.* **2018**, *161*, 381–391. [CrossRef] [PubMed]
34. Thinh, N.X.; Bräuer, A.; Teucher, V. Introduction into work package urban flooding of the bmbf megacity research project TP. Ho Chi Minh. *Environ. Inform. Ind. Environ. Prot. Concepts Methods Tools* **2009**, *2009*, 223–232.
35. Van Leeuwen, C.J.; Dan, N.P.; Dieperink, C. The challenges of water governance in Ho Chi Minh City. *Integr. Environ. Assess. Manag.* **2016**, *12*, 345–352. [CrossRef]
36. ETCCDI Climate Change Indices. RClimdex and RHtests Toolkits. Available online: http://etccdi.pacificclimate.org/software.shtml (accessed on 2 September 2018).
37. Alexander, L.V.; Zhang, X.; Peterson, T.C.; Caesar, J.; Gleason, B.; Klein Tank, A.M.G.; Haylock, M.; Collins, D.; Trewin, B.; Rahimzadeh, F.; et al. Global observed changes in daily climate extremes of temperature and precipitation. *J. Geophys. Res.* **2006**, *111*, D05109. [CrossRef]
38. Khoi, D.N.; Trang, H.T. Analysis of changes in precipitation and extremes events in Ho Chi Minh City, Vietnam. *Procedia Eng.* **2016**, *142*, 228–234. [CrossRef]
39. Mann, H.B. Nonparametric tests against trend. *Econometrica* **1945**, *13*, 245. Available online: https://www.jstor.org/stable/1907187?origin=crossref (accessed on 11 October 2020). [CrossRef]
40. Kendall, M.G. Further contributions to the theory of paired comparisons. *Biometrics* **1955**, *11*, 43. Available online: https://www.jstor.org/stable/3001479?origin=crossref (accessed on 11 October 2020). [CrossRef]
41. Sen, P.K. Robustness of some nonparametric procedures in linear models. *Ann. Math. Stat.* **1968**, *39*, 1913–1922. Available online: http://projecteuclid.org/euclid.aoms/1177698021 (accessed on 11 October 2020). [CrossRef]
42. Zhang, Y.; Huang, G.; Wang, X.; Liu, Z. Observed changes in temperature extremes for the Beijing-Tianjin-Hebei region of China. *Meteorol. Appl.* **2017**, *24*, 74–83. [CrossRef]
43. von Storch, H. Misuses of Statistical Analysis in Climate Research. In *Analysis of Climate Variability: Applications of Statistical Techniques*; von Storch, H., Navarra, A., Eds.; Springer: Berlin/Heidelberg, Germany, 1999; pp. 11–26.
44. Gocic, M.; Trajkovic, S. Analysis of changes in meteorological variables using Mann-Kendall and Sen's slope estimator statistical tests in Serbia. *Glob. Planet. Chang.* **2013**, *100*, 172–182. Available online: http://linkinghub.elsevier.com/retrieve/pii/S0921818112002032 (accessed on 11 October 2020). [CrossRef]
45. Hamed, K.H.; Rao, A.R. A modified Mann-Kendall trend test for autocorrelated data. *J. Hydrol.* **1998**, *204*, 182–196. [CrossRef]
46. Semenov, M.A.; Stratonovitch, P. Use of multi-model ensembles from global climate models for assessment of climate change impacts. *Clim. Res.* **2010**, *41*, 1–14. [CrossRef]
47. Bian, H.; Lü, H.; Sadeghi, A.M.; Zhu, Y.; Yu, Z.; Ouyang, F.; Su, J.; Chen, R. Assessment on the Effect of Climate Change on Streamflow in the Source Region of the Yangtze River, China. *Water* **2017**, *9*, 70.
48. Kumar, D.; Arya, D.S.; Murumkar, A.R.; Rahman, M.M. Impact of climate change on rainfall in Northwestern Bangladesh using multi-GCM ensembles. *J. Climatol.* **2014**, *34*, 1395–1404. [CrossRef]
49. King, L.M.M.; Irwin, S.; Sarwar, R.; McLeod, A.I.A.; Simonovic, S.P.P. The Effects of Climate Change on Extreme Precipitation Events in the Upper Thames River Basin: A Comparison of Downscaling Approaches. *Can. Water Resour. J. Rev. Can. Ressour. Hydr.* **2012**, *37*, 253–274. Available online: https://www.tandfonline.com/doi/full/10.4296/cwrj2011-938 (accessed on 11 October 2020). [CrossRef]

50. Lu, Y.; Qin, X.S.; Mandapaka, P.V. A combined weather generator and K-nearest-neighbour approach for assessing climate change impact on regional rainfall extremes. *Int. J. Climatol.* **2015**, *35*, 4493–4508. [CrossRef]
51. Khoi, D.N.; Sam, T.T.; Nhi, P.T.T.; Quan, N.T.; Hung, B.V.; Phung, N.K.; Van Thinh, N. Uncertainty assessment for climate change impact on streamflow and water quality in the dong nai river basin, vietnam. In *World Environmental and Water Resources Congress*; American Society of Civil Engineers: Reston, VA, USA, 2019; pp. 366–373. Available online: http://ascelibrary.org/doi/10.1061/9780784482346.037 (accessed on 11 October 2020).
52. Semenov, M.A. Simulation of extreme weather events by a stochastic weather generator. *Clim. Res.* **2008**, *35*, 203–212. [CrossRef]
53. Semenov, M.A.; Stratonovitch, P. Adapting wheat ideotypes for climate change: Accounting for uncertainties in CMIP5 climate projections. *Clim. Res.* **2015**, *65*, 123–139. [CrossRef]
54. Sam, T.T.; Khoi, D.N.; Thao, N.T.T.; Nhi, P.T.T.; Quan, N.T.; Hoan, N.X.; Nguyen, V.T. Impact of climate change on meteorological, hydrological and agricultural droughts in the Lower Mekong River Basin: A case study of the Srepok Basin, Vietnam. *Water Environ. J.* **2018**. [CrossRef]
55. Willmott, C.J. On the validation of models. *Phys. Geogr.* **1981**, *2*, 184–194. [CrossRef]
56. Moriasi, D.N.; Gitau, M.W.; Pai, N.; Daggupati, P. Hydrologic and Water Quality Models: Performance Measures and Evaluation Criteria. *Trans. ASABE* **2015**, *58*, 1763–1785.
57. Legates, D.R.; McCabe, G.J. Evaluating the use of "goodness-of-fit" Measures in hydrologic and hydroclimatic model validation. *Water Resour. Res.* **1999**, *35*, 233–241. [CrossRef]
58. Hadi, S.J.; Tombul, M. Long-term spatiotemporal trend analysis of precipitation and temperature over Turkey. *Meteorol. Appl.* **2018**, *25*, 445–455. [CrossRef]
59. MONRE. *Climate Change and Sea Level Rise Scenarios for Vietnam: Summary for Policymakers*; Ministry of Natural Resources and Environment: Vietnam, Asia, 2016; p. 39.
60. Ngo-Duc, T.; Tangang, F.T.; Santisirisomboon, J.; Cruz, F.; Trinh-Tuan, L.; Nguyen-Xuan, T.; Phan-Van, T.; Juneng, L.; Narisma, G.; Singhruck, P.; et al. Performance evaluation of RegCM4 in simulating extreme rainfall and temperature indices over the CORDEX-Southeast Asia region. *Int. J. Climatol.* **2017**, *37*, 1634–1647. [CrossRef]
61. Trinh-Tuan, L.; Matsumoto, J.; Tangang, F.; Juneng, L.; Cruz, F.; Narisma, G.; Santisirisomboon, J.; Phan-Van, T.; Gunawan, D.; Aldrian, E.; et al. Application of quantile mapping bias correction for mid-future precipitation projections over Vietnam. *SOLA* **2019**, *15*, 1–6. [CrossRef]
62. Tangang, F.; Supari, S.; Chung, J.X.; Cruz, F.; Salimun, E.; Ngai, S.T.; Juneng, L.; Santisirisomboon, J.; Santisirisomboon, J.; Ngo-Duc, T.; et al. Future changes in annual precipitation extremes over Southeast Asia under global warming of 2 °C. *APN Sci. Bull.* **2018**, *8*, 3–8. [CrossRef]
63. Guo, D.; Zheng, F.; Gupta, H.; Maier, H.R. An inverse approach to perturb historical rainfall data for scenario-neutral climate impact studies. *J. Hydrol.* **2018**, *556*, 877–890. [CrossRef]
64. Jianga, C.; Xionga, L.; Wangb, D.; Liua, P.; Guoa, S.; Xua, C.Y. Separating the impacts of climate change and human activities on runoff using the Budyko-type equations with time-varying parameters. *J. Hydrol.* **2015**, *522*, 326–338. [CrossRef]
65. Kim, H.; David, W.; Marcouillerb, D.W.; Woosnam, K.M. Rescaling social dynamics in climate change: The implications of cumulative exposure, climate justice, and community resilience. *Geoforum* **2018**, *96*, 129–140. [CrossRef]

water

MDPI

Article

Climatic Controls on Mean and Extreme Streamflow Changes Across the Permafrost Region of Canada

Rajesh R. Shrestha *, Jennifer Pesklevits, Daqing Yang, Daniel L. Peters and Yonas B. Dibike

Watershed Hydrology and Ecology Research Division, Environment and Climate Change Canada, University of Victoria, 2472 Arbutus Rd., Victoria, BC V8N 1V8, Canada; jennifer.pesklevits@canada.ca (J.P.); daqing.yang@canada.ca (D.Y.); daniel.peters@Canada.ca (D.L.P.); yonas.dibike@canada.ca (Y.B.D.)
* Correspondence: rajesh.shrestha@canada.ca

Abstract: Climatic change is affecting streamflow regimes of the permafrost region, altering mean and extreme streamflow conditions. In this study, we analyzed historical trends in annual mean flow (Q_{mean}), minimum flow (Q_{min}), maximum flow (Q_{max}) and Q_{max} timing across 84 hydrometric stations in the permafrost region of Canada. Furthermore, we related streamflow trends with temperature and precipitation trends, and used a multiple linear regression (MLR) framework to evaluate climatic controls on streamflow components. The results revealed spatially varied trends across the region, with significantly increasing (at 10% level) Q_{min} for 43% of stations as the most prominent trend, and a relatively smaller number of stations with significant Q_{mean}, Q_{max} and Q_{max} timing trends. Temperatures over both the cold and warm seasons showed significant warming for >70% of basin areas upstream of the hydrometric stations, while precipitation exhibited increases for >15% of the basins. Comparisons of the 1976 to 2005 basin-averaged climatological means of streamflow variables with precipitation and temperature revealed a positive correlation between Q_{mean} and seasonal precipitation, and a negative correlation between Q_{mean} and seasonal temperature. The basin-averaged streamflow, precipitation and temperature trends showed weak correlations that included a positive correlation between Q_{min} and October to March precipitation trends, and negative correlations of Q_{max} timing with October to March and April to September temperature trends. The MLR-based variable importance analysis revealed the dominant controls of precipitation on Q_{mean} and Q_{max}, and temperature on Q_{min}. Overall, this study contributes towards an enhanced understanding of ongoing changes in streamflow regimes and their climatic controls across the Canadian permafrost region, which could be generalized for the broader pan-Arctic regions.

Keywords: climatic controls; multiple linear regression; permafrost region; streamflow extremes; trend analysis; variable importance analysis

Citation: Shrestha, R.R.; Pesklevits, J.; Yang, D.; Peters, D.L.; Dibike, Y.B. Climatic Controls on Mean and Extreme Streamflow Changes Across the Permafrost Region of Canada. *Water* **2021**, *13*, 626. https://doi.org/10.3390/w13050626

Academic Editor: Enrique Morán Tejeda

Received: 22 January 2021
Accepted: 23 February 2021
Published: 27 February 2021

1. Introduction

The Arctic and subarctic regions of the world—mostly underlain by continuous/discontinuous permafrost—are affected by a range of extreme streamflow conditions, which include low to no flows in winter under ice-cover; spring floods due to river-ice breakup and snowmelt; snowmelt driven peak flows during spring/early summer; and in rare instances, rainfall driven peak flows in late summer. These extreme conditions are influenced by various climatological and landscape drivers and controls, and changes in these factors affect the magnitude, timing and duration of such events. Particularly, enhanced warming in the northern latitudes (which is almost twice the global average temperature increase [1–3]) and amplified poleward moisture transport to the region [4,5] (which propagates into increased precipitation) are affecting different components of the regional hydro-climatic systems. Prominent changes include reductions in the magnitude, duration and extent of snow cover [6–8], enhanced permafrost thaw [9,10] and changes in the snowfall–rainfall balance [11], all with the potential to alter the mean and extreme streamflow conditions across the permafrost region.

For instance, while warmer temperatures lead to an earlier snowmelt-driven peak flow [12,13], the magnitude of peak flow is mainly determined by the precipitation amount [14]. Furthermore, increased winter precipitation implies greater water input and larger snowpack storage in the basin, and thus, the potential for higher peak flows. In contrast, warmer temperatures tend to moderate peak flows by reducing the fraction of winter precipitation falling as snow and shortening the snow accumulation period [15]. On the other hand, spring rainfall and rain-on-snow events—which appear to be increasing in the recent years [16–19]—could enhance peak flow events. At the same time, permafrost degradation associated with the warming temperatures could lead to increases in subsurface flow pathways, soil moisture volume and groundwater recharge [20,21], which could lead to winter low flow increases. Additionally, permafrost thawing augments a basin's water storage capacity, and could lead to decreases in the annual maximum discharge [22] as well as the maximum-to-minimum discharge ratio [23].

Changes in the mean streamflow state and extremes could make areas within permafrost region vulnerable. For instance, although current infrastructure in the permafrost region is limited, the development (e.g., hydroelectric water resources) is progressing at a rapid pace [24]. Thus, infrastructure design and operation need to take into account the changes in extremes. For instance, while increased low flow could augment the water supply in the winter, the potential increase in peak flow could make the built hydraulic structures inadequate, and result in more frequent and widespread flooding. Higher peak flows and flooding could also threaten infrastructure and operational services, such as mines, roads, railways, bridges and settlements located nearby [25].

Given these potential impacts, it is important to enhance our understanding of changes in streamflow regimes in the permafrost region beyond the magnitude and direction of trends. For instance, understanding the role of climatic drivers on streamflow changes could provide valuable insights, not only for process understanding, but also to diagnose future directions of change. In this context, while a number of previous studies have analyzed pan-Arctic wide streamflow trends [26–28], to our knowledge, neither the analyses of streamflow trends across the entire permafrost region of Canada nor evaluation of streamflow trends in relation to climatic drivers (i.e., precipitation and temperature) have been undertaken.

In this study, we address the aforementioned knowledge gaps first by analyzing trends in key streamflow variables across the continental-scale permafrost region of Canada. Secondly, we conduct a novel analysis of climatic controls on streamflow changes by relating spatial variability and trends in streamflow, air temperature and precipitation in a statistical framework. We also use a multiple linear regression framework (MLR) to evaluate the temporal relationships of seasonal precipitation and temperature with mean and extreme streamflow conditions for selected rivers. We focus on select streamflow variables including annual mean flow, minimum flow, and maximum flow and its timing, and assess the streamflow trends in the context of pan-Arctic wide trends. Further, we discuss the potential future changes in mean and extreme streamflow conditions.

2. Data and Methods

2.1. Streamflow Data

We used the permafrost database based on Heginbottom et al. [29] and published by Natural Resources Canada [30] as a base map, which consists of a number of permafrost zones (continuous, extensive, sporadic, isolated), for the selection of streamflow stations. We selected 83 hydrometric stations that lie within these zones from the Water Survey of Canada hydrometric station network [31], with the criteria of basin area > 2000 km^2 and year-round observations. Additionally, a United States Geological Survey station (15356000 Yukon River at Eagle AK [32]) was included because only a limited number of years (1983 to 2017) of data are available for the corresponding station in the Canadian side (09ED001).

The selected hydrometric stations are unregulated (natural) except for the Mackenzie River affected by the upstream Peace River regulation, whose effect can be assumed relatively minor on selected three downstream mainstem stations [33]. We extracted four streamflow variables from the daily time series data: (i) annual mean flow (Q_{mean}), (ii) annual minimum flow (Q_{min}), (iii) annual maximum flow (Q_{max}), and (iv) timing of annual maximum flow (Q_{max} timing). For each of the four streamflow variables, the stations with more than 40 years of records were first selected, out of which stations with >30% missing values were discarded. This left 83, 83, 79, and 78 stations for Q_{max}, Q_{max} timing, Q_{min} and Q_{mean}, respectively. Additionally, the datasets cover different periods between 1945 and 2018, with missing values in different years, while some stationed have been discontinued. However, further constraining the station selection over a consistent period covering > 30 years of record would result in elimination of over a third of stations, and a lack of representative stations over some regions. Thus, as in the case of St. Jacques and Sauchyn [34], varying periods of records were used and trend analysis results provide a general direction of historical changes rather than changes over a specific time period.

2.2. Climate Data

We used two observation-based gridded climate data products consisting of the 1945 to 2012 Pacific Northwest North American meteorological (PNWNAmet) data [35] for western Canada and the 1950 to 2016 Natural Resources Canada meteorological (NRCANmet) data [36] for rest of Canada. Although NRCANmet is available for the entirety Canada, PNWNAmet was preferred for western Canada (where it is available) as it showed improved performance for climate means, extremes and variability compared to NRCANmet [35]. Both datasets consist of daily precipitation, maximum and minimum temperature, with spatial resolutions of ~7 km (0.0625 degrees) and ~10 km (0.0833 degrees) at 60 °N latitude, respectively, for PNWNAmet and NRCANmet. From the two datasets, basin averaged precipitation and temperature were extracted using the National hydrometric network basin polygons [37] for the respective hydrometric stations. For comparison with the streamflow variables, the time periods of precipitation and temperature data were matched with the streamflow records of each station. Given the delayed snowmelt-driven streamflow response in the region, water years (October to September) were considered for all analyses.

2.3. Analyse

Trends in streamflow, precipitation and temperature variables were analyzed using the R "zyp" package [38] by employing a non-parametric Mann–Kendall test [39] with iterative pre-whitening [40]. The iterative pre-whitening method, employed to remove the effects of serial correlation, was used because it has been found to be robust in handling Type-I error and produce conservative estimates of p-values, thus limiting false trend detection [41]. Besides accounting for serial correlations for each variable and at each individual station, the effect of spatial correlation among stations (field significance) was evaluated by using the false discovery rate technique suggested by Wilks [42]. Results of the analyses are presented for 10% ($p < 0.10$) and 5% ($p < 0.05$) significance levels, with the former considered as the level of statistically significant trend, while the latter signifies a stronger trend.

We analyzed the spatial relationships amongst climatological means of the temperature, precipitation and streamflow variables as well as their trends using the Kendall's τ correlation. This method was preferred to the more common Pearson correlation because Kendall's τ uses a rank-based procedure, is more resistant to outliers, and measures all monotonic (linear and nonlinear) correlations [43]. All streamflow values and trend values were normalized by their respective drainage areas when comparing them with precipitation and temperature values/trends.

It should be noted that, in most cases, the drainage areas of most hydrometric stations extend beyond a particular permafrost zone, and in some cases, to the permafrost-free zone.

Therefore, analyses specific to different permafrost zones were not undertaken. However, we considered trends for seven (out of 11) of the Water Survey of Canada's drainage regions [37], which include: (3) Labrador, Northwestern Quebec; (4) Albany; (5) Nelson; (6) Arctic, Hudson Bay; (7) Upper Mackenzie; (8–9) Pacific and Yukon; and (10) Arctic, Lower Mackenzie and Queen. Region 8 has only one station and its results were combined with region 9.

For selected stations, we characterized temporal sensitivities of streamflow variables to climatic variables using the MLR framework. Precipitation, temperature and their interaction were considered as the driving variables and streamflow components were considered as the response variables, given by:

$$\Delta Q = \beta_1 + \beta_2 \Delta P + \beta_3 \Delta T + \beta_4 (\Delta P \Delta T) \quad (1)$$

where β_1, β_2, β_3 and β_4 are the coefficients of MLR. ΔQ, ΔP and ΔT are the anomalies of streamflow, precipitation and temperature variables with respect to their mean values over the period 1976 to 2005, with ΔQ and ΔP considered as the percentage change and ΔT considered as the absolute change. $\Delta P \Delta T$ represents the interactions between precipitation and temperature. Lehner et al. [44] and Chegwidden et al. [45] used a similar concept in their evaluations using MLR models, although the latter did not use the interaction term. Further, we used the R "Caret" package [46] variable importance (VI) metric, which is based on the absolute t-statistic of each model coefficient, to assess the relative sensitivity of each input variable with respect to the output variable of the MLR model.

3. Results and Discussion

3.1. Spatial Variations of Streamflow Trends

Figure 1 depicts the trends in Q_{mean}, Q_{min} and Q_{max} along with the Q_{max} timing at 84 hydrometric stations across the permafrost region of Canada. The results generally corroborate with previous studies of six large Arctic flowing rivers (Appendix A Table A1), suggesting trends in the Canadian permafrost region are consistent with those across the pan-Arctic. Specifically, similar to the trends across the pan-Arctic, there are larger fractions of stations with increasing Q_{mean} (62%) and Q_{min} (68%) trends than decreases, although some of the trends are not significant. The larger fraction of stations (44%) with significant increases ($p < 0.10$) in Q_{min} is also consistent with the previous studies. Overall, Q_{min} increases represent the most prominent trend in this study, with 38% of trends field significant (Table 1); thus, most trends are globally significant.

Table 1. Number of stations with significant trends ($p < 0.1$) using the Mann–Kendall test with iterative pre-whitening, and the number of field significant trends with false detection rate method.

Variable	Significant Increasing/Decreasing Trends/Total Stations	Field significant Increasing/Decreasing Trends/Total Stations
Q_{mean}	12/5/79	6/0/79
Q_{min}	35/2/79	30/1/79
Q_{max}	9/7/83	3/1/83
Q_{max} timing	8/7/83	0/0/83
Q_{max}/Q_{min}	2/33/79	1/19/79
O–M_T	67/0/84	63/0/84
A–S_T	60/0/84	55/0/84
O–M_P	23/3/84	11/0/84
A–S_P	13/1/84	2/0/84

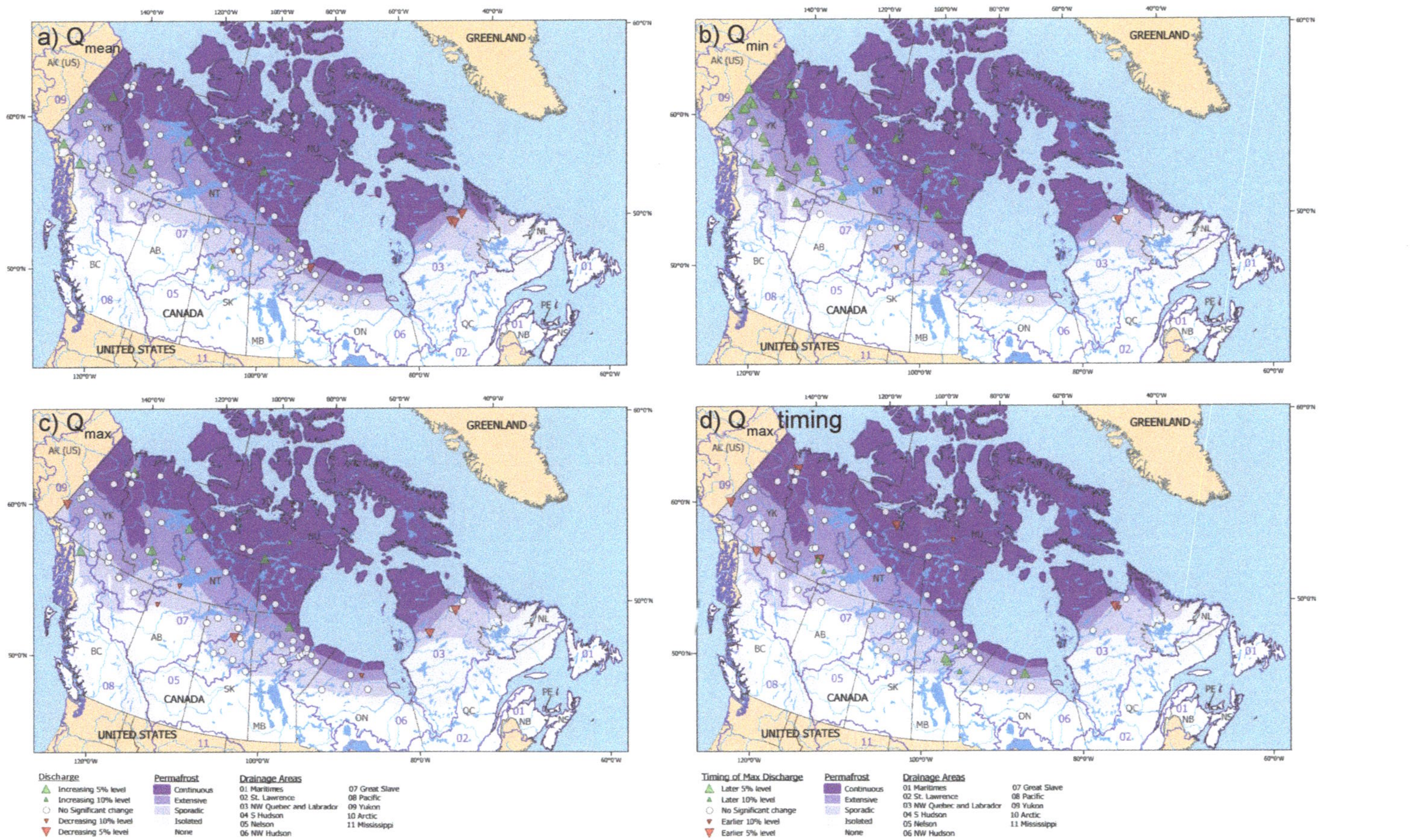

Figure 1. Historical trends in streamflow variables for the permafrost region of Canada: (**a**) Mean flow (Q_{mean}); (**b**) Minimum flow (Q_{min}); (**c**) maximum flow (Q_{max}) and (**d**) Q_{max} timing. Significant increasing and decreasing trends at 5% and 10% levels are shown.

Most stations with increasing Q_{mean} trends in this study are located in north-western Canada, while the southern and eastern regions show no trends, or even a few decreasing trends. Likewise, while there are few stations with significant trends in Q_{max} and Q_{max} timing in north-western Canada, the majority of stations in other regions show no significant trends. Furthermore, number of stations with field significant trends are even smaller, implying the lack of globally significant trends in these variables. This is especially the case for Q_{max} timing, for which none of the trends are field significant, thus, these trends are practically insignificant. Nevertheless, the effect of increasing Q_{min} is reflected in declining Q_{max} to Q_{min} ratios for some of the stations (Table 1). Specifically, 33 out of 79 stations exhibit significant downward trends, with field significant trends for 19 stations.

Considering the trends for the seven drainage regions of Canada, it can be seen that the bulk of the stations are located in western Canada, with only a limited number of stations in eastern Canada (Table 2). Given the limited number of stations, field significance testing was not done at the regional level. Considering the temporal trends, regions 8, 9 and 10 have larger fractions of stations with trends in Q_{min} and Q_{max}/Q_{min}, while regions 3, 4, 5, 6 and 7 have smaller fractions of stations with trends. Some of the trends in region 3 are opposite to other regions, e.g., while there are declining Q_{mean} and Q_{max} trends in region 3, the declining trends are less prevalent in other regions. However, given the limited number of stations, generalization/extrapolation of patterns for these data-poor regions (regions 3 to 5) is considered problematic. There are also regional differences in western Canada; e.g., in the upper Mackenzie (region 7), there is a larger fraction of stations with increasing Q_{min} trends compared to the lower Mackenzie (region 10).

Table 2. Number of stations in seven regions with significant trends ($p < 0.1$) using the Mann–Kendall test with iterative pre-whitening. Summarized trends include significant increasing/ decreasing trends/ total stations.

Variable	Region 3	Region 4	Region 5	Region 6	Region 7	Region 8 & 9	Region 10
Q_{mean}	0/3/5	0/0/5	0/0/6	4/1/15	0/0/9	4/0/14	2/1/25
Q_{min}	0/1/5	0/0/6	2/0/7	4/1/15	2/0/9	11/0/13	14/1/24
Q_{max}	0/2/5	0/1/5	0/0/7	2/1/15	1/2/9	1/1/14	4/1/28
Q_{max} timing	0/0/5	2/0/5	3/0/7	1/0/15	0/0/9	0/2/14	2/5/28
Q_{max}/Q_{min}	1/3/5	0/0/6	0/2/7	1/1/15	0/6/9	0/10/13	2/9/24
O–M_T	4/0/5	4/0/6	4/0/7	10/0/15	8/0/9	10/0/14	27/0/28
A–S_T	3/0/5	6/0/6	5/0/7	11/0/15	5/0/9	9/0/14	21/0/28
O–M_P	0/0/5	1/0/6	2/0/7	5/0/15	3/0/9	0/0/14	12/3/28
A–S_P	0/0/5	0/0/6	0/0/7	0/0/15	3/0/9	0/0/14	10/1/28

Seven regions are defined according to the Water Survey of Canada drainage regions, which include: (3) Labrador, Northwestern Quebec; (4) Albany; (5) Nelson; (6) Arctic, Hudson Bay; (7) Upper Mackenzie; (8) and (9) Pacific and Yukon; and (10) Arctic, Lower Mackenzie and Queen.

Regions 9 and 10 consist of the Yukon and Mackenzie rivers, respectively, along with their tributaries. The streamflow trends at the most downstream stations and upstream stations are generally consistent. Specifically, the significantly increasing Q_{min} trend in the Yukon Eagle station (1951 to 2018) is consistent with 9 out of 11 upstream stations. The Yukon Eagle station does not have significant trend in Q_{mean}, Q_{max} and Q_{max} timing, which is in general agreement with the lack of trends in 8 (12), 10 (12) and 10 (12) upstream stations, respectively (numbers in the parentheses are the total number of stations). Similar results were obtained when comparing the 1973 to 2016 trends for the most downstream station on the Mackenzie River (Mackenzie River at Arctic Red River) with upstream stations. Specifically, the absence of significant trends in Q_{mean}, Q_{max} and Q_{max} timing are in agreement with 14 (15), 14 (18) and 15 (18) upstream stations, respectively. However, the downstream station exhibits no significant trend in Q_{min}, while 11 (15) upstream stations show increasing trends. It is also noteworthy that for the Mackenzie River streamflow variables, the lack of significant trends over 1973 to 2016 are in agreement with Yang et al. [47] over 1973 to 2011. For the Yukon Eagle station, significant increases in Q_{min} and insignificant change in Q_{max} for the period 1951 to 2018 agree with Bennett et al. [48], in which the 1954/1964 to 2013 periods were used. However, insignificant change in the Yukon Eagle Q_{mean} for the period 1951 to 2018 differs from the significant increasing trend for the period 1975 to 2016 according to Box et al. [27].

Given that the Q_{min} increase is the most prominent trend in this study, as well as other studies across the pan-Arctic (see Appendix A), an important question arises on its link to permafrost degradation. In this respect, while a majority of stations in the discontinuous permafrost zones in western Canada show increasing Q_{min} trends, only a few stations in the discontinuous permafrost zones in Saskatchewan, Manitoba and Ontario show increasing Q_{min} (Figure 1). Furthermore, Smith et al. [21] found greater Q_{min} increases in the non-permafrost zone than permafrost zone of Eurasia and speculated that decreased seasonal freezing of soils caused by warmer winters and/or deeper snowpack may be causing the Q_{min} increases. Such processes could also be contributing to Q_{min} increases in this study, and consequently, the decline of the Q_{max}-to-Q_{min} ratio.

3.2. Spatial Streamflow and Climatic Trends Relationships

Figure 2 shows the results of the Mann–Kendall trend analyses for the cold season (October to March, i.e., O–M) and warm season (April to September, i.e., A–S) temperature (T) and precipitation (P), with both calculated as the basin-average values for the areas that drain to hydrometric stations. The results show distinct patterns in temperature and precipitation changes. While the majority of O–M_T and A–S_T show increasing trends for the entire permafrost region, a smaller fraction of O–M_P and A–S_P show increasing trends. The differences become more distinct when the trends for the seven regions are considered (Figure 2, Table 2). Specifically, while the stations with temperature increases are distributed across all regions, the bulk of stations with increasing precipitation trends are located in region 10 followed by region 7. Some stations in regions 5 and 6 also show an increasing trend in O–M_P.

Next, we present correlations among the 1976 to 2005 climatological means of streamflow components and driving precipitation and temperature (Figure 3a), with Q_{mean}, Q_{min} and Q_{max} values normalized by the basin areas (hereafter referred to as the q_{mean}, q_{min} and q_{max}). The correlations specify the strength of spatial relationships, such as the significant positive correlations among q_{mean}, q_{min} and q_{max} suggest that the basins with higher values in one flow component are also likely to have higher values in the other two components. Furthermore, as precipitation is the main driver of annual flow, q_{mean} for the basins are positively correlated with both O–M_P and A–S_P. The negative correlation of q_{mean} with A–S_T is likely due to higher evapotranspiration losses in warmer basins. q_{max} is negatively correlated with its timing, which is probably due to the wet-to-dry precipitation gradient, and warm-to-cold temperature gradient from south to north in the permafrost region [49]. Thus, the warmer southern basins are likely to produce higher q_{max}, and earlier timing than the colder northern basins. On the other hand, the negative correlation between q_{min} and q_{max} could be linked to snowpack and baseflow generation. For instance, the relatively colder basins are likely to have lower baseflow and larger snowpack than warmer basins with a similar precipitation amount, thus producing relatively smaller q_{min} and larger q_{max} in colder basins. Besides the spatial variability of precipitation and temperature, the seasonality also plays a role on streamflow responses. Specifically, the positive correlations of q_{min} and q_{max} with O–M_P are in line with the expectation that the basins with higher winter precipitation also produce higher flow rates. q_{min} also has a positive correlation with O–M_T. Therefore, the higher the winter temperature in the basin, the higher the low-flow response, likely due to higher rainfall-to-total-precipitation ratio and higher subsurface flow. Furthermore, the negative correlation of q_{max} with A–S_T is likely related to spring snow accumulation and melt, which is constrained by spring temperature; thus, the higher the A–S_T, the earlier the snowmelt and the smaller the q_{max}. However, although warmer basins are generally expected to have earlier Q_{max} timings, there is no significant correlation between Q_{max} timing and A–S_T.

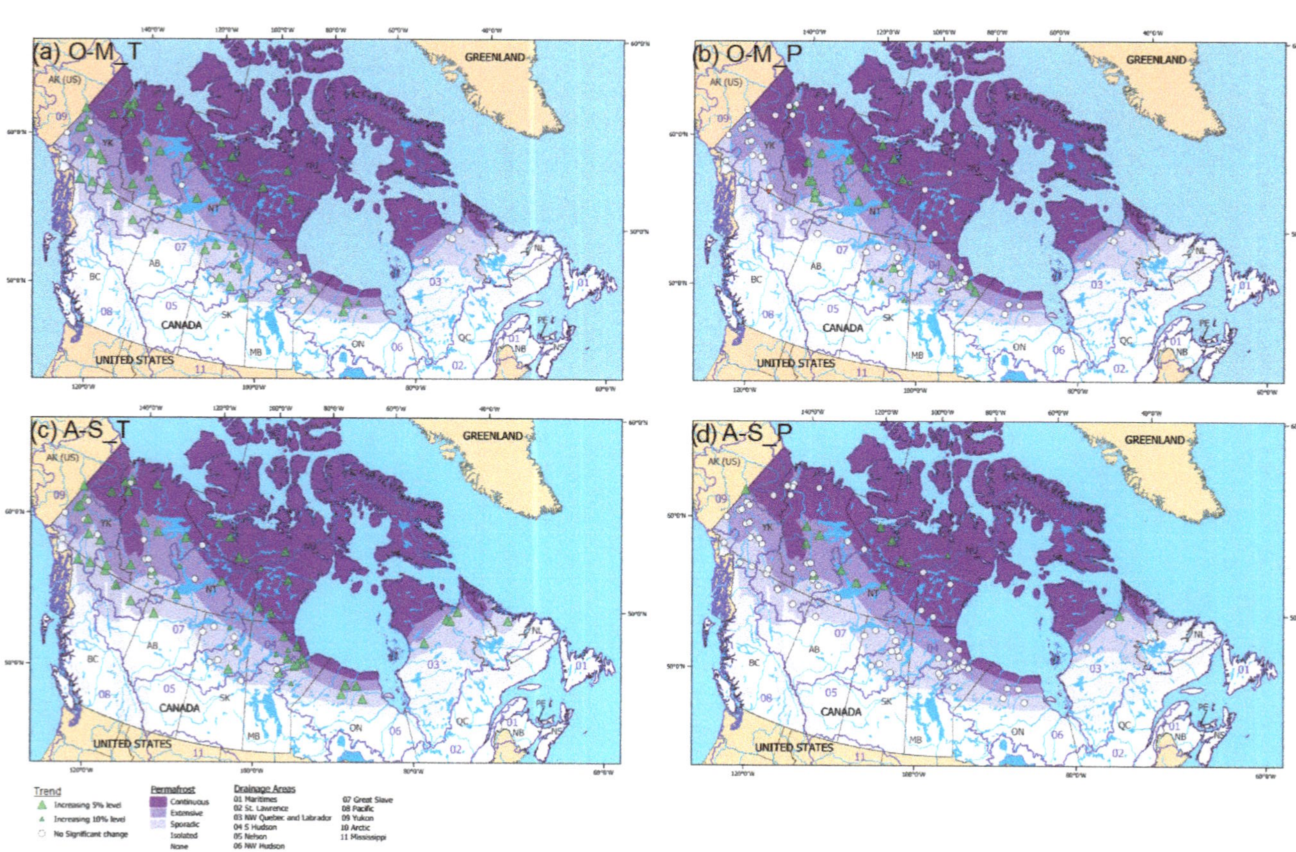

Figure 2. Historical trends in basin-averaged temperature and precipitation for the permafrost region of Canada: (**a**) October–March mean temperature (O–M_T); (**b**) October–March total precipitation (O–M_P); (**c**) April–September mean temperature (A–S_T); (**d**) April–September total precipitation (A–S_P). Significant increasing and decreasing trends at 5 and 10% levels are shown.

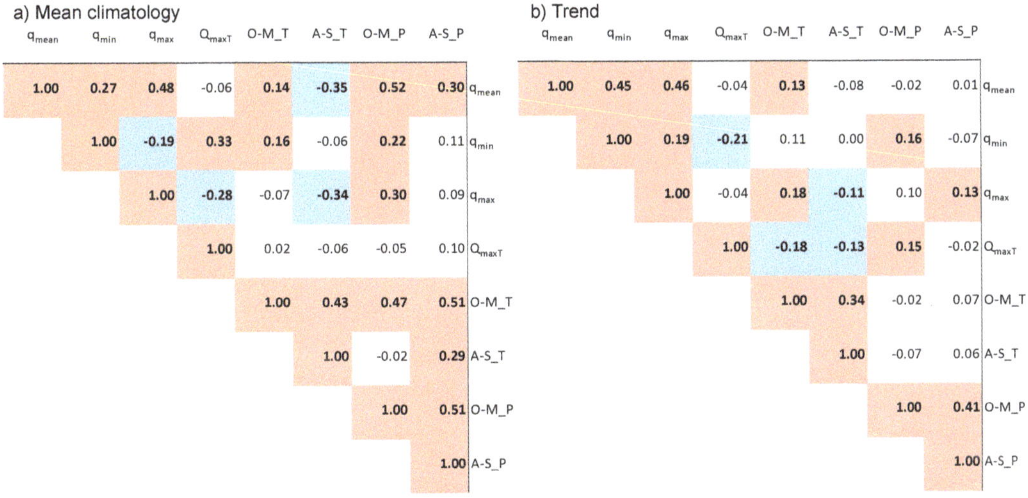

Figure 3. Correlogram of climatic and streamflow variables: (**a**) 1976 to 2005 mean climatology; (**b**) trends for available period. Both climatic and streamflow variables are averaged over the basins. The numbers on the plots are Kendall's τ correlations, with significant correlations ($p < 0.1$) indicated by bold numbers.

The correlations among streamflow and climate variable trend values (Sen's slope) signify the relationships in the magnitude and direction of changes (Figure 3b). The correlogram reveals significant but mostly weak correlations (Kendall's $\tau < 0.2$) for some of the variables. The positive correlations amongst q_{mean}, q_{min} and q_{max} trends imply that the increase in one of the streamflow variables is likely accompanied by increases in other variables, although the trends for individual stations may not always be significant. With reference to the driving variables, q_{max} and q_{mean} trends show weak positive correlations with O–M_T trends. Such unidirectional flow and temperature trends appear counterintuitive, as an increased temperature can normally be expected to lead to decreased snowpack and consequently decreased flows. In this case, other factors that accompanied increasing temperature trends may have played a role, such as precipitation trends, although the correlations are not significant. The positive correlation of q_{min} with O–M_P trends indicate increasing low-flow with increasing winter precipitation. The significant negative correlations of Q_{max} timing trends with A–S_T and O–M_T trends are in line with the expectation of earlier timing with increasing temperature, which cause smaller snowpack and earlier snowmelt. Furthermore, the negative correlations between q_{min} and Q_{max} timing trends are likely linked to temperature and snowpack changes, because increasing winter temperatures lead to larger q_{min} and smaller snowpack, with earlier snowmelt leading to earlier Q_{max} timing. However, the expected linkage between q_{min} increases with temperature increases were not found to be significant.

Furthermore, a closer examination of the temperature, precipitation and streamflow changes reveals expected patterns. For instance, 31 of the 35 significantly increasing q_{min} trends coincide with the O–M_T increasing trend. However, only a few q_{mean} and q_{max} results showed significant trends, and these trends do not always line up with the significant O–M_P and A–S_P trends. A number of factors affect the presence/absence of the relationships between streamflow and climate variable trends. Firstly, streamflow trends reflect the cumulative effects of precipitation, temperature and snow trends. Therefore, trends in one of the driving variables (temperature or precipitation) may not lead to trends in streamflow variables. Secondly, given the complex interactions between the climatic drivers, the trend values may not change monotonically for different variables, thus resulting in weak or no correlations with the streamflow trends.

There are also a number of sources of uncertainty that could lead to discrepancies and inconsistencies. A major source of uncertainty is the representativeness of the precipitation and temperature in the PNWMAmet and NRCanmet dataset, particularly due to the sparse station network in the high-latitude region [50]. Additionally, uncertainties arise from the quality of the observed discharge data, especially during the ice-covered low-flow period [51], as well as missing values in the discharge data. There are also inconsistencies in the length of the records. Given that some of the Q_{min}, Q_{mean}, Q_{max} and Q_{max} timing records have gaps, the compared lengths are not always consistent with temperature and precipitation records. Nevertheless, the analyses of climatological means and trends of the driving and resultant variables generally provide physically plausible explanations.

3.3. Climatic Controls on Streamflow Variables

We focused our temporal analyses on stations with the longest available flow records, which included: (i) 1947 to 2012 (with few years of missing records in between) period for the Liard Upper Crossing (Liard-UC) and (ii) 1951 to 2012 (complete) period for the Yukon Eagle (Yukon-E), Alaska stations (Appendix A Figure A1). Note that the flow records were trimmed at the year 2012 to match the precipitation and temperature records in the PNWNAmet datasets. Both the Liard-UC and Yukon-E stations flow records have a significantly increasing Q_{min} trend ($p < 0.05$, Liard-UC: slope 5.1%/decade; Yukon-E: slope 4.1%/decade). Q_{mean} and Q_{max} trends for both stations are not significant at $p < 0.1$. In the case of basin-averaged temperature, both A–S_T and O–M_T are increasing significantly ($p < 0.05$) for Liard-LC (decadal trends, A–S_T: 0.2 °C, O–M_T: 0.4 °C) and Yukon-E (decadal trends, A–S_T: 0.3 °C, O–M_T: 0.5 °C). There were no significant precipitation trends (for both A–S_P and O–M_P) for either basin.

The sensitivities of precipitation and temperature changes on Q_{mean}, Q_{min} and Q_{max} anomalies relative to 1976 to 2005 together with the results of MLR analyses are shown in Figure 4. Note that Q_{max} timing results are not shown because of the overall lack of relationships with climatic drivers. The relative controls of precipitation and temperature on the three streamflow variables are depicted in terms of VI scores, with R^2 summarizing the MLR model fits. We also explored different combinations of the precipitation and temperature seasons in the MLR model setup. The results that yielded the best R^2 values were depicted, which included annual values (Ann_P and Ann_T) for Q_{mean}, October to March values (O–M_P and O–M_T) for Q_{min}, and October to July (O–J_P and O–J_T) values for Q_{max} for both stations. These best combinations of months are in agreement with the analyses of climatic controls on these three streamflow variables using the historical and projected future temperature and precipitation for the Liard River basin [14]. The relative controls of temperature, precipitation and their interaction term are depicted by VI_1, VI_2 and VI_3, respectively.

The MLR model results for Q_{mean} indicate the dominant controls of Ann_P ($VI_2 > 0.75$) for both basins (Figure 4a,b). While Ann_T has a small influence on Q_{mean} for Liard-LC, its influence for Yukon-E is negligible. The controls of precipitation–temperature interactions on Q_{mean} for the two basins are non-existent to negligible. Furthermore, as illustrated by the stratified patterns of Q_{mean} anomalies relative to Ann_P anomalies, the sensitivities of Ann_P on Q_{mean} are positive for both basins; i.e., the larger the Ann_P, the larger the Q_{mean}. Overall, the model performance for Q_{mean} is good for both basins ($R^2 > 0.7$), i.e., Ann_P and Ann_T anomalies can explain over 70% of variance in Q_{mean} anomalies.

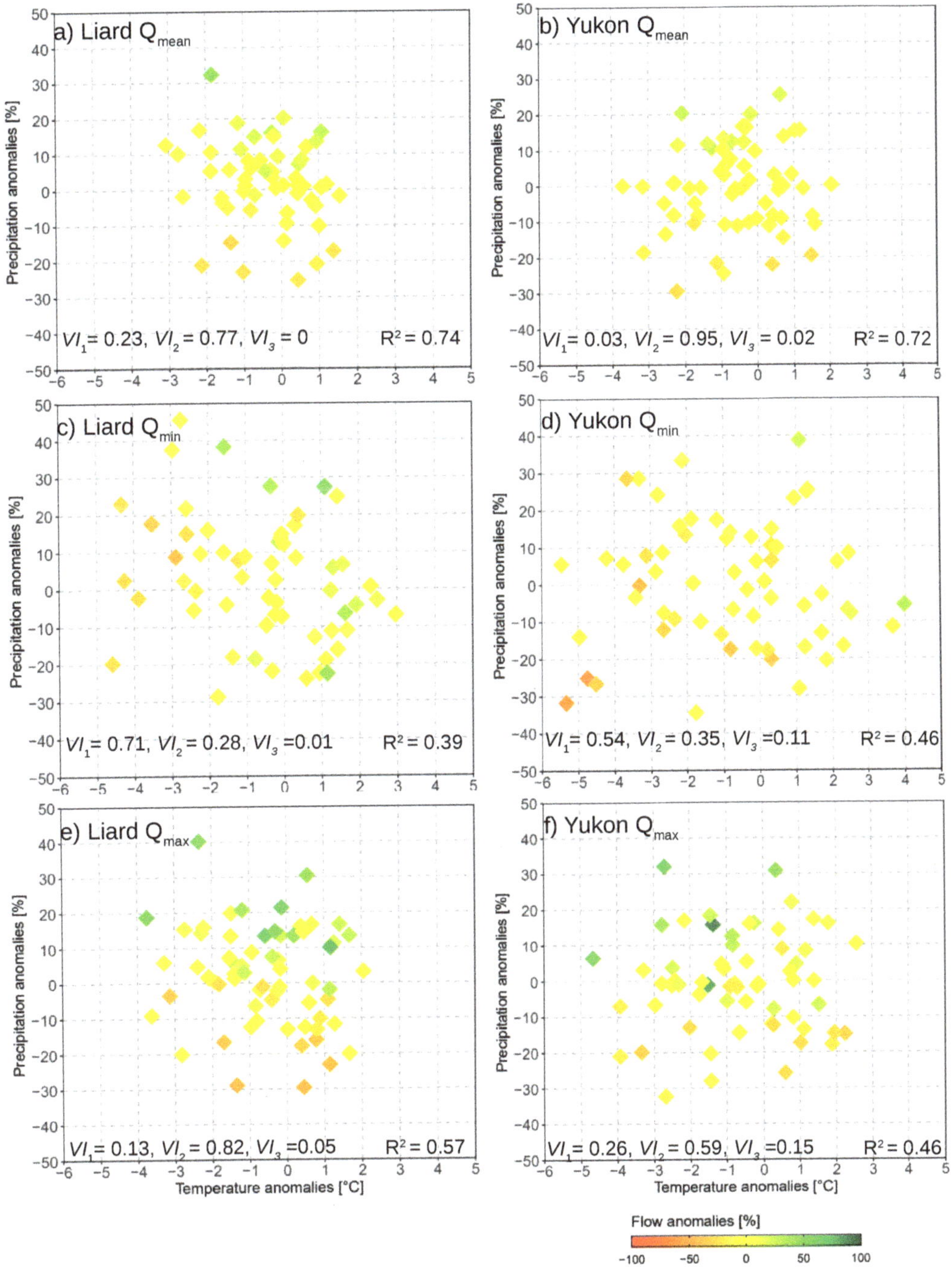

Figure 4. Sensitivity plots of streamflow anomalies against annual/seasonal temperature (°C) and precipitation (%) anomalies with respect to 1976–2005 period. Illustrated results for Liard and Yukon Rivers include: (**a**,**b**) Q_{mean} against Ann_T and Ann_P; (**c**,**d**) Q_{min} against October to March temperature (O–M_T) and precipitation (O–M_P); and (**e**,**f**) Q_{max} against October–June temperature (O–J_T) and precipitation (O–J_P) anomalies.

In the case of Q_{min}, temperature plays a larger role, with VI_1 for O–M_T > 0.5 for both basins. Precipitation has a smaller control, and the precipitation–temperature interaction has negligible to minor influence (Figure 4c,d). The higher temperature control on Q_{min} is also illustrated by stratified patterns along O–M_T axis, i.e., larger Q_{min} with higher O–M_T, especially in the case of Liard-LC basin. The MLR model predictability for Q_{min} as given by R^2 range between 0.39 and 0.46; thus, the models can only explain <50% of the variance of Q_{min}. Nevertheless, the higher temperature control on low flow—which in this region is mainly contributed by baseflow from groundwater [52]—align with the expectation of higher groundwater recharge and subsurface flow and higher Q_{min}, as temperature increases [20].

In contrast to Q_{min}, precipitation has higher control ($VI_2 \geq 0.59$) on Q_{max}, especially in the case of Liard-LC. This is also illustrated by a stratified positive relationship between Q_{max} and O–J_P, i.e., higher Q_{max} for larger precipitation (Figure 4e,f). The temperature control on Q_{max} is ≤ 0.26, indicating the smaller role of O–J_T change on Q_{max} change. In this case also, the precipitation–temperature interaction plays a minor role on Q_{max}. The MLR model R^2 range between 0.46 and 0.57, indicating a moderate predictability of Q_{max} with O–J_P and O–J_T.

Overall, the evaluation of precipitation and temperature controls on Q_{mean}, Q_{min} and Q_{max} provide further insights into the effect of these variables on streamflow response. The dominant control of precipitation on Q_{mean} and Q_{max}, and temperature on Q_{min} from this observation-based study is consistent with the modelling study that include historical and projected future simulations [14]. This provides some confidence in using the MLR model for diagnosing the future direction of streamflow change with respect to the projected precipitation and temperature changes. As such, future warming can be generally expected to lead to higher Q_{min}, while precipitation increase can be generally expected to lead to increased Q_{mean} and Q_{max}. However, the MLR models, with the R^2 predictability scores ranging between 0.39 and 0.74, have limited skills in explaining the variability of streamflow components. Furthermore, given the lack of physical representation, the MLR model is considered not suitable for extrapolation problems, such as projecting future changes. Nevertheless, the MLR-based VI analysis contributes to a process understanding with regard to dominant driving variables affecting streamflow responses.

4. Discussion on Future Changes in Extremes

The preceding investigation on the relationships of precipitation and temperature with Q_{mean}, Q_{min} and Q_{max}, based on the historical climate, leads to a subsequent question about the potential future changes in these variables in the context of enhanced warming and increased precipitation across the pan-Arctic region [53]. A number of other factors are also associated with the higher temperature and precipitation are that directly or indirectly affect these streamflow variables. For instance, with warmer temperatures, there will be a reduction in the snowfall fraction and an increase in the rainfall fraction [11]. The changing precipitation phase could lead to an increased frequency and areal extent of rain-on-snow events [16,18,54]. Future projections of snow indicate a general decline in the snow cover extent [7,55] and highly variable maximum snow–water equivalent changes, with increases, decreases or no change depending on the region, and climate model used [7,56–58]. Furthermore, in response not only to warming, but also to changes in snow cover, which exerts a control on the underlying soil, Arctic permafrost is projected to undergo substantial degradation [59].

Given that the projected changes in temperature, precipitation, snow and permafrost are generally a continuation of historical trends, future changes in streamflow variables can be expected to follow the historical changes. For instance, previous global-scale or pan-Arctic-wide studies indicate future increases in Q_{mean} and Q_{min}, and earlier Q_{max} timings [60–62]. Similar changes in Q_{mean} and Q_{min} were also projected for the Liard River basin [14]. However, the potential future changes in Q_{max} over the northern region, which is conditional on the evolution of precipitation, temperature, snowpack and permafrost changes and their interactions, remain less clear. Nevertheless, there are indications of changes to the nonstationary streamflow extremes (i.e., alterations in return periods and return levels). For instance, Hirabayashi et al. [63] projected future increases in the frequencies of the historical 100-year return period floods for the Yenisei, Lena, Mackenzie and Yukon Rivers, and a decreased frequency for the Ob River.

Summer rainfall-driven peak flow events are also related to rainfall/snowfall balance, which are historically very rare in large Arctic rivers, and typically occur only in the small coastal catchments [64–66]. For instance, an extreme summer flood event occurred in the Upper Kuparuk River catchment (drainage area 142 km^2) on the northern slope of Alaska in response to a 50-h, 80 mm rainfall event in July 1999, with the peak flow exceeding snowmelt peak discharge for 1993 to 2001 [64]. Additionally, a closer examination of the Yukon-E streamflow revealed occurrences of multiple summer peaks, with secondary peaks in late summers of 2010 and 2012 (not shown). Given that portions of the Yukon Basin are experiencing extensive melting of alpine glaciers and perennial snow fields [67,68], these summer peak flow events may have been influenced by glacier melt. In this respect, the projected glacier mass loss has the potential to modify the mean and extreme streamflow states of basins with glaciers across the pan-Arctic, including changes in the summer peak flows [69].

5. Summary and Conclusions

This study contributes towards understanding the historical changes in streamflow regimes and the role of climatic drivers on streamflow changes across the continental scale permafrost region of Canada. We analyzed historical trends in several streamflow components (i.e., Q_{mean}, Q_{min}, Q_{max} and Q_{max} timing) across 84 streamflow stations in the permafrost region of Canada. Overall, the trends are generally consistent with previous studies across the pan-Arctic (Appendix A Table A1), with increasing Q_{min} for 43% of stations ($p < 0.10$) as the most prominent trend. Q_{min} trends are also field significant for 38% of stations, and the majority of the stations with Q_{min} increases are located in western Canada. Similar to previous studies (Appendix A Table A1), the number of stations with significant trends in Q_{mean}, Q_{max} and Q_{max} timing are small, with increasing or decreasing trends of less than 15%, and even smaller fractions with field significant trends. Temperatures over the basins that drain to hydrometric stations show widespread increases, with significant trends in October to March and April to September temperatures for 80 and 71% of the basins, respectively; the majority of trends are field significant. Upward precipitation trends were obtained for some of the basins, especially for the months of October to March, with 27% of basins showing increases and 13% field significant.

The analysis of climatic controls (precipitation and temperature) provides key insights into their relationships with the streamflow changes. For instance, the Kendall's τ correlations of the basin-averaged runoff with precipitation and temperature indicate weak to moderate associations, including positive correlations of q_{mean} with O–M_P and A–S_P, and q_{min} with O–M_P and O–M_T. These associations reinforce the role of higher basin precipitation in producing higher runoffs. In contrast, the negative correlations of A–S_T with q_{mean} and q_{max} is indicative of the temperature influence on evaporation, snow accumulation and melt; therefore, the higher the temperatures, the lower the runoffs. The correlations of streamflow trends with precipitation and temperature trends are usually weak, although directions of trends are generally in line with the expectation, i.e., the positive correlation between q_{min} and O–M_P trends, and the negative correlations between q_{max} timing and A–S_T and O–M_T trends.

The MLR-based variable importance analysis captured the sensitivity of climatic drivers on streamflow components and provided plausible explanations on driver-response relationships. The results revealed the dominant control of precipitation on Q_{mean} and Q_{max}, with increasing wetness generally leading to higher discharge. On the other hand, the dominant temperature control on Q_{min} reinforces the impact of climate warming on increasing the Q_{min} trend, apparently by increasing the fraction of precipitation falling as rain and promoting subsurface flow with increasing temperature. Overall, the insights gained from this study, i.e., the effect of climatic drivers on streamflow changes, could be generalized for understanding streamflow changes across the permafrost region of the world. Furthermore, with a careful selection of climatic drivers, the methodologies used in this study could be applied to evaluate streamflow changes in the permafrost-free regions.

The historical streamflow trends and their relationships with climatic drivers lead to a subsequent question on future streamflow changes. In this respect, previous studies suggest future increases in Q_{mean}, Q_{min} and earlier Q_{max} timings across the pan-Arctic [14,60–62], while the direction of Q_{max}—which depends on the interactions of precipitation, temperature and snow—remain uncertain for most regions. Furthermore, given that most of the impact studies over the northern region are large-scale studies, and not designed to represent the finer details of the streamflow extremes, there is a need to develop basin-scale hydrologic models by incorporating key processes (such as permafrost degradation) for an improved understanding of the ongoing and future hydrologic changes. Furthermore, there is a need to improve the ground-based streamflow observation network as well as leverage remotely sensed products to better evaluate and attribute the streamflow changes, especially in the areas with sparse coverage as identified in this study. Overall, the results of this study, including the enhanced understanding of ongoing streamflow changes and climatic controls affecting these changes, provide valuable insights into the hydrologic processes in the permafrost region of Canada and beyond.

Author Contributions: Conceptualization, R.R.S., D.Y., D.L.P. and Y.B.D.; data curation, analyses and figures, J.P. and R.R.S.; draft manuscript, R.R.S.; review and editing, D.Y., D.L.P., Y.B.D., J.P.; All authors have read and agreed to the published version of the manuscript.

Funding: This study was conducted with internal funding from Environment and Climate Change Canada.

Institutional Review Board Statement: Not applicable.

Informed Consent Statement: Not applicable.

Data Availability Statement: Streamflow data used in the study is available through the Government of Canada, Historical Hydrometric Data Archive, http://wateroffice.ec.gc.ca/ (accessed on 26 February 2021). Additional streamflow data for Yukon River at Eagle station is obtained from the United States Geological Survey, https://waterdata.usgs.gov/usa/nwis/uv?15356000 (accessed on 26 February 2021). Gridded precipitation and temperature data used in this study is available through the Pacific Climate Impacts Consortium data portal, https://data.pacificclimate.org/portal/gridded_observations/map/ (accessed on 26 February 2021).

Acknowledgments: We acknowledge Natural Resources Canada for permafrost database and Water Survey of Canada, Environment and Climate Canada, for hydrometric network basin polygons used in this study.

Conflicts of Interest: The authors declare no conflict of interest.

Appendix A. Review of Streamflow Trends for Major Pan-Arctic Rivers

We compiled trends for Q_{mean}, Q_{min}, Q_{max} and Q_{max} timing from previous studies (most recent study for each river) for six large Arctic flowing rivers that include the Kolyma, Lena, Yenisey and Ob rivers in Eurasia, and the Mackenzie and Yukon rivers in North America (Figure A1; Table A1). Also included are results for multiple stations within these basins, summarized as a count of stations with significant trends out of the total number of stations. Note that these published studies present trends for different time periods, thus reflecting the responses to different periods of long-term climate forcing (e.g., temperature and precipitation) and short-term variability (i.e., Arctic Oscillation, Pacific Decadal Oscillation and El Niño Southern Oscillation). Some of the compiled results are for different stations (upstream or downstream), and/or flow conditions (naturalized or observed), which could produce different trends. Furthermore, the studies use different trend analysis methods that could also cause differences on the reported results.

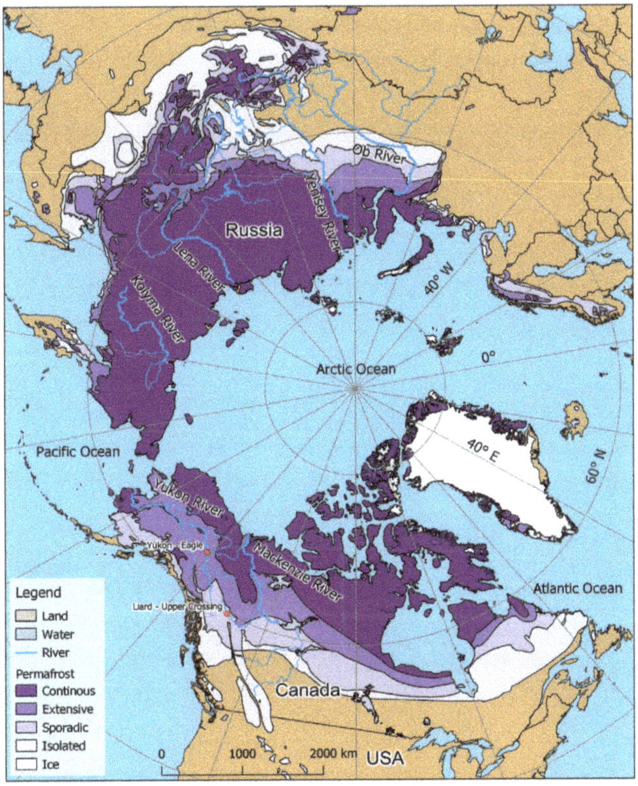

Figure A1. Map showing the six large Arctic flowing rivers summarized in Table 1 (Kolyma, Lena, Yenisey, Ob, Mackenzie, and Yukon). Red dots indicate the location of two hydrometric stations: (**1**) Yukon River at Eagle (Yukon-E) and (**2**) Liard River at Upper Crossing (Liard-UC); for which detailed analyses of climatic controls on streamflow extremes were conducted.

The review of streamflow trends across the Pan-Arctic generally suggests consistent patterns. For instance, in the case of Q_{mean}, most studies found an increasing trend or no trend ($p < 0.05$ or $p < 0.10$), with no study showing significantly decreasing trends. Furthermore, studies that considered aggregated discharge from several stations also found generally increasing discharges. For instance, Zhang et al. [4] found an increasing trend for the combined annual discharge from Lena, Yenisey and Ob Rivers from 1948 to 2008. Likewise, Durocher et al. [26] found a general increase in freshwater flow to the Arctic Ocean for the period 1975 to 2015, with the increase being more prominent in the Eurasian rivers than in the North American rivers. Ahmed et al. [28] found that the combined annual mean flow from the Mackenzie, Lena, Yenisey and Ob rivers to the Arctic Ocean has increased by 14% during in period 1980 to 2009. The increasing trends in Q_{mean} are also consistent with the annual cold season and warm season precipitation increases across the Pan-Arctic [27].

From the few available studies on Q_{min}, the trends are mostly increasing, some of which are significant [21,70]. Additionally, winter (e.g., December to February or January to March) flows are also significantly increasing [28,34]. In the case of Q_{max}, the changes are mostly insignificant, except for the tributaries of the Yukon and Lena rivers, which show either increases or decreases [48,70]. However, besides hydro-climatic controls, other factors may contribute to these trends. For example, the increasing trend for the Lena River is also influenced by reservoir filling that reduced its outflows in the 1960s [71,72]. From the very few available studies, it appears that Q_{max} timings are shifting earlier [47,72] although the trends are generally not significant. Nevertheless, there is a general expectation of earlier Q_{max} timing, along with the earlier onset of snowmelt and earlier centroid of annual flow in a warmer climate [73].

Table A1. Summary of previous studies on streamflow trends for major pan-Arctic Rivers.

River/Region	Reference	Study Period	Trends			
			Q_{mean}	Q_{min}	Q_{max}	Q_{max} Timing
Kolyma (Observed)	Box et al. [27]	1971–2015	++1 (1)			
Lena (Observed)	Box et al. [27]	1971–2015	++1 (1)			
Lena (Observed)	Ahmed et al. [28]	1936–2009	++1 (1) annual volume	+NS10 (1) December–February volume	−− (1)	
Lena (multiple stations)	Tananaev et al. [70]	Varying periods between 1925–2013	++29 (100); −−2 (100)	++30 (55); −−2 (55)	++6 (105); −−3 (105)	
Lena + Eastern Siberia (multiple natural tributaries)	Smith et al. [21]	1958–1989		+46 (212); −17 (212)		
Yenisey (Observed)	Box et al. [27]	1971–2015	+NS10 (1)			
Yenisey (Observed)	Ahmed et al. [28]	1980–2009	++1 (1) annual volume	++1 (1) December–February volume	−NS10 (1)	
Ob (Observed)	Box et al. [27]	1971–2015	−NS10 (1)			
Ob (Observed)	Ahmed et al. [28]	1936–2009	+NS10 (1) annual volume	++1 (1) December–February volume	+NS10 (1)	
Mackenzie	Box et al. [27]	1973–2015	+NS10 (1)			
Mackenzie	Yang et al. [47]	1973–2011	+NS10 (1)	+NS10 (1)	−NS10 (1)	−NS10 (1)
Mackenzie and tributaries	St. Jacques and Sauchyn [34]	Varying periods between 1939–2007	+9 (23)	+20 (23) January–March mean flow		
Yukon	Box et al. [27]	1975–2016	++1 (1)			
Yukon (Canadian Portion)	Déry et al. [74]	1964–2013	+NS05 (1)			
Yukon and other rivers in Alaska	Bennett et al. [48]	1954/1964–2013		+3 (8)	−3 (8)	

++ (−−) indicate increasing (decreasing) trends at 5% significance level; + (−) indicate increasing (decreasing) trends at 10% significance level. NS05 and NS10 signify no significant trends at the 5% and 10% significance levels, respectively. Blank cells mean no analysis for that variable is performed in the cited study. The numbers in the trend columns are the number of stations with statistically significant trends (out of total number of stations).

References

1. Serreze, M.C.; Barry, R.G. Processes and impacts of Arctic amplification: A research synthesis. *Glob. Planet. Chang.* **2011**, *77*, 85–96. [CrossRef]
2. Cohen, J.; Screen, J.A.; Furtado, J.C.; Barlow, M.; Whittleston, D.; Coumou, D.; Francis, J.; Dethloff, K.; Entekhabi, D.; Overland, J.; et al. Recent Arctic amplification and extreme mid-latitude weather. *Nat. Geosci.* **2014**, *7*, 627–637. [CrossRef]
3. Zhang, X.; Flato, G.; Kirchmeier-Young, M.; Vincent, L.; Wan, H.; Wang, X.; Rong, R.; Fyfe, J.; Li, G.; Kharin, V.V. Changes in temperature and precipitation across Canada; Chapter 4. In *Canada's Changing Climate Report*; Bush, E., Lemmen, D.S., Eds.; Government of Canada: Ottawa, ON, Canada, 2019; pp. 112–193.
4. Zhang, X.; He, J.; Zhang, J.; Polyakov, I.; Gerdes, R.; Inoue, J.; Wu, P. Enhanced poleward moisture transport and amplified northern high-latitude wetting trend. *Nat. Clim. Chang.* **2012**, *3*, 47–51. [CrossRef]
5. Vihma, T.A.; Screen, J.; Tjernström, M.; Newton, B.W.; Zhang, X.; Popova, V.; Deser, C.; Holland, M.M.; Prowse, T.D. The atmospheric role in the Arctic water cycle: A review on processes, past and future changes, and their impacts. *J. Geophys. Res. Biogeosci.* **2016**, *121*, 586–620. [CrossRef]
6. Derksen, C.; Brown, R.; Mudryk, L.; Luojus, K. Arctic: Terrestrial Snow. State of the Climate in 2014. *Bull. Am. Meteorol. Soc.* **2015**, *96*, 133–135.

7. Mudryk, L.R.; Derksen, C.; Howell, S.; Laliberté, F.; Thackeray, C.; Sospedra-Alfonso, R.; Vionnet, V.; Kushner, P.J.; Brown, R. Canadian snow and sea ice: Historical trends and projections. *Cryosphere* **2018**, *12*, 1157–1176. [CrossRef]
8. Derksen, C.; Burgess, D.; Duguay, C.; Howell, S.; Murdyk, L.; Smith, S.; Thackeray, C.; Kirchmeier-Young, M. Changes in snow, ice, and permafrost across Canada. In *Changes in Snow, Ice, and Permafrost across Canada*; Bush, E., Lemmen, D.S., Eds.; Government of Canada: Ottawa, ON, Canada, 2019; pp. 194–260.
9. Romanovsky, V.E.; Smith, S.L.; Christiansen, H.H. Permafrost thermal state in the polar Northern Hemisphere during the international polar year 2007-2009: A synthesis. *Permafr. Periglac. Process.* **2010**, *21*, 106–116. [CrossRef]
10. Grosse, G.; Goetz, S.; McGuire, A.D.E.; Romanovsky, V.; Schuur, E.A.G. Changing permafrost in a warming world and feedbacks to the Earth system. *Environ. Res. Lett.* **2016**, *11*, 040201. [CrossRef]
11. Bintanja, R.; Andry, R.B.O. Towards a rain-dominated Arctic. *Nat. Clim. Chang.* **2017**, *7*, 263–267. [CrossRef]
12. Barnett, T.P.; Adam, J.C.; Lettenmaier, D.P. Potential impacts of a warming climate on water availability in snow-dominated regions. *Nat. Cell Biol.* **2005**, *438*, 303–309. [CrossRef]
13. Barnett, T.P.; Pierce, D.W.; Hidalgo, H.G.; Bonfils, C.; Santer, B.D.; Das, T.; Bala, G.; Wood, A.W.; Nozawa, T.; Mirin, A.A.; et al. Human-Induced Changes in the Hydrology of the Western United States. *Science* **2008**, *319*, 1080–1083. [CrossRef] [PubMed]
14. Shrestha, R.R.; Cannon, A.J.; Schnorbus, M.A.; Alford, H. Climatic Controls on Future Hydrologic Changes in a Subarctic River Basin in Canada. *J. Hydrometeorol.* **2019**, *20*, 1757–1778. [CrossRef]
15. Shrestha, R.R.; Cannon, A.J.; Schnorbus, M.A.; Zwiers, F.W. Projecting future nonstationary extreme streamflow for the Fraser River, Canada. *Clim. Chang.* **2017**, *145*, 289–303. [CrossRef]
16. Rennert, K.J.; Roe, G.; Putkonen, J.; Bitz, C.M. Soil Thermal and Ecological Impacts of Rain on Snow Events in the Circumpolar Arctic. *J. Clim.* **2009**, *22*, 2302–2315. [CrossRef]
17. Liston, G.E.; Hiemstra, C.A. The Changing Cryosphere: Pan-Arctic Snow Trends (1979–2009). *J. Clim.* **2011**, *24*, 5691–5712. [CrossRef]
18. Hansen, B.B.; Isaksen, K.; Benestad, R.E.; Kohler, J.; Pedersen, Å.Ø.; Loe, L.E.; Coulson, S.J.; Larsen, J.O.; Varpe, Ø. Warmer and wetter winters: Characteristics and implications of an extreme weather event in the High Arctic. *Environ. Res. Lett.* **2014**, *9*, 114021. [CrossRef]
19. Bokhorst, S.; Pedersen, S.H.; Brucker, L.; Anisimov, O.; Bjerke, J.W.; Brown, R.D.; Ehrich, D.; Essery, R.L.H.; Heilig, A.; Ingvander, S.; et al. Changing Arctic snow cover: A review of recent developments and assessment of future needs for observations, modelling, and impacts. *Ambio* **2016**, *45*, 516–537. [CrossRef]
20. Bense, V.F.; Ferguson, G.; Kooi, H. Evolution of shallow groundwater flow systems in areas of degrading permafrost. *Geophys. Res. Lett.* **2009**, *36*. [CrossRef]
21. Smith, L.C.; Pavelsky, T.M.; Macdonald, G.M.; Shiklomanov, A.I.; Lammers, R.B. Rising minimum daily flows in northern Eurasian rivers: A growing influence of groundwater in the high-latitude hydrologic cycle. *J. Geophys. Res. Space Phys.* **2007**, *112*. [CrossRef]
22. Matti, B.; Dahlke, H.E.; Lyon, S.W. On the variability of cold region flooding. *J. Hydrol.* **2016**, *534*, 669–679. [CrossRef]
23. Ye, B.; Yang, D.; Zhang, Z.; Kane, D.L. Variation of hydrological regime with permafrost coverage over Lena Basin in Siberia. *J. Geophys. Res. Space Phys.* **2009**, *114*. [CrossRef]
24. Cherry, J.E.; Knapp, C.; Trainor, S.; Ray, A.J.; Tedesche, M.; Walker, S. Planning for climate change impacts on hydropower in the Far North. *Hydrol. Earth Syst. Sci.* **2017**, *21*, 133–151. [CrossRef]
25. Instanes, A.; Kokorev, V.; Janowicz, R.; Bruland, O.; Sand, K.; Prowse, T. Changes to freshwater systems affecting Arctic infrastructure and natural resources. *J. Geophys. Res. Biogeosci.* **2016**, *121*, 567–585. [CrossRef]
26. Durocher, M.; Requena, A.I.; Burn, D.H.; Pellerin, J. Analysis of trends in annual streamflow to the Arctic Ocean. *Hydrol. Process.* **2019**, *33*, 1143–1151. [CrossRef]
27. Box, J.; Colgan, W.T.; Christensen, T.R.; Schmidt, N.M.; Lund, M.; Parmentier, F.-J.W.; Brown, R.; Bhatt, U.S.; Euskirchen, E.S.; Romanovsky, V.E.; et al. Key indicators of Arctic climate change: 1971–2017. *Environ. Res. Lett.* **2019**, *14*, 045010. [CrossRef]
28. Ahmed, R.; Prowse, T.; Dibike, Y.; Bonsal, B.; O'Neil, H. Recent Trends in Freshwater Influx to the Arctic Ocean from Four Major Arctic-Draining Rivers. *Water* **2020**, *12*, 1189. [CrossRef]
29. Heginbottom, J.A.; Dubreuil, M.A.; Harker, P.A. *Canada—Permafrost, National Atlas of Canada, National Atlas Information Service*; Natural Resources Canada, MCR: Ottawa, ON, Canada, 1995; p. 4177.
30. Natural Resources Canada Index of Permafrost Database. Available online: https://ftp.maps.canada.ca/pub/nrcan_rncan/archive/vector/geology/Permafrost/ (accessed on 16 February 2021).
31. Water Survey of Canada Water Level and Flow - Environment Canada. Available online: https://wateroffice.ec.gc.ca/ (accessed on 16 February 2021).
32. United States Geological Survey USGS Current Conditions for USGS 15356000 YUKON R AT EAGLE AK. Available online: https://waterdata.usgs.gov/usa/nwis/uv?15356000 (accessed on 16 February 2021).
33. De Rham, L.P.; Prowse, T.D.; Beltaos, S.; Lacroix, M.P. Assessment of annual high-water events for the Mackenzie River basin, Canada. *Hydrol. Process.* **2008**, *22*, 3864–3880. [CrossRef]
34. Jacques, J.-M.S.; Sauchyn, D.J. Increasing winter baseflow and mean annual streamflow from possible permafrost thawing in the Northwest Territories, Canada. *Geophys. Res. Lett.* **2009**, *36*. [CrossRef]

35. Werner, A.T.; Schnorbus, M.A.; Shrestha, R.R.; Cannon, A.J.; Zwiers, F.W.; Dayon, G.; Anslow, F. A long-term, temporally consistent, gridded daily meteorological dataset for northwestern North America. *Sci. Data* **2019**, *6*, 180299. [CrossRef]
36. Hutchinson, M.F.; McKenney, D.W.; Lawrence, K.; Pedlar, J.H.; Hopkinson, R.F.; Milewska, E.; Papadopol, P. Development and Testing of Canada-Wide Interpolated Spatial Models of Daily Minimum–Maximum Temperature and Precipitation for 1961–2003. *J. Appl. Meteorol. Clim.* **2009**, *48*, 725–741. [CrossRef]
37. Environment and Climate Change Canada National Hydrometric Network Basin Polygons-Open Government Portal. Available online: https://open.canada.ca/data/en/dataset/0c121878-ac23-46f5-95df-eb9960753375 (accessed on 16 February 2021).
38. Bronaugh, D.; Werner, A. Zyp: Zhang + Yue-Pilon Trends Package. 2019. Available online: https://cran.r-project.org/web/packages/zyp/zyp.pdf. (accessed on 27 February 2021).
39. Kendall, M.G. *Rank Correlation Methods*; Charles Griffin: London, UK, 1955.
40. Zhang, X.; Zwiers, F.W. Comment on "Applicability of prewhitening to eliminate the influence of serial correlation on the Mann-Kendall test" by Sheng Yue and Chun Yuan Wang. *Water Resour. Res.* **2004**, *40*, 03805. [CrossRef]
41. Bürger, G. On trend detection. *Hydrol. Process.* **2017**, *31*, 4039–4042. [CrossRef]
42. Wilks, D.S. On "Field Significance" and the False Discovery Rate. *J. Appl. Meteorol. Clim.* **2006**, *45*, 1181–1189. [CrossRef]
43. Helsel, D.R.; Hirsch, R.M. *Statistical Methods in Water Resources*; US Geological Survey: Reston, VA, USA, 2002; Volume 323.
44. Lehner, F.; Wood, A.W.; Vano, J.A.; Lawrence, D.M.; Clark, M.P.; Mankin, J.S. The potential to reduce uncertainty in regional runoff projections from climate models. *Nat. Clim. Chang.* **2019**, *9*, 926–933. [CrossRef]
45. Chegwidden, O.S.; Rupp, D.E.; Nijssen, B. Climate change alters flood magnitudes and mechanisms in climatically-diverse headwaters across the northwestern United States. *Environ. Res. Lett.* **2020**, *15*, 094048. [CrossRef]
46. Kuhn, M.; Wing, J.; Weston, S.; Williams, A.; Keefer, C.; Engelhardt, A.; Cooper, T.; Mayer, Z.; Kenkel, B. Caret: Classification and Regression Training. 2018. Available online: https://ui.adsabs.harvard.edu/abs/2015ascl.soft05003K/abstract (accessed on 26 February 2021).
47. Yang, D.; Shi, X.; Marsh, P.D. Variability and extreme of Mackenzie River daily discharge during 1973–2011. *Quat. Int.* **2015**, *380–381*, 159–168. [CrossRef]
48. Bennett, K.; Cannon, A.; Hinzman, L. Historical trends and extremes in boreal Alaska river basins. *J. Hydrol.* **2015**, *527*, 590–607. [CrossRef]
49. Yang, D.; Shrestha, R.R.; Park, H. Heat Flux from 15 Canadian Northern Rivers Draining to Arctic Ocean and Hudson/James Bay. *Glob. Planet. Chang.* under review.
50. Mekis, É.; Vincent, L.A. An Overview of the Second Generation Adjusted Daily Precipitation Dataset for Trend Analysis in Canada. *Atmosphere-Ocean.* **2011**, *49*, 163–177. [CrossRef]
51. Hamilton, S. Sources of Uncertainty in Canadian Low Flow Hydrometric Data. *Can. Water Resour. J. / Rev. Can. des ressources hydriques* **2008**, *33*, 125–136. [CrossRef]
52. Woo, M.-K.; Thorne, R. Winter Flows in the Mackenzie Drainage System. *Arct.* **2014**, *67*, 238–256. [CrossRef]
53. IPCC. Annex I: Atlas of Global and Regional Climate Projections. In *Climate Change 2013*; Van Oldenborgh, G.J., Collins, M., Arblaster, J., Christensen, J.H., Marotzke, J., Power, S.B., Rummukainen, M., Zhou, T., Stocker, T., Qin, D., Eds.; Cambridge University Press: Cambridge, UK; New York, NY, USA, 2013; pp. 1311–1393.
54. Jeong, D.I.; Sushama, L. Rain-on-snow events over North America based on two Canadian regional climate models. *Clim. Dyn.* **2017**, *50*, 303–316. [CrossRef]
55. Mudryk, L.; Santolaria-Otín, M.; Krinner, G.; Ménégoz, M.; Derksen, C.; Brutel-Vuilmet, C.; Brady, M.; Essery, R. Historical Northern Hemisphere snow cover trends and projected changes in the CMIP6 multi-model ensemble. *Cryosphere* **2020**, *14*, 2495–2514. [CrossRef]
56. Callaghan, T.V.; Johansson, M.; Brown, R.D.; Groisman, P.Y.; Labba, N.; Radionov, V.; Barry, R.G.; Bulygina, O.N.; Essery, R.L.H.; Frolov, D.M.; et al. The Changing Face of Arctic Snow Cover: A Synthesis of Observed and Projected Changes. *Ambio* **2011**, *40*, 17–31. [CrossRef]
57. Diffenbaugh, N.S.; Scherer, M.; Ashfaq, M. Response of snow-dependent hydrologic extremes to continued global warming. *Nat. Clim. Chang.* **2012**, *3*, 379–384. [CrossRef] [PubMed]
58. Shrestha, R.R.; Bonsal, B.R.; Bonnyman, J.M.; Cannon, A.J.; Najafi, M.R. Heterogeneous snowpack response and snow drought occurrence across river basins of northwestern North America under 1.0 °C to 4.0 °C global warming. *Clim. Chang.* **2021**, *164*, 1–21. [CrossRef]
59. Collins, M.; Knutti, R.; Arblaster, J.; Dufresne, J.; Fichefet, T.; Friedlingstein, P.; Gao, X.; Gutowski, W.; Johns, T.; Krinner, G.; et al. Long-term Climate Change: Projections, Commitments and Irreversibility. In *Climate Change 2013: The Physical Science Basis. Contribution of Working Group I to the Fifth Assessment Report of the Intergovernmental Panel on Climate Change*; Stocker, T., Qin, D., Plattner, G., Tignor, M., Allen, S., Boschung, J., Nauels, A., Xia, Y., Bex, V., Midgley, P., Eds.; Cambridge University Press: Cambridge, UK, 2013; pp. 1029–1136.
60. Koirala, S.; Yeh, P.J.-F.; Hirabayashi, Y.; Kanae, S.; Oki, T. Global-scale land surface hydrologic modeling with the representation of water table dynamics. *J. Geophys. Res. Atmos.* **2014**, *119*, 75–89. [CrossRef]
61. Schewe, J.; Heinke, J.; Gerten, D.; Haddeland, I.; Arnell, N.W.; Clark, D.B.; Dankers, R.; Eisner, S.; Fekete, B.M.; Colón-González, F.J.; et al. Multimodel assessment of water scarcity under climate change. *Proc. Natl. Acad. Sci. USA* **2014**, *111*, 3245–3250. [CrossRef] [PubMed]

62. Bring, A.; Shiklomanov, A.; Lammers, R.B. Pan-Arctic river discharge: Prioritizing monitoring of future climate change hot spots. *Earth's Futur.* **2017**, *5*, 72–92. [CrossRef]
63. Hirabayashi, Y.; Mahendran, R.; Koirala, S.; Konoshima, L.; Yamazaki, D.; Watanabe, S.; Kim, H.; Kanae, S. Global flood risk under climate change. *Nat. Clim. Chang.* **2013**, *3*, 816–821. [CrossRef]
64. Kane, D.L.; McNamara, J.P.; Yang, D.; Olsson, P.Q.; Gieck, R.E. An Extreme Rainfall/Runoff Event in Arctic Alaska. *J. Hydrometeorol.* **2003**, *4*, 1220–1228. [CrossRef]
65. Kane, D.L.; Hinzman, L.D.; Gieck, R.E.; McNamara, J.P.; Youcha, E.K.; Oatley, J.A. Contrasting extreme runoff events in areas of continuous permafrost, Arctic Alaska. *Hydrol. Res.* **2008**, *39*, 287–298. [CrossRef]
66. Dugan, H.A.; Lamoureux, S.F.; Lafrenière, M.J.; Lewis, T. Hydrological and sediment yield response to summer rainfall in a small high Arctic watershed. *Hydrol. Process.* **2009**, *23*, 1514–1526. [CrossRef]
67. Striegl, R.G.; Dornblaser, M.M.; Aiken, G.R.; Wickland, K.P.; Raymond, P.A. Carbon export and cycling by the Yukon, Tanana, and Porcupine rivers, Alaska, 2001-2005. *Water Resour. Res.* **2007**, *43*. [CrossRef]
68. Aiken, G.R.; Spencer, R.G.M.; Striegl, R.G.; Schuster, P.F.; Raymond, P.A. Influences of glacier melt and permafrost thaw on the age of dissolved organic carbon in the Yukon River basin. *Glob. Biogeochem. Cycles* **2014**, *28*, 525–537. [CrossRef]
69. Bliss, A.; Hock, R.; Radić, V. Global response of glacier runoff to twenty-first century climate change. *J. Geophys. Res. Earth Surf.* **2014**, *119*, 717–730. [CrossRef]
70. Tananaev, N.I.; Makarieva, O.M.; Lebedeva, L.S. Trends in annual and extreme flows in the Lena River basin, Northern Eurasia. *Geophys. Res. Lett.* **2016**, *43*, 10–764. [CrossRef]
71. Yang, D.; Ye, B.; Kane, D.L. Streamflow changes over Siberian Yenisei River Basin. *J. Hydrol.* **2004**, *296*, 59–80. [CrossRef]
72. Shiklomanov, A.I.; Lammers, R.B.; Rawlins, M.A.; Smith, L.C.; Pavelsky, T.M. Temporal and spatial variations in maximum river discharge from a new Russian data set. *J. Geophys. Res. Space Phys.* **2007**, *112*. [CrossRef]
73. Tan, A.; Adam, J.C.; Lettenmaier, D.P. Change in spring snowmelt timing in Eurasian Arctic rivers. *J. Geophys. Res. Space Phys.* **2011**, *116*. [CrossRef]
74. Déry, S.J.; Stadnyk, T.A.; Macdonald, M.K.; Gauli-Sharma, B. Recent trends and variability in river discharge across northern Canada. *Hydrol. Earth Syst. Sci.* **2016**, *20*, 4801–4818. [CrossRef]

water

MDPI

Article

Multiple Indicators of Extreme Changes in Snow-Dominated Streamflow Regimes, Yakima River Basin Region, USA

Anna M. Wagner [1,*], Katrina E. Bennett [2], Glen E. Liston [3], Christopher A. Hiemstra [4] and Dan Cooley [5]

[1] US Army Cold Regions Research and Engineering Laboratory, Fort Wainwright, AK 99703, USA
[2] Los Alamos National Laboratory, Earth and Environmental Sciences, Los Alamos, NM 87545, USA; kbennett@lanl.gov
[3] Cooperative Institute for Research in the Atmosphere, Colorado State University, Fort Collins, CO 80523, USA; Glen.Liston@colostate.edu
[4] USDA Forest Service, Geospatial Management Office, Salt Lake City, UT 84138, USA; Christopher.Hiemstra@usda.gov
[5] Department of Statistics, Colorado State University, Fort Collins, CO 80523, USA; cooleyd@rams.colostate.edu
* Correspondence: Anna.M.Wagner@erdc.dren.mil

Abstract: Snow plays a major role in the hydrological cycle. Variations in snow duration and timing can have a negative impact on water resources. Excluding predicted changes in snowmelt rates and amounts could result in deleterious infrastructure, military mission, and asset impacts at military bases across the US. A change in snowpack can also lead to water shortages, which in turn can affect the availability of irrigation water. We performed trend analyses of air temperature, snow water equivalent (SWE) at 22 SNOTEL stations, and streamflow extremes for selected rivers in the snow-dependent and heavily irrigated Yakima River Basin (YRB) located in the Pacific Northwest US. There was a clear trend of increasing air temperature in this study area over a 30 year period (water years 1991–2020). All stations indicated an increase in average air temperatures for December (0.97 °C/decade) and January (1.12 °C/decade). There was also an upward trend at most stations in February (0.28 °C/decade). In December–February, the average air temperatures were 0.82 °C/decade. From these trends, we estimate that, by 2060, the average air temperatures for December–February at most (82%) stations will be above freezing. Furthermore, analysis of SWE from selected SNOTEL stations indicated a decreasing trend in historical SWE, and a shift to an earlier peak SWE was also assumed to be occurring due of the shorter snow duration. Decreasing trends in snow duration, rain-on-snow, and snowmelt runoff also resulted from snow modeling simulations of the YRB and the nearby area. We also observed a shift in the timing of snowmelt-driven peak streamflow, as well as a statistically significant increase in winter maximum streamflow and decrease in summer maximum and minimum streamflow trends by 2099. From the streamflow trends and complementary GEV analysis, we show that the YRB basin is a system in transition with earlier peak flows, lower snow-driven maximum streamflow, and higher rainfall-driven summer streamflow. This study highlights the importance of looking at changes in snow across multiple indicators to develop future infrastructure and planning tools to better adapt and mitigate changes in extreme events.

Keywords: climate change; extreme events; hydrology; snow; trends; snow water equivalent; Yakima River basin

Citation: Wagner, A.M.; Bennett, K.E.; Liston, G.E.; Hiemstra, C.A.; Cooley, D. Multiple Indicators of Extreme Changes in Snow-Dominated Streamflow Regimes, Yakima River Basin Region, USA. *Water* **2021**, *13*, 2608. https://doi.org/10.3390/w13192608

Academic Editors: Rajesh R. Shrestha and Mohammad Reza Najafi

Received: 16 August 2021
Accepted: 14 September 2021
Published: 22 September 2021

1. Introduction

About one-sixth of the world's population is dependent on seasonal snowpacks and glaciers for water resources [1]. There is mounting evidence that the Northern Hemisphere is experiencing changes in snowpack characteristics [2–7]. More recent studies indicate that snowpacks are continuing to change in subarctic, Arctic, alpine, and mid-latitude regions [8–10]. For the period of 1980 to 2018, Pulliainen et al. [11] reported a decreasing trend of annual maximum snow mass for the Northern Hemisphere. Additionally,

Coupled Model Intercomparison project 6 (CMIP6) models project strong negative snow extent and mass trends in the Northern Hemisphere [10]. In the western US, changes in snowpack could result in major consequences for the overall hydrology [3,5,12–17]. For example, in western watersheds, snow is a critical surface water resource; 50% to 80% of total runoff is from spring/summer runoff [18–20], leading to a significant loss in water availability. Several studies show evidence that snowmelt is occurring earlier in the spring (e.g., [12,18,21,22]). For example, in the western US, there is an indication that snowmelt is taking place up to 26 days earlier [6]. Moreover, springtime snow water equivalent (SWE) has been declining since 1925 [12]. According to a more recent study by Mote et al. [23], a majority of the sites continue to show a decreasing trend. Not only is the timing of snowmelt changing, but snow duration is also declining [24]. Future predictions indicate a reduction in both total snow amount and snow duration for the western US (e.g., [18,25–31]). In the western US mountain regions, [32] projects that snow duration will decrease by about 2 months by the mid-21st century, and there will be a 30% reduction in areal extent of wintertime snow-dominated area. Similarly, Lute et al. [33] estimated a decline in both the total number of days with snowfall and the SWE in the same period.

Spring runoff occurs earlier in snowmelt-dominated rivers, with a shift of up to 3 weeks earlier [34]. Even with no change in precipitation intensity, predictions indicate that this trend will continue, and that, by 2050, the maximum peak will take place approximately 1 month earlier [1]. Additionally, Casola et al. [35] indicated that a warming climate will result in a loss of about 20% of the 1 April snowpack in the Washington Cascade Mountains [35].

The Columbia River, located in the Pacific Northwest region of the western US, is the fourth largest river in the US, when comparing annual flow. The Columbia is mainly fed by mountain snowmelt, and approximately half of the annual flow is stored for flood control, hydropower, and irrigation [19]. A tributary of the Columbia River, the Yakima River, in the south central and eastern Washington State, houses important US military investments, such as Hanford and Yakima Training Center. To help with future land management planning in the Yakima River basin (YRB) region, Washington (WA), USA, a Strategic Environmental Research and Development Program (SERDP) study was funded to examine future impacts to snowpack at this site [36].

In this paper, we present the results of a 3 year SERDP study. This study included an assessment of our hypothesis that there are multiple indicators of a change in climate in the Yakima River region, with a specific focus on impacts on the US Army's Yakima Training Center (YTC). The paper considers trends in air temperature, snow accumulation, SWE arrival and departure dates, SWE peak, and streamflow within YRB and close-by watersheds as indicators of climate change.

2. Study Area, Data, and Methods

2.1. Study Area

The YRB is located in the northwestern corner of the Columbia Basin, Washington (Figure 1). Within the YRB is YTC, a 327,000 acre maneuver and live-fire training area [37]. The Yakima River is a main branch of Columbia River with 15,941 km^2 to its drainage outlet, and the upper basin above Union Gap is 9018 km^2. The mean annual precipitation varies from 180 mm at lower elevations to 3000 mm at higher elevations [38]. Most of the runoff from the basin is generated from the snowpack [39]. The Bureau of Reclamation initiated the Yakima Project for irrigation efforts in the YRB in 1910. In 1932, the basin was also developed for hydropower but the power is presently mostly generated for agricultural use [40]. Several streams in YRB are regulated, where excess water is stored in reservoirs and released during low flow. From this process, Yakima River and its tributaries provide approximately 180,000 hectares of irrigation water to agriculturalists operating in the region [39]. Vaccaro [41] estimated that regulation of flow reduced the mean annual flow by about 20% at Union Gap and 48% near Parker.

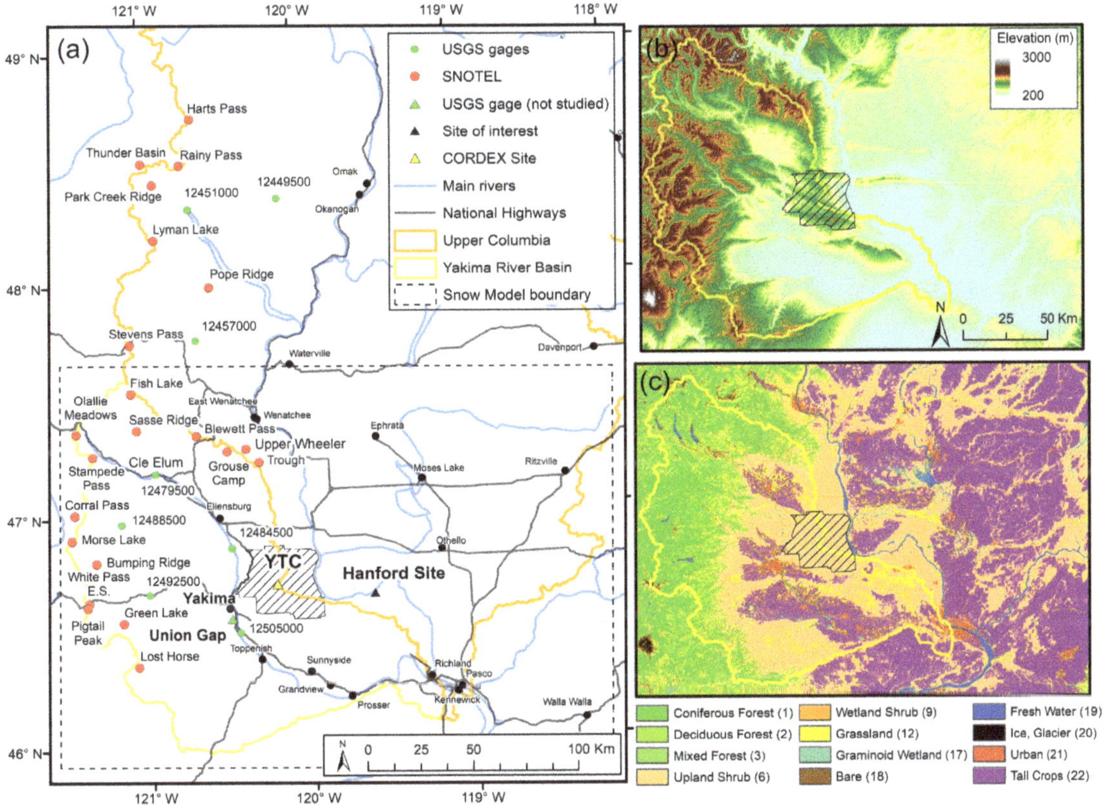

Figure 1. (**a**) Map of the study area and river basins, showing the YTC, selected cities including the town of Yakima (black dots), the snow modeling boundary (dashed line), hydrological unit code (HUC) 8 boundaries for the Yakima and Upper Columbia river basins, USGS gages (green solid dots, see Table S2), and SNOTEL stations (red solid dots, see Table S1). (**b**) Elevation and (**c**) land cover from 2013 National Land Cover Data [42] for the snow modeling area.

2.2. Data

2.2.1. SNOTEL Sites and Climate Data

In the US, repeated manual measurements of snowpack along staked out snow courses started in early 1900s. By the 1970s, the seasonal snowpack in the US was manually measured along snow courses at close to 2000 stations [19]. In the mid-1960s, a new fully automated and unattended snow telemetry (SNOTEL) network to measure snow depth was initiated. The network is operated by the US Department of Agriculture (USDA) Natural Resources Conversation Service (NRCS) and currently includes over 900 stations. Snow water equivalent (SWE) measurements are made using snow pillows, and other measurements include snow depths, minimum and maximum air temperatures, and precipitation.

For this work, we used SWE and air temperature from 22 SNOTEL stations, where seven stations were located within the YRB and the remaining stations were in close-by watersheds (Figure 1). The selected stations had a record of at least 30 years and included data from water year (WY) WY1991–WY2020 (Table S1). The stations varied in elevation (el.) from 1100 m at Fish Lake, located in the Headwaters Cle Elum River watershed, to 2000 m at Harts Pass, located in the Upper Canyon Creek watershed. Air temperature and precipitation data for the Yakima Airport site were extracted from the Natural Resources Conservation Service (NRCS) database.

2.2.2. Streamflow Data

Streamflow gage information was compiled for eight US Geological Survey (USGS) gages within our study area (Table S2). Historical USGS streamflow gage and naturalized, modeled streamflow information was generated by the University of Washington's study of the Columbia River basin (Chegwidden, et al. [43], referred to as UW17). The UW17 dataset includes output from several hydrologic models, run with different parameterizations (P2 was selected for this work). For the purposes of this work, we selected output from the Variable Infiltration Capacity (VIC) hydrologic model that matched well with historical gage data [44]. We additionally included gages outside the Yakima River basin for comparison with the Yakima gages, which are heavily impacted by dams, diversion structures, and water withdrawals. The UW17 data also included future model projections, as described below.

2.2.3. Climate Scenarios

For the streamflow analysis, we utilized future projections from two different regional climate models (RCMs) and earth system models (ESMs)—Canadian Earth System Model v2, CanESM2 (CanRCM4) and Global Fluid Dynamical Lab Environmental System Model v2 Weather and Research Forecast model, GFDL-ESM2M (WRF)—that represented high (wet) and low (dry) scenarios of change as reported in the aforementioned SERDP study [36]. To project future streamflow changes, we focused on representative concentration pathway (RCP) 8.5 regional and earth system models scenarios, extracted from the Coordinated Regional Downscaling Experiment (CORDEX) [45]. Additional analysis of precipitation comparison between 2006 and 2099 for the CORDEX RCP 8.5 models that support the selection of our two future models is presented in Supplemental Figure S1.

Future streamflow predictions were extracted from the University of Washington's Columbia River Basin study and utilized the CanESM2 and GFDL-ESM2M models, based on multivariate adaptive constructed analogs (MACA) downscaled climate data and the VIC hydrology model. More details on the datasets can be found in the Supplementary Materials [46]. Future streamflow was obtained at eight USGS gages (ESMs; see Table S2). CanESM2 (CanRCM4) simulates regional climate model feedbacks and responses using the Canadian Earth System Model as boundary conditions and the Canadian Regional Climate Model for internal energy and water simulations [47]. The GFDL-ESM2M Earth System Model provides boundary conditions to the WRF model [47].

2.2.4. SnowModel Data

We present snow modeling simulation results derived from our SERDP study [36] to illustrate the spatial representation of the snow coverage and trends of our study area. In that study, we used SnowModel [48,49], a physically based model. We applied a 3 h time step for 36 years starting on 1 September 1979 to 2015. We used a 300 m grid increment, and the simulation domain was approximately 50,600 km^2 (Figure 1b,c). The dominant vegetation is cropland with upland shrubs and coniferous forest at higher elevations. We used topography from the National Elevation Database (NED) and land-cover data from the National Land Cover Dataset [42]. The meteorological data were from the National Aeronautics and Space Administration's (NASA's) North American Land Data Assimilation System (NLDAS, accessed at https://disc.gsfc.nasa.gov/datasets?keywords=NLDAS, accessed on: 5 December 2015). For this work, the spatial changes in snow duration, rain-on-snow (ROS), and SWE runoff are presented to illustrate the general basin trends.

2.3. Methods

2.3.1. Air Temperature and SWE Trends

For each SNOTEL station and the Yakima Airport, we calculated both the average monthly air temperatures for each month from December through February and the average air temperature for the winter months (December–February) for WY1991–WY2020.

From these averages, we performed a linear fit analysis and calculated the yearly increase in monthly air temperatures over our studied time period. We recognize that assuming a linear trend in time may underestimate temperature changes. If we were to fit a more complex model to the future project climate data, there would be considerable feedbacks that may lead to stronger increases in temperature. Thus, we believe that, by using a simple linear approximation for modeling, our extrapolation of the air temperature is an appropriate approach, particularly in combination with our other multiple (and nonlinear) methods applied in our work. We used this linear increase to estimate the future year when the average air temperatures would be above freezing. This was used to calculate the percentage of stations that would be above freezing (or already at positive temperatures) by 2020, 2040, 2060, 2080, and 2100.

We also identified maximum peak SWE and 1 April SWE for each of the years available for the SNOTEL stations. For each station, we performed a linear fit trend analysis from the time series (WY1991–WY2020) of maximum peak SWE and 1 April SWE. These results were used to determine the historical SWE change. Additionally, we investigated the trends of the SWE accumulation and snow-free dates for all years in the time period. These dates were chosen from longest period of continuous snow cover for each WY following the definition by Liston and Hiemstra [4].

2.3.2. Streamflow Trends and Generalized Extreme Value (GEV)

We analyzed the weekly, seasonal, and annual trends of maximum and minimum streamflow trends for the eight USGS gages using GFDL-ESM2M and CanESM2. We consider a historical time period (1991–2010) and a future time period (2070–2099), along with the entire duration of time (1991–2099). Both trends and the GEV approach examine the annual (YR) and seasonal (December, January, February (DJF); March, April, May (MAM); June, July, August (JJA); September, October, November (SON)) time series trends. Nonparametric trends were calculated using R's zyp package, using Sen's slope to estimate the magnitude of the trend and the Mann–Kendall to estimate the trend significance; significance was determined at a p-value ≤ 0.05.

We utilized a GEV distribution to further consider the type of changes projected in streamflow. GEV is based on the theory of extremal limits, which states that "a sufficiently long time series of block maxima will approach the GEV distribution asymptotically at large sample sizes" [50]. We applied a technique that allowed us to consider the best fit form of nonstationarity for the streamflow. For this component of the work, we considered only the maximum streamflow and examined the time period from 1991–2099. Additionally, for the GEV, we focused only on GFDL-ES2M streamflow projections, because of the strong similarities found between CanESM2 and GFDL-ES2M trend results (see Figures S2 and S3, Table S3).

The GEV approach in this study used the R-project's Generalized Extreme Values conditional density estimation network (GEVcdn) package [51,52], which was described in detail in Bennett et al. [53]. For this application, we followed the same criteria to select the 'best' model on the basis of the minimized Akaike information criteria, corrected for small sample sizes (AICc) [54]. We considered five candidate models, including a stationary (S), a linear nonstationary (LNS), and three nonlinear nonstationary models with changing the number of nodes from one to three. In the S model, the parameters were not allowed to vary in time, whereas location and scale parameters were allowed to vary in time in the other candidate models.

3. Results

3.1. Air Temperature at SNOTEL Sites and Yakima Airport

To investigate regions vulnerable to changes in water storage and availability, as well as future increases in air temperatures, monthly and seasonal air temperature trends for each SNOTEL site and Yakima Airport were calculated for WY 1991–2020. From these trends, we projected air temperatures for each SNOTEL site and the years when temperatures above freezing would occur (Figure 2a). Such an increase in air temperature to above freezing indicates a lesser snowpack in that year, which would most likely result in a decrease in water storage. As expected, the air temperatures were lower at higher elevations. Surprisingly, the air temperatures in February increased at a lower rate than during other months. In fact, in February, three stations (Trough, Corral Pass, and Harts Pass) did not result in an increasing trend in air temperature, indicating that temperatures at these stations will continue to be below freezing during this century. We found a significant increasing trend in air temperatures that varied both temporally and spatially. The monthly average air temperature across stations was 0.97 °C/decade (0.59 to 1.42 °C/decade) in December, 1.19 °C/decade (0.81 to 1.82 °C/decade) in January, and 0.28 °C/decade (no increase to 0.73 °C/decade) in February. The average air temperature increase for December–February was between 0.51 and 1.32 °C/decade (average of 0.82 °C/decade). Trends in yearly average air temperatures for 1991 to 2019 ranged from 0.33 to 1.03 °C/decade.

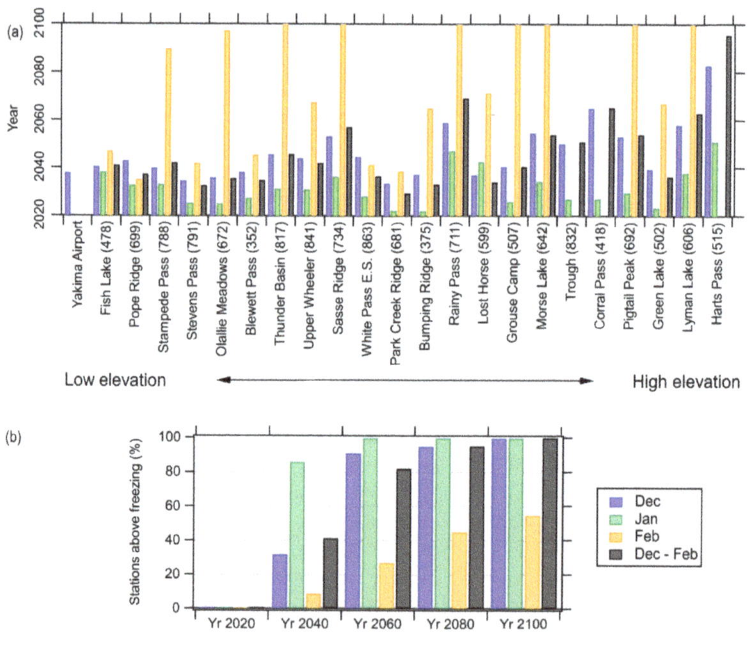

Figure 2. (**a**) Projected years when mean air temperatures will be above freezing for Yakima Airport and the 22 SNOTEL stations. Note that the current mean air temperature at the Yakima Airport was above freezing in 2020 for all months but December. Furthermore, there is no projection for the mean air temperature trends at Trough, Corral Pass, and Harts Pass during February because of a decreasing air temperature trend. Lowest to highest elevation stations are listed from left to right. (**b**) Projection of stations above freezing for winter months starting December through February and average for December through February.

From our linear trend analyses, we also estimated the percentage of studied SNOTEL stations where positive air temperatures were encountered for every 20 years starting in 2020 through 2100, to see if similar trends continued (Figure 2b). In year 2040, several stations were already at above freezing air temperatures. In fact, more than one-third of the stations (41%) experienced above freezing air temperatures in the period from December through February. This percentage increased to 82% in 2060, and, by 2080, only one station (Harts Pass) was below freezing temperatures. In 2100, below freezing was only encountered at about half of the stations in the month of February.

3.2. SWE Analysis

SWE observations at the 22 SNOTEL stations displayed high variability across both years and elevation (Figure 3). The SWE ranged from a minimum of zero (several stations) to close to a 2700 mm maximum at Pigtail Peak. The highest average 30 year SWE was at Lyman Lake (el. 1823 m), the second highest elevated SNOTEL station studied. The stations with the lowest average maximum SWE were, surprisingly, not at the lowest elevation stations but at stations with various elevations (Trough SWE = 255 mm, el. 1670 m; Upper Wheeler SWE = 303 mm, el. 1320 m; Blewett Pass SWE = 325 mm, el. 1292 m; Pope Ridge SWE = 424 mm, el. 1094 m; Grouse Camp SWE = 475 mm, el. 1643 m).

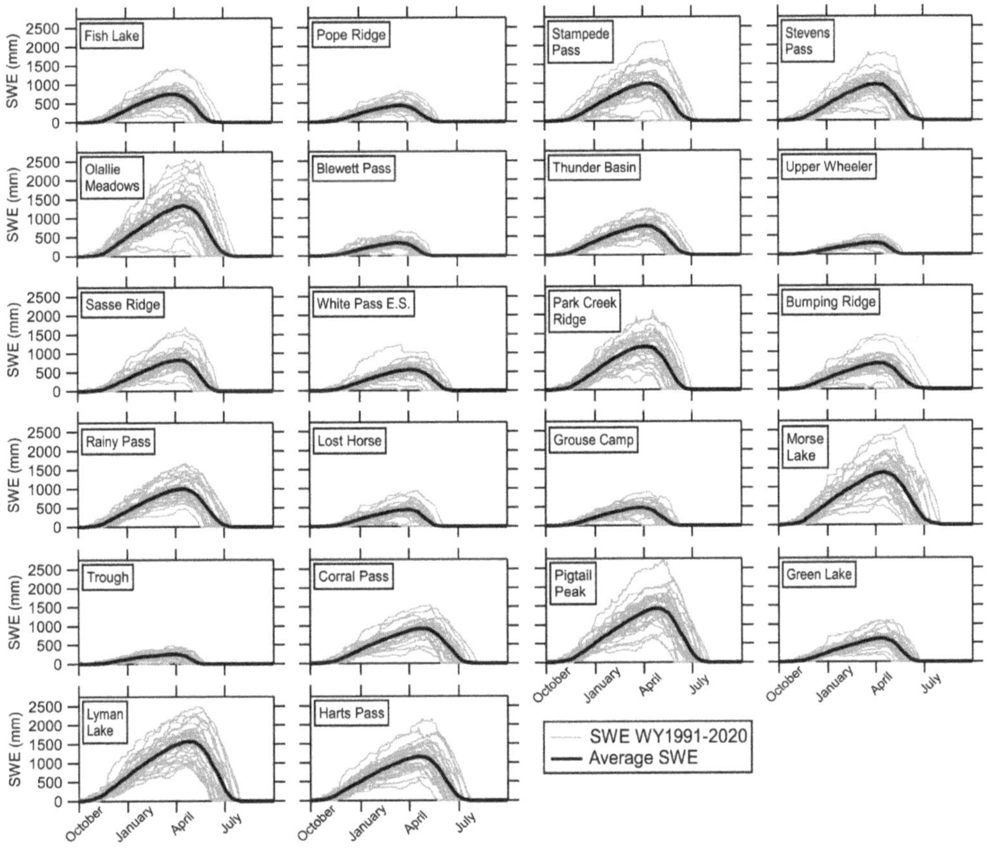

Figure 3. SWE for WY1991–WY2020 at 22 SNOTEL stations (see information in Figure 1 and Table S1). Gray lines indicate SWE for each WY and the black solid line is the average SWE for the WY1991-WY2020. The marks on the x-axis indicate the first day of each month.

From the SWE dataset (Figure 3), we calculated the 30 year linear trends of SWE accumulation (Figure 4a) and snow-free date (Figure 4b) for all stations. Most stations were experiencing a later SWE start date and an earlier SWE end date. There was a strong negative trend in SWE starting date for stations with low to high elevation, where the longest delay in start date was at the lower elevation stations. The longest delay in start date was 18 days (Sasse Ridge). Opposite to this trend, some of the highest elevation stations were experiencing an earlier start date (Pigtail Peak, Lyman Lake, and Harts Pass). Green Lake, also a high-elevation station, was experiencing a later SWE start date of 11 days. A positive trend was seen in SWE end dates where the end date was earlier at lower elevations compared to higher elevations. The earliest SWE end date was 19 days earlier, and this occurred at Blewett Pass. This station was also the station experiencing the greatest difference in the number of total SWE days, with 29 fewer days in WY 2020 compared to WY 1991.

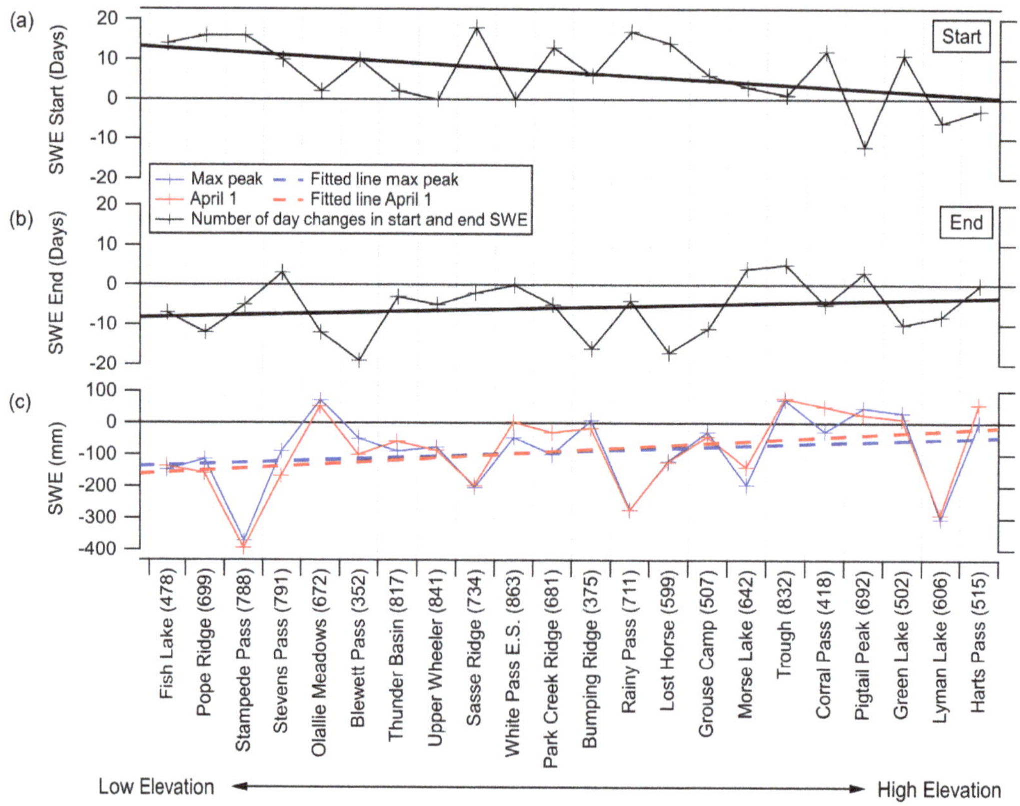

Figure 4. Trends in SWE start where a positive number indicates a later start date (**a**), and trends in SWE end where a negative number indicates an earlier end date (**b**); change in SWE for WY1991–WY2020 (**c**).

We also calculated the trend in peak SWE; there was a clear decreasing trend in peak SWE for the 30 year period (Figure 4c). The greatest decrease in peak SWE was 393 mm found at Stampede Pass. Among the 22 stations, this station was the third lowest elevation station. Even though it was one of the lowest elevation stations, it had a high SWE (max 1 April SWE peak of 2042 mm). In addition to investigating the trends in maximum SWE on 1 April, which is typically greatest and also commonly used by water resource planners, we also analyzed trends in maximum peak (Figure 4c). In general, the trends were similar; however, for the lower stations, the 1 April SWE was lower than the SWE peak.

3.3. Snow Model Simulations

Snow model simulations were performed to develop a spatial coverage of snow duration, ROS, and snowmelt runoff averages and trends from 1 September 1979–2015 for the studied area. Figure 5a–c shows the snow duration and trends during the core snow season. Following Liston and Hiemstra [4], we defined the core season as the longest period of continuous snow cover in each year (see the inset in Figure 5a). The average snow duration at lower elevations was less than 90 days during the 36 year period. For higher elevations, snow duration was over 300 days. Trends in snow-cover duration varied between ±40 days/decade. In general, lower elevations showed an increase in snow trends and, at intermediate and higher elevations, snow-cover duration increased. When averaging the yearly snow duration for the study area, it showed a declining trend of 20 days/decade for the 36 year period (Figure 5c). The number of ROS events was low at lower elevations and up to 50 days/year at higher elevations where snowpacks are long-lived (Figure 5d). Most of the studied area indicated almost no change in ROS trends (Figure 5e). Overall, there was a decreasing trend of ROS events (Figure 5f). As expected, higher elevations corresponded to higher snowmelt runoff values (400 cm) compared to much smaller values at lower elevations (Figure 5g). The entire modeling domain showed mostly negative snowmelt runoff trends, with −10 cm/decade at the highest elevations not uncommon. A decline in average snowmelt runoff was, therefore, no surprise (Figure 5i).

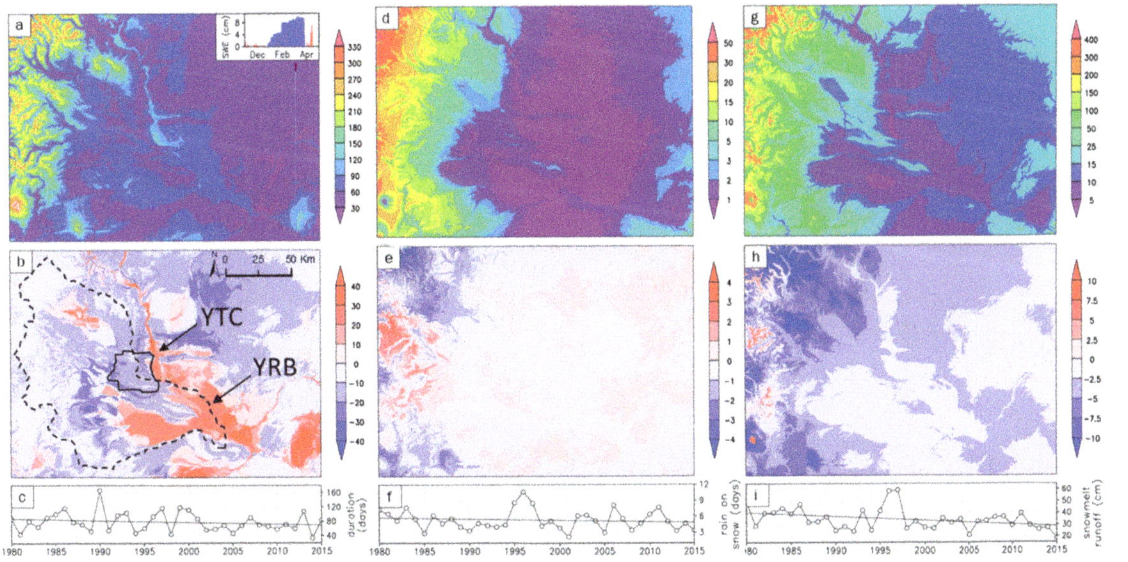

Figure 5. Snow model results for simulations (1979 to 2015): (**a**) average 36 year snow duration, (**b**) snow duration trend per decade, (**c**) yearly snow duration area average, (**d**) average 36 year ROS events, (**e**) ROS trend per decade, (**f**) yearly ROS area average, (**g**) average 36 year snowmelt runoff, (**h**) snowmelt runoff trend per decade, and (**i**) yearly snowmelt runoff area average (modified from [36]).

3.4. Streamflow Patterns and Trends

We analyzed historical streamflow and two ESMs at eight USGS gages (Table S2) for the mean, maximum, and minimum streamflow historical and future streamflow patterns (Figure 6). The historical streamflow was lowest in August–September and increased gradually until peak flow in late spring or early summer, with several systems experiencing higher flows in October–December. In all but the Tieton River watershed, peak flow occurred close to mid-June to mid-July, with higher-elevation gages peaking later in the year. Future streamflow patterns exhibit large shifts, with streamflow peaking much earlier in the season (i.e., January through May) in the Yakima River basin, with the higher-elevation systems (Wenatchee, Stehekin, Methow) peaking approximately 1 month earlier. Watersheds ranged in size from the Yakima River USGS station at Parker to the smaller American River at Nile station; hence, flows also varied on the basis of this factor.

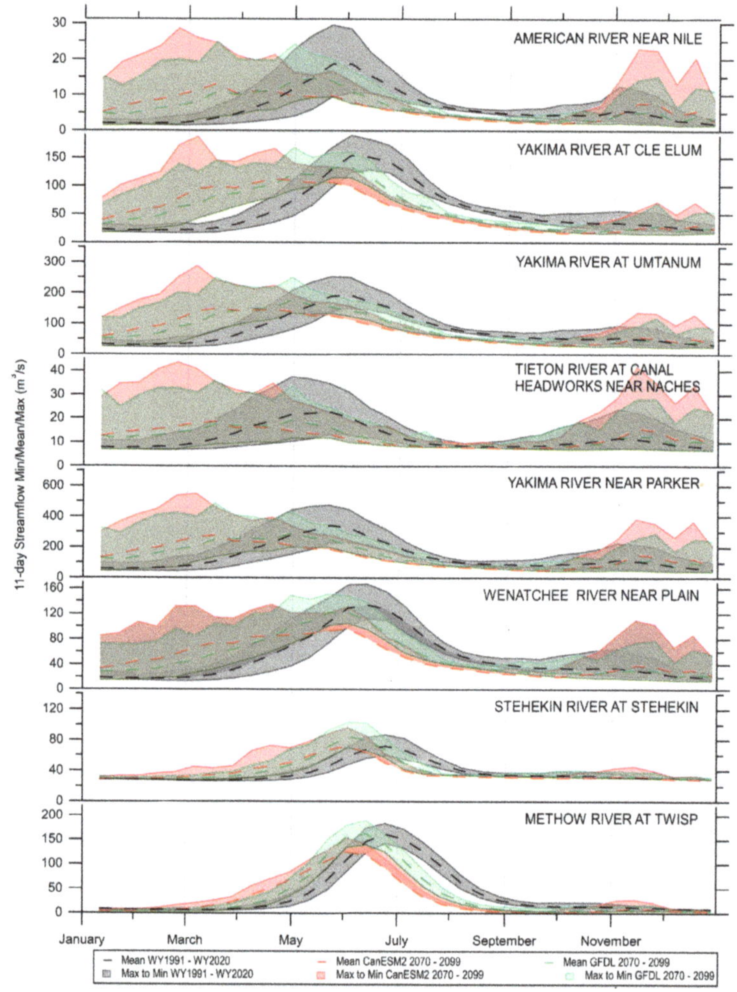

Figure 6. Streamflow trends for the eight USGS stations.

Over the 109 year period, minimum streamflow trends are projected to decrease significantly in summer and fall at almost all stations with the exception of Tieton River and Stehekin River, where only small declines are projected. Spring minimum streamflow trends are expected to increase at all stations, with the exception of Methow River, where streamflow trends are projected to decrease in all seasons and annually. Maximum streamflow is expected to increase significantly in winter at all stations with the exception of Methow River. Furthermore, for maximum streamflow, trends for all stations are projected to significantly decrease in summer, again with the exception of Stehekin River, where summer streamflow is expected to increase (not significant). Annually, maximum streamflow patters are projected to increase over the 109 year period, while minimum streamflow trends are expected to decrease overall at most stations.

3.5. GEV Analysis

The GEV results illustrate how the maximum winter and maximum summer streamflow is projected to change in the future for the Yakima River basin region. For the GFDL-ESM2M data, the *LNS* model was minimized for the annual, winter, and summer streamflow maximums (Figures S4 and S5). The linear, nonstationary responses indicate that shifts in the systems will occur linearly with strong trends occurring in the systems over time, indicative of the overarching impacts of snowfall reductions and rainfall increases observed in the region. The lack of nonlinear responses in these systems reinforces that the changes are projected to be unidirectional under the snow-to-rain transition in regime. The GEV revealed strongly increasing maximum streamflow in winter (DJF) over the historical-to-future periods, while summer (JJA) maximum streamflow decreased over the period at all stations, further supporting the trend analysis results for the region. Annually, the three systems north of Yakima (Wenatchee, Stehekin, and Methow) are projected to exhibit some slight downward shift in trends by the 2050s, indicating a potential change in the systems, possibly indicative of the shift toward a loss of snowpack, transitioning toward a rainfall-dominant system by this time.

4. Discussion

Our study presented historical and future trends in multiple indicators (air temperature, precipitation (snow and rain), and timing of minimum, average, and maximum streamflow) for YRB, a snow-dominated region in the western US. We present three decades of air temperature trends, SWE measurements, and snow duration trends, in addition to simulated streamflow trends and GEV analysis from present to 2100. This study offers insight into how changes in climate affect this region, and how these changes impact streamflow. Even though hydrological simulations in a heavily irrigated watershed like the YRB can be difficult [55], results as presented in this study across multiple indicators represent the best estimate of how changes in climate could impact the region.

The air temperature trends for a 30 year historical period indicates that air temperatures are increasing at the majority of YRB stations and during most winter months except for February. Confirming these colder trends is the study by Kapnick and Hall [56], which reported lower warming trends for February for 12 western US states compared to other months at more isolated locations. As such, our results indicate that YRB is in a vulnerable transitional phase between precipitations falling as snow or rain. Hamlet and Lettenmaier [57] reported that, in a transient watershed, any change in temperature could have a great impact on runoff. Therefore, these increasing trends in air temperature will affect future water storage in the basin. For our study period (WY1991–WY2020), the highest increase in monthly air temperature observed was 1.82 °C/decade (January), recorded at a mid-elevation SNOTEL location (Park Ridge). We saw an increase in winter temperature (December–February) of 0.82 °C/decade as an average between the 22 stations. Similar warming was reported by Hu and Nolin [58], who indicated a warming trend of 0.4–1.2 °C/decade for the Interior West. Other studies also estimated warming rates of about 0.5 °C/decade [56,59,60], which are similar to the trends in yearly average air temperatures reported in this study (0.33 to 1.03 °C/decade). This extremely high rate of warming across the region that is already being observed historically is indicative of the rate of warming that has already occurred across the YRB.

As expected, associated with the increasing trend in air temperature, our analyses showed a decrease in both peak SWE and 1 April SWE over the 30 year period (WY1991–WY2020). Moreover, trends at most stations have experienced a delay in SWE start date and an earlier SWE end date, resulting in a shorter snow duration. A trend in shorter snow duration over the period of WY1980–WY2015 for the snow modeling area was also evident. Considering these changes, as recommended by other studies (e.g., [32,61,62]), a different date for peak SWE than 1 April should be considered for water resource planning.

Previous studies in the Washington area showed similar outcomes to our analyses, where major shifts in snowpack and precipitation transitions from snow to rain were identified. Klos, Link, and Abatzoglou [32] projected changes in winter precipitation (December–February) from late 20th century (1979–2012) for HUC4 watersheds, and Yakima indicated a 16% decrease in snow-dominated extent and a 25% increase in rain-dominated extent by the mid-21st century (2036–2050). Previous work predicted that snowmelt-dominated watersheds in the region will be reduced by 2080 [63]. Yakima River is one of three western US watersheds predicted to be almost entirely depleted in 1 April SWE [30] by the 2080s. The authors also reported that, by 2080, the Yakima watershed will experience a shift from a transient rain–snow basin to a rain-dominated basin [30].

In a larger part of the US, the major driver for severe spring flooding is snow meltwater [64,65]. ROS events are a large cause of peak flow events [64,66]. At higher elevations in snow-dominated areas, the projected future increase in ROS makes flood risk assessments complicated [67,68]. Our snow modeling simulations for the historical period of WY1980–WY2015 showed a similar outcome, where an increase of up to 4 days/decade was seen in some higher elevation areas.

Patterns in streamflow indicate a system in transition between the historical periods of 1970–1999, as well as into the future to 2099. Under a slightly drier future climate scenario, all streamflow at most sites is projected to experience declining summer minimum and maximum streamflow, with one high-elevation station (Methow) experiencing declines across all seasons and annually over the 1991–2099 period. Many systems are also projected to experience increasing spring minimum streamflow, due to higher winter streamflow influenced by increased rain in winter. On the other hand, winter streamflow is projected to increase strongly across the region. These finding have strong implications for increasing wintertime flooding events and drought scenarios in summer, highlighting the potential for increased extreme events of opposing direction that may occur throughout the water year.

When considering changes using the GEV approach, we observed a linear and strongly decreasing summer streamflow and increasing winter flows, again reinforcing the trends and the transition in the systems toward a rainfall-dominant winter streamflow pattern, with a shift in snowmelt streamflow peaks to an earlier and more drawn out melt, leading to lower streamflow availability during the summer. The lack of nonlinearity in responses highlights the combined impact of increasing temperature, reductions in snow, and increases in rainfall on the systems observed across the region.

In a study where the focus was to assess hydrology implications from climate change of Washington state, Elsner et al. [30] projected a similar shift to that reported in this study in future streamflow peak from summer to winter for Yakima River at Parker (see location in Figure 1) when compared to historical measurements. Our results are also comparable to a study by Vano et al. [69], who performed a hydrological seasonal sensitivity study to climate variations for several regions in the Pacific Northwest. They reported similar shifts in streamflow for Yakima River near Parker. In their study, out of the five basins studied, the Yakima River basin was the most sensitive to warming. The sensitivity is highlighted here in the observed trends in streamflow projected to occur, with similar responses noted for both the irrigated Yakima system and the adjacent, unimpacted river basins of Wenatchee, Stehekin, and Methow. These findings have implications for other systems adjacent to the sites we studied, including those systems draining into the Columbia River basin.

The changes observed in the multiple indicators have broad-reaching implications for flow. Water managers, agriculturalists, irrigation managers, and urban planners may require adjustments to planning and water retention scheduling to accommodate and retain winter rainfall storage for drier summer months in lieu of the reduced natural water storage in snowpack. Importantly, YTC will want to take into account the drier summer conditions, which may affect military missions. Hanford may want to consider the possibility of increased movement of solutes and sediments posed by the increase in winter runoff, as well as the overall increase in water moving through the surface, subsurface, and soils in the region. Additionally, the declines in summer streamflow will likely have an impact on other important components of the ecology and economy in the region, such as fisheries. Studies in the YRB indicate that a warming climate will decrease the salmonid habitat due to both increased streamflow during winter months and higher water temperatures [70].

5. Summary and Conclusions

Future climate predictions indicate that temperatures will continue to increase, and that this can negatively impact snow-dominated watersheds in regions that rely on the snowpack for water resources. Therefore, it is greatly important to investigate changes in climate during the design phase of new infrastructure. Moreover, military operations and missions can be affected by alterations in climate, making it necessary to account for these changes during future land management. This study focused on historical and future trends in climate, as well as their effect on streamflow for the Yakima River Basin, a snow-dominated area located in the western US, in the same region as a large military installation (YTC). Our analyses included air temperature and SWE data from a total of 23 stations over a 30 year period. From the air temperature trends, we also projected by what year winter temperatures would be above freezing and would lead to a depleting snowpack. Additionally, we analyzed streamflow trends across five stations in the YRB and, because this watershed is heavily regulated, we also ran trend analyses for three close-by stations.

Our air temperature trend analyses agree with previous studies showing increasing air temperatures. The 30 year historical record of air temperature showed an increasing trend for almost each month starting in December through February, as well as for the 3 month average (December through February) air temperatures. Surprisingly, a few stations showed no increase in air temperature during the month of February, indicating this as the coldest of the months analyzed. More interesting, more than one-third of the stations (41%) are predicted to have above freezing air temperatures by 2040 in December through February. Most of these stations are located at lower elevations, but some stations are at higher elevations (i.e., Lost Horse and Green Lake). By 2060, a majority of the stations (82%) are projected to show air temperatures above freezing during this time period. The increase in air temperature will result in a shift in snow arrival and departure date, SWE peak flow, and maximum spring streamflow. The streamflow trend and GEV analyses indicated that the YRB basin is a system in transition, where streamflow is changing to earlier peak flows as a result of increasing temperatures. The findings from this study provide valuable input for planning infrastructure, water, and land management in the YRB region and locations with similar climate. Furthermore, this study highlights the local variability in climate, which proves the need for and importance of incorporating spatial differences during the design phase of hydraulic structures in order to avoid negative impacts.

Supplementary Materials: The following are available online at https://www.mdpi.com/article/10.3390/w13192608/s1, Table S1. Characteristics for the 22 automated SNOTEL stations from Natural Resources Conservation Service, including SNOTEL site name and number, latitude, longitude, elevation, and basin location. Table S2. Characteristics for the eight Washington state streamflow gages and sites analyzed in this study. Site name and abbreviated description, USGS ID, latitude, longitude, basin area, elevation of USGS gage, and basin location. Table S3. Seasonal and annual trends for minimum and maximum streamflow (m3/s) for 1991–2099 for GFDL-ES2M. Trends are values occuring over the 109 year time period. Bold values indicate significant trends (p-value ≤ 0.05). Figure S1. Precipitation comparison between 2006-2016 (current/historical) and 2070-2099 (future) for the CORDEX RCP 8.5 models that support the selection of our two future models. The CanESM2.CanRCM4 model shows the wettest winter and spring for the current climate compared to future projections and the GFDL-ESM2M.WRF model is the driest (winter and spring) in comparison to the future climate. Figure S2. Left, GFDL-ESM2M for 1991–2099 maximum streamflow. Right, CanESM2 for 1991–2099 maximum streamflow. Darker shades of blue represent p-value ≤ 0.01, while lighter shades of blue represent p-value ≤ 0.05 for statistical significance. Figure S3. Left, GFDL-ESM2M for 1991–2099 minimum streamflow trends. Right, CanESM2 for 1991–2099 minimum streamflow trends. Darker shades of blue represent p-value ≤ 0.01, while lighter shades of blue represent p-value ≤ 0.05 for statistical significance. Figure S4. GEV results for eight stations for the maximum annual streamflow from 1991–2099 for the GFDL- ESM2M model. Figure S5. GEV results for eight stations for the maximum DJF (top) and JJA (bottom) streamflow from 1991–2099 for the GFDL-ESM2M model.

Author Contributions: Conceptualization, methodology, and validation, A.M.W. and K.E.B.; formal analysis, A.M.W., K.E.B., and G.E.L.; investigation, A.M.W., K.E.B., and C.A.H.; data curation, A.M.W., K.E.B., and G.E.L.; writing—original draft preparation, A.M.W. and K.E.B.; writing—review and editing, A.M.W., K.E.B., C.A.H., G.E.L., and D.C.; visualization, A.M.W., K.E.B., and G.E.L. All authors have read and agreed to the published version of the manuscript.

Funding: This study was supported by the Strategic Environmental Research and Development Program (SERDP) Resource Conservation (RC) and Resilience project RC-2515.

Institutional Review Board Statement: Not applicable.

Informed Consent Statement: Not applicable.

Data Availability Statement: The data are available upon reasonable request to the corresponding author.

Acknowledgments: This study was supported by the Strategic Environmental Research and Development Program (SERDP) Resource Conservation (RC) and Resilience project RC-2515. The authors would also like to thank the anonymous reviewers that helped improve this manuscript. Permission to publish was granted by the Director of Cold Regions Research and Engineering Laboratory.

Conflicts of Interest: The authors declare no conflict of interest. The funders had no role in the design of the study; in the collection, analyses, or interpretation of data; in the writing of the manuscript, nor in the decision to publish the results.

References

1. Barnett, T.P.; Adam, J.C.; Lettenmaier, D.P. Potential impacts of a warming climate on water availability in snow-dominated regions. *Nature* **2005**, *438*, 303–309. [CrossRef] [PubMed]
2. Edwards, A.C.; Scalenghe, R.; Freppaz, M. Changes in the seasonal snow cover of alpine regions and its effect on soil processes: A review. *Quat. Int.* **2007**, *162*, 172–181. [CrossRef]
3. Stewart, I.T. Changes in snowpack and snowmelt runoff for key mountain regions. *Hydrol Process* **2009**, *23*, 78–94. [CrossRef]
4. Liston, G.E.; Hiemstra, C.A. The Changing Cryosphere: Pan-Arctic Snow Trends (1979–2009). *J. Clim.* **2011**, *24*, 5691–5712. [CrossRef]
5. Derksen, C.; Brown, R. Spring snow cover extent reductions in the 2008–2012 period exceeding climate model projections. *Geophys. Res. Lett.* **2012**, *39*, L19504. [CrossRef]
6. Hall, D.K.; Crawford, C.J.; DiGirolamo, N.E.; Riggs, G.A.; Foster, J.L. Detection of earlier snowmelt in the Wind River Range, Wyoming, using Landsat imagery, 1972–2013. *Remote Sens. Environ.* **2015**, *162*, 45–54. [CrossRef]
7. Hori, M.; Sugiura, K.; Kobayashi, K.; Aoki, T.; Tanikawa, T.; Kuchiki, K.; Niwano, M.; Enomoto, H. A 38-year (1978–2015) Northern Hemisphere daily snow cover extent product derived using consistent objective criteria from satellite-borne optical sensors. *Remote Sens. Environ.* **2017**, *191*, 402–418. [CrossRef]
8. Hock, R.; Rasul, G.; Adler, C.; Cáceres, B.; Gruber, S.; Hirabayashi, Y.; Jackson, M.; Kääb, A.; Kang, S.; Kutuzov, S. High Mountain Areas. In *IPCC Special Report on the Ocean and Cryosphere in a Changing Climate*; NOAA: Washington, DC, USA, 2019.
9. Meredith, M.; Sommerkorn, M.; Cassotta, S.; Derksen, C.; Ekaykin, A.; Hollowed, A.; Kofinas, G.; Mackintosh, A.; Melbourne-Thomas, J.; Muelbert, M. Polar Regions. Chapter 3. In *IPCC Special Report on the Ocean and Cryosphere in a Changing Climate*; NOAA: Washington, DC, USA, 2019.
10. Mudryk, L.; Santolaria-Otín, M.; Krinner, G.; Ménégoz, M.; Derksen, C.; Brutel-Vuilmet, C.; Brady, M.; Essery, R. Historical Northern Hemisphere snow cover trends and projected changes in the CMIP6 multi-model ensemble. *Cryosphere* **2020**, *14*, 2495–2514. [CrossRef]
11. Pulliainen, J.; Luojus, K.; Derksen, C.; Mudryk, L.; Lemmetyinen, J.; Salminen, M.; Ikonen, J.; Takala, M.; Cohen, J.; Smolander, T. Patterns and trends of Northern Hemisphere snow mass from 1980 to 2018. *Nature* **2020**, *581*, 294–298. [CrossRef]
12. Mote, P.W.; Hamlet, A.F.; Clark, M.P.; Lettenmaier, D.P. Declining mountain snowpack in western North America. *Bull. Am. Meteorol. Soc.* **2005**, *86*, 39–49. [CrossRef]
13. Yang, D.Q.; Zhao, Y.Y.; Armstrong, R.; Robinson, D.; Brodzik, M.J. Streamflow response to seasonal snow cover mass changes over large Siberian watersheds. *J. Geophys. Res. Earth Surf.* **2007**, *112*. [CrossRef]
14. Clow, D.W. Changes in the timing of snowmelt and streamflow in Colorado: A response to recent warming. *J. Clim.* **2010**, *23*, 2293–2306. [CrossRef]
15. Pederson, G.T.; Gray, S.T.; Ault, T.; Marsh, W.; Fagre, D.B.; Bunn, A.G.; Woodhouse, C.A.; Graumlich, L.J. Climatic Controls on the Snowmelt Hydrology of the Northern Rocky Mountains. *J. Clim.* **2011**, *24*, 1666–1687. [CrossRef]
16. Gan, T.Y.; Barry, R.G.; Gizaw, M.; Gobena, A.; Balaji, R. Changes in North American snowpacks for 1979–2007 detected from the snow water equivalent data of SMMR and SSM/I passive microwave and related climatic factors. *J. Geophys. Res. Atmos.* **2013**, *118*, 7682–7697. [CrossRef]
17. Li, D.; Wrzesien, M.L.; Durand, M.; Adam, J.; Lettenmaier, D.P. How much runoff originates as snow in the western United States, and how will that change in the future? *Geophys. Res. Lett.* **2017**, *44*, 6163–6172. [CrossRef]

18. Stewart, I.T.; Cayan, D.R.; Dettinger, M.D. Changes in snowmelt runoff timing in western North America under abusiness as usual'climate change scenario. *Clim. Change* **2004**, *62*, 217–232. [CrossRef]
19. Serreze, M.C.; Clark, M.P.; Armstrong, R.L.; McGinnis, D.A.; Pulwarty, R.S. Characteristics of the western United States snowpack from snowpack telemetry (SNOTEL) data. *Water Resour. Res.* **1999**, *35*, 2145–2160. [CrossRef]
20. McCabe, G.J.; Clark, M.P.; Hay, L.E. Rain-on-snow events in the western United States. *Bull. Am. Meteorol. Soc.* **2007**, *88*, 319–328. [CrossRef]
21. Cayan, D.R.; Kammerdiener, S.A.; Dettinger, M.D.; Caprio, J.M.; Peterson, D.H. Changes in the onset of spring in the western United States. *Bull. Am. Meteorol. Soc.* **2001**, *82*, 399–416. [CrossRef]
22. Gergel, D.R.; Nijssen, B.; Abatzoglou, J.T.; Lettenmaier, D.P.; Stumbaugh, M.R. Effects of climate change on snowpack and fire potential in the western USA. *Clim. Change* **2017**, *141*, 287–299. [CrossRef]
23. Mote, P.W.; Li, S.; Lettenmaier, D.P.; Xiao, M.; Engel, R. Dramatic declines in snowpack in the western US. *Npj Clim. Atmos. Sci.* **2018**, *1*, 1–6. [CrossRef]
24. Knowles, N. Trends in snow cover and related quantities at weather stations in the conterminous United States. *J. Clim.* **2015**, *28*, 7518–7528. [CrossRef]
25. Gleick, P.H. The development and testing of a water balance model for climate impact assessment: Modeling the Sacramento basin. *Water Resour. Res.* **1987**, *23*, 1049–1061. [CrossRef]
26. Lettenmaier, D.P.; Gan, T.Y. Hydrologic sensitivities of the Sacramento-San Joaquin River basin, California, to global warming. *Water Resour. Res.* **1990**, *26*, 69–86. [CrossRef]
27. Dettinger, M.D.; Cayan, D.R.; Meyer, M.K.; Jeton, A.E. Simulated hydrologic responses to climate variations and change in the Merced, Carson, and American River basins, Sierra Nevada, California, 1900–2099. *Clim. Change* **2004**, *62*, 283–317. [CrossRef]
28. Knowles, N.; Cayan, D.R. Elevational dependence of projected hydrologic changes in the San Francisco estuary and watershed. *Clim. Change* **2004**, *62*, 319–336. [CrossRef]
29. Lemke, P.; Ren, J.; Alley, R.B.; Allison, I.; Carrasco, J.; Flato, G.; Fujii, Y.; Kaser, G.; Mote, P.; Thomas, R.H.; et al. Observations: Changes in Snow, Ice and Frozen Ground. In *Climate Change 2007: The Physical Science Basis. Contribution of Working Group I to the Fourth Assessment Report of the Intergovernmental Panel on Climate Change*; Solomon, S., Qin, D., Manning, M., Chen, Z., Marquis, M., Averyt, K.B., Tignor, M., Miller, H.L., Eds.; Cambridge University Press: Cambridge, UK; New York, NY, USA, 2007.
30. Elsner, M.M.; Cuo, L.; Voisin, N.; Deems, J.S.; Hamlet, A.F.; Vano, J.A.; Mickelson, K.E.; Lee, S.-Y.; Lettenmaier, D.P. Implications of 21st century climate change for the hydrology of Washington State. *Clim. Change* **2010**, *102*, 225–260. [CrossRef]
31. Rupp, D.E.; Mote, P.W.; Bindoff, N.L.; Stott, P.A.; Robinson, D.A. Detection and attribution of observed changes in Northern Hemisphere spring snow cover. *J. Clim.* **2013**, *26*, 6904–6914. [CrossRef]
32. Klos, P.Z.; Link, T.E.; Abatzoglou, J.T. Extent of the rain-snow transition zone in the western US under historic and projected climate. *Geophys. Res. Lett.* **2014**, *41*, 4560–4568. [CrossRef]
33. Lute, A.; Abatzoglou, J.; Hegewisch, K. Projected changes in snowfall extremes and interannual variability of snowfall in the western U nited S tates. *Water Resour. Res.* **2015**, *51*, 960–972. [CrossRef]
34. Stewart, I.T.; Cayan, D.R.; Dettinger, M.D. Changes toward earlier streamflow timing across western North America. *J. Clim.* **2005**, *18*, 1136–1155. [CrossRef]
35. Casola, J.H.; Cuo, L.; Livneh, B.; Lettenmaier, D.P.; Stoelinga, M.T.; Mote, P.W.; Wallace, J.M. Assessing the impacts of global warming on snowpack in the Washington Cascades. *J. Clim.* **2009**, *22*, 2758–2772. [CrossRef]
36. Wagner, A.M.; Hiemstra, C.A.; Liston, G.E.; Bennett, K.B.; Cooley, D.S.; Gelvin, A.B. *Changes in Climate and its Effect on Timing of Snowmelt and Intensity-Duration-Frequency Curves*; ERDC/CRREL TR-21-8; U.S. Army Engineer Research and Development Center, Cold Regions Researech and Engineering Laboratory: Hanover, NH, USA, 2021; p. 147.
37. Warren, S.D.; Ruzycki, T.S.; Vaughan, R.; Nissen, P.E. Validation of the Unit Stream Power Erosion and Deposition (USPED) Model at Yakima Training Center, Washington. *Northwest Sci.* **2019**, *92*, 338–345.
38. Hillman, B.; Douglas, E.M.; Terkla, D. An analysis of the allocation of Yakima River water in terms of sustainability and economic efficiency. *J. Environ. Manag.* **2012**, *103*, 102–112. [CrossRef]
39. Vano, J.A.; Scott, M.J.; Voisin, N.; Stöckle, C.O.; Hamlet, A.F.; Mickelson, K.E.; Elsner, M.M.; Lettenmaier, D.P. Climate change impacts on water management and irrigated agriculture in the Yakima River Basin, Washington, USA. *Clim. Change* **2010**, *102*, 287–317. [CrossRef]
40. Mitchell, C.J.; Lawrence, E.; Chu, V.R.; Harris, M.J.; Landis, W.G.; von Stackelberg, K.E.; Stark, J.D. Integrating metapopulation dynamics into a Bayesian network relative risk model: Assessing risk of pesticides to Chinook salmon (Oncorhynchus tshawytscha) in an ecological context. *Integr. Environ. Assess. Manag.* **2021**, *17*, 95–109. [CrossRef]
41. Vaccaro, J. *Comparison of Unregulated and Regulated Streamflow for the Yakima River at Union Gap and Near Parker, Washington*; US Department of the Interior, Geological Survey: Tacoma, WA, USA, 1986; Volume 82.
42. Yang, L.; Jin, S.; Danielson, P.; Homer, C.; Gass, L.; Bender, S.M.; Case, A.; Costello, C.; Dewitz, J.; Fry, J. A new generation of the United States National Land Cover Database: Requirements, research priorities, design, and implementation strategies. *ISPRS J. Photogramm. Remote Sens.* **2018**, *146*, 108–123. [CrossRef]
43. Chegwidden, O.; Nijssen, D.; Rupp, D.; Mote, P. *Hydrologic Response of the Columbia River System to Climate Change*; Zenodo: Geneva, Switzerland, 2017.

44. Queen, L.E.; Mote, P.W.; Rupp, D.E.; Chegwidden, O.; Nijssen, B. Ubiquitous increases in flood magnitude in the Columbia River basin under climate change. *Hydrol. Earth Syst. Sci.* **2021**, *25*, 257–272. [CrossRef]
45. Mearns, L.; McGinnis, S.; Korytina, D.; Arritt, R.; Biner, S.; Bukovsky, M.; Chang, H.; Christensen, O.; Herzmann, D.; Jiao, Y. *The NA-CORDEX Dataset, Version 1.0*; NCAR Climate data Gateway: Boulder, CO, USA, 2017. [CrossRef]
46. Hamlet, A.F.; Elsner, M.M.; Mauger, G.S.; Lee, S.-Y.; Tohver, I.; Norheim, R.A. An overview of the Columbia Basin Climate Change Scenarios Project: Approach, methods, and summary of key results. *Atmos. Ocean* **2013**, *51*, 392–415. [CrossRef]
47. Scinocca, J.; Kharin, V.; Jiao, Y.; Qian, M.; Lazare, M.; Solheim, L.; Flato, G.; Biner, S.; Desgagne, M.; Dugas, B. Coordinated global and regional climate modeling. *J. Clim.* **2016**, *29*, 17–35. [CrossRef]
48. Liston, G.E.; Elder, K. A distributed snow-evolution modeling system (SnowModel). *J. Hydrometeorol.* **2006**, *7*, 1259–1276. [CrossRef]
49. Liston, G.E.; Itkin, P.; Stroeve, J.; Tschudi, M.; Stewart, J.S.; Pedersen, S.H.; Reinking, A.K.; Elder, K. A Lagrangian snow-evolution system for sea-ice applications (SnowModel-LG): Part I—Model description. *J. Geophys. Res. Ocean.* **2020**, *125*, e2019JC015913. [CrossRef] [PubMed]
50. Coles, S.; Bawa, J.; Trenner, L.; Dorazio, P. *An Introduction to Statistical Modeling of Extreme Values*; Springer: Berlin/Heidelberg, Germany, 2001; Volume 208.
51. Cannon, A.J. A flexible nonlinear modelling framework for nonstationary generalized extreme value analysis in hydroclimatology. *Hydrol. Process.* **2010**, *24*, 673–685. [CrossRef]
52. Cannon, A.J. GEVcdn: An R package for nonstationary extreme value analysis by generalized extreme value conditional density estimation network. *Comput. Geosci.* **2011**, *37*, 1532–1533. [CrossRef]
53. Bennett, K.E.; Cannon, A.J.; Hinzman, L. Historical trends and extremes in boreal Alaska river basins. *J. Hydrol.* **2015**, *527*, 590–607. [CrossRef]
54. Burnham, K.P.; Anderson, D.R.; Huyvaert, K.P. AIC model selection and multimodel inference in behavioral ecology: Some background, observations, and comparisons. *Behav. Ecol. Sociobiol.* **2011**, *65*, 23–35. [CrossRef]
55. Qiu, J.; Yang, Q.; Zhang, X.; Huang, M.; Adam, J.C.; Malek, K. Implications of water management representations for watershed hydrologic modeling in the Yakima River basin. *Hydrol. Earth Syst. Sci.* **2019**, *23*, 35–49. [CrossRef]
56. Kapnick, S.; Hall, A. Causes of recent changes in western North American snowpack. *Clim. Dyn.* **2012**, *38*, 1885–1899. [CrossRef]
57. Hamlet, A.F.; Lettenmaier, D.P. Effects of 20th century warming and climate variability on flood risk in the western US. *Water Resour. Res.* **2007**, *43*. [CrossRef]
58. Hu, J.M.; Nolin, A.W. Widespread warming trends in storm temperatures and snowpack fate across the Western United States. *Environ. Res. Lett.* **2020**, *15*, 034059. [CrossRef]
59. Harpold, A.; Brooks, P.; Rajagopal, S.; Heidbuchel, I.; Jardine, A.; Stielstra, C. Changes in snowpack accumulation and ablation in the intermountain west. *Water Resour. Res.* **2012**, *48*. [CrossRef]
60. Harpold, A.A.; Brooks, P.D. Humidity determines snowpack ablation under a warming climate. *Proc. Natl. Acad. Sci. USA* **2018**, *115*, 1215–1220. [CrossRef]
61. Milly, P.; Betancourt, J.; Falkenmark, M.; Hirsch, R.M.; Kundzewicz, Z.W.; Lettenmaier, D.P.; Stouffer, R.J. Stationarity is dead: Whither water management? *Earth* **2008**, *4*, 20. [CrossRef]
62. Pederson, G.T.; Gray, S.T.; Woodhouse, C.A.; Betancourt, J.L.; Fagre, D.B.; Littell, J.S.; Watson, E.; Luckman, B.H.; Graumlich, L.J. The unusual nature of recent snowpack declines in the North American Cordillera. *Science* **2011**, *333*, 332–335. [CrossRef]
63. Mantua, N.; Tohver, I.; Hamlet, A. Climate change impacts on streamflow extremes and summertime stream temperature and their possible consequences for freshwater salmon habitat in Washington State. *Clim. Change* **2010**, *102*, 187–223. [CrossRef]
64. Berghuijs, W.R.; Woods, R.A.; Hutton, C.J.; Sivapalan, M. Dominant flood generating mechanisms across the United States. *Geophys. Res. Lett.* **2016**, *43*, 4382–4390. [CrossRef]
65. Li, D.; Lettenmaier, D.P.; Margulis, S.A.; Andreadis, K. The role of rain-on-snow in flooding over the conterminous United States. *Water Resour. Res.* **2019**, *55*, 8492–8513. [CrossRef]
66. Villarini, G. On the seasonality of flooding across the continental United States. *Adv. Water Resour.* **2016**, *87*, 80–91. [CrossRef]
67. Musselman, K.N.; Lehner, F.; Ikeda, K.; Clark, M.P.; Prein, A.F.; Liu, C.; Barlage, M.; Rasmussen, R. Projected increases and shifts in rain-on-snow flood risk over western North America. *Nat. Clim. Change* **2018**, *8*, 808–812. [CrossRef]
68. Jeong, D.I.; Sushama, L. Rain-on-snow events over North America based on two Canadian regional climate models. *Clim. Dyn.* **2018**, *50*, 303–316. [CrossRef]
69. Vano, J.A.; Nijssen, B.; Lettenmaier, D.P. Seasonal hydrologic responses to climate change in the Pacific Northwest. *Water Resour. Res.* **2015**, *51*, 1959–1976. [CrossRef]
70. Hatten, J.R.; Waste, S.M.; Maule, A.G. Assessing climate-change risks to cultural and natural resources in the Yakima River Basin, Washington, USA. *Clim. Change* **2014**, *124*, 363–370. [CrossRef]

water

MDPI

Article

Assessing Climatic Drivers of Spring Mean and Annual Maximum Flows in Western Canadian River Basins

Yonas B. Dibike [1,2,*], Rajesh R. Shrestha [1], Colin Johnson [1], Barrie Bonsal [3] and Paulin Coulibaly [2]

1 Environment and Climate Change Canada, Watershed Hydrology and Ecology Research Division, University of Victoria, 2472 Arbutus Rd., Victoria, BC V8N 1V8, Canada; rajesh.shrestha@canada.ca (R.R.S.); colinjohnson@uvic.ca (C.J.)
2 Civil Engineering Department and the School of Earth Environment and Society, McMaster University, Hamilton, ON L8S 4K1, Canada; couliba@mcmaster.ca
3 Environment and Climate Change Canada, Watershed Hydrology and Ecology Research Division, National Hydrology Research Centre, Saskatoon, SK S7N 3H5, Canada; barrie.bonsal@canada.ca
* Correspondence: yonas.dibike@canada.ca

Abstract: Flows originating from cold and mountainous watersheds are highly dependent on temperature and precipitation patterns, and the resulting snow accumulation and melt conditions, affecting the magnitude and timing of annual peak flows. This study applied a multiple linear regression (MLR) modelling framework to investigate spatial variations and relative importance of hydroclimatic drivers of annual maximum flows (AMF) and mean spring flows (MAMJflow) in 25 river basins across western Canada. The results show that basin average maximum snow water equivalent (SWEmax), April 1st SWE and spring precipitation (MAMJprc) are the most important predictors of both AMF and MAMJflow, with the proportion of explained variance averaging 51.7%, 44.0% and 33.5%, respectively. The MLR models' abilities to project future changes in AMF and MAMJflow in response to changes to the hydroclimatic controls are also examined using the Canadian Regional Climate Model (CanRCM4) output for RCP 4.5 and RCP8.5 scenarios. The results show considerable spatial variations depending on individual watershed characteristics with projected changes in AMF ranging from −69% to +126% and those of MAMJflow ranging from −48% to +81% by the end of this century. In general, the study demonstrates that the MLR framework is a useful approach for assessing the spatial variation in hydroclimatic controls of annual maximum and mean spring flows in the western Canadian river basins. However, there is a need to exercise caution in applying MLR models for projecting changes in future flows, especially for regulated basins.

Keywords: peak flows; multiple linear regression; predictor; predictand; snow water equivalent; annual maximum flow; climate change; western Canada

Citation: Dibike, Y.B.; Shrestha, R.R.; Johnson, C.; Bonsal, B.; Coulibaly, P. Assessing Climatic Drivers of Spring Mean and Annual Maximum Flows in Western Canadian River Basins. *Water* **2021**, *13*, 1617. https://doi.org/10.3390/w13121617

Academic Editor: Thomas Meixner

Received: 21 April 2021
Accepted: 5 June 2021
Published: 8 June 2021

Publisher's Note: MDPI stays neutral with regard to jurisdictional claims in published maps and institutional affiliations.

1. Introduction

Streamflows originating from cold and mountainous regions are significantly affected by increasing air temperature and changes in precipitation patterns associated with global warming. A warming climate results in a shift in precipitation from snow towards rain, affecting the snowpack volume and snowmelt timing [1]. The magnitude and timing of peak streamflow events are also affected, often exacerbating flood events and causing significant damages [2–4]. Western Canada consists of a diverse region spanning mid- to high-latitudes with highly contrasting topography (Figure 1) and hydroclimatic regimes [5,6]. Water availability over the majority of the region is largely controlled by snowmelt, especially from alpine areas in the headwaters of many of the river basins [7]. Hence, the snowmelt-driven spring freshet is the dominant hydrological event for most of the rivers [8]. While winter temperature and precipitation affect snowfall amount and late-winter snowpack, spring temperatures affect the rate and timing of spring snowmelt, directly influencing spring runoff volumes and peak flows [9]. Some cold-season high flows in coastal regions are also associated with long-duration rainfalls resulting from persistent storms, while intense short-duration rainstorms are often responsible for high flow events in relatively small drainage basins [10]. Peak flows generated by such intense and/or excessive rainfall typically occur in late spring and summer, when atmospheric convective precipitation is more common [11].

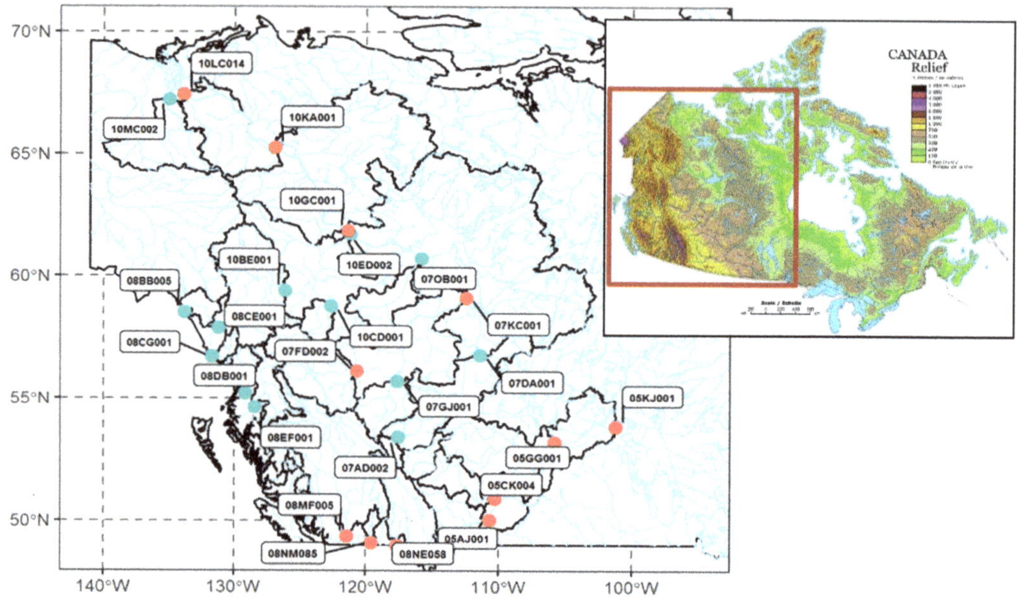

Figure 1. The western Canadian study area including the relief map, the drainage basins and the location and regulation status of hydrometric stations. Red and blue points depict regulated and unregulated stations, respectively (source of the relief map: Natural Resources Canada).

Several studies indicate that western Canadian snowpack is diminishing, especially in the southern regions, reducing the amount of water stored over the winter months and affecting the amount of runoff produced in the spring and summer [12–15]. Both glacial retreat and shrinking snowpacks have been accompanied by changes in runoff patterns and streamflow timing, two factors that can have substantial effects on aquatic ecosystems and urban water systems [16–18]. Studies also suggested that the changes in the timing and magnitude of hydrologic extremes may be one of the most significant consequences of climate change in Canada [2,19]. Future projections over cold region watersheds indicate continued changes in the different components of the hydrologic cycle, such as temperature, precipitation, snow accumulation and melt, with the potential to further impact local and regional hydrological regimes. In many cases, such projected changes are also expected to cause changes in the magnitude and timing of the spring freshet and peak flow events [15,20]. However, peak flow prediction is a challenging endeavor due to the different mechanisms involved and the nonlinear, nonstationary nature of the underlying hydrological processes.

Statistical modelling techniques such as multiple linear regression (MLR) have been widely used in hydrology, e.g., for establishing predictor–predictand relationships and identifying predictors' relative importance such as in spring freshet and peak flow prediction [21–24]. Such statistical predictions of hydrologic time series mostly depend on historic observations and are based on the correlations between the predictand and predictor variables that manifest the influence of large-scale climate on the hydrologic regime [21]. In this context, for snow-dominated regions like western Canada, where peak streamflow is highly dependent on snowpack [10], the inclusion of snow storage in the regression model could provide a potential pathway for improving the linear regression model. If acceptable levels of correlations are found, an MLR framework can be useful to build models for predicting the spring freshet and peak flows. There may also be a possibility to extend the MLR approach to projecting flows for future climate. However, this could be challenging if projected changes in the temperature and precipitation distributions are large, thus shifting the distribution of streamflow extremes beyond the range of historical observations. Therefore, there is a need to evaluate carefully the ability of the MLR models to simulate future streamflow changes. Such evaluations can provide insights that could be useful for other snow-dominated regions of the world. Besides, there is also a need to analyze the model's predictive ability in basins affected by direct human impacts, such as regulation and diversion that could significantly alter the streamflow regime and affect the predictability of streamflow response.

Given the aforementioned knowledge gaps, the first objective of this study is an application of the MLR modelling framework to assess the relative importance of different climatic drivers on the spatial and temporal variation in annual maximum flow (AMF) and mean spring flows (comprising March, April, May and June, hereafter MAMJflow). The application is conducted over 25 western Canadian river basins where the relative importance of a number of predictors, including the annual maximum snow water equivalent (SWEmax) or April 1st SWE, mean spring precipitation (MAMJpcp) and temperature (MAMJtemp), is analyzed over the 1980 to 2012 historical period. The second objective of the study is to investigate the applicability of the MLR model to project changes in the AMF and MAMJflow in the region, using CanRCM4 projected climatic drivers corresponding to RCP4.5 and RCP 8.5 scenarios over the 21st century. The study also examines differences in MLR model performances between regulated and unregulated basins.

2. Study Area and Data Sets

This study assesses all major river basins across western Canada with a total area of around 2.5 million square kilometers (Figure 1). The region's physiography is dominated by the Western Cordillera mountains, which are the hydrologic apex of major western North American rivers that drain to the Pacific and Arctic oceans [17]. The major watersheds in the region include the Mackenzie, Yukon, Fraser, Colombia and Saskatchewan, and the flow in each of these river systems is heavily dependent on from mountain snowpack and glaciers [7]. Twenty-five Water Survey of Canada (WSC) hydrometric stations each with drainage areas of more than 7500 square kilometers are utilized. The threshold for the drainage area was chosen to ensure an ample number of grids for the predictor variables described below. Most of the selected stations are located in British Columbia (10), Alberta (6) and the Northwest Territories (7), with the remaining two in Manitoba (1) and Saskatchewan (1). Daily streamflow data from the Water Survey of Canada (WSC) HYDAT database over the 1980–2012 historical period were used to extract AMF and MAMJflow at each of the 25 stations. Basin boundary delineation for each station was obtained from the National Hydrometric Network basin polygons online database [25] and verified with in-house delineations generated by the GRASS GIS tools [26].

Some of the rivers in the study region, such as the Peace River in the Mackenzie basin, tributaries of the Fraser basin and a large part of the Colombia River, are regulated for hydropower production. Similarly, most parts of the Saskatchewan and Okanagan rivers have a system of reservoirs and diversions for irrigation and hydroelectricity [7]. Therefore, while data from 13 of the selected stations represent unregulated natural flows, the 12 remaining stations are designated as regulated with major upstream structure(s) altering the natural flow regime. The level of flow alterations in the regulated stations varies, with the Fraser River at Hope having minor flow alteration, while stations in the Peace and Columbia basins are affected by major flow alterations from upstream dams [27]. Flows in the Saskatchewan and Okanagan basins are also affected by water withdrawal. Nonetheless, the effect of upstream regulation on hydrologic regime usually diminishes with distance downstream of the regulation location. Therefore, regulated stations were included in the study to explore whether the annual peak flows in these locations are influenced by antecedent climatic conditions and if the MLR can still capture the prevailing relationship between the climatic drivers and mean spring and annual peak flows in such basins. Since major regulations in each of those 12 basins started before 1980, the study employed the 1980–2012 historical streamflow data for better consistency. The 25 stations and their drainage areas and geographic coordinates are provided in Table 1.

Table 1. Hydrometric stations across western Canada used for annual maximum and mean spring flow analysis and MLR modelling.

Number	Station Name	Area (km^2)	Lat	Lon	Regulation
10LC014	Mackenzie River at Arctic Red River	1,680,000	67.4560	−133.7533	Yes
10KA001	Mackenzie River at Norman Wells	1,590,000	65.2720	−126.8500	Yes
10GC001	Mackenzie River at Fort Simpson	1,300,000	61.8684	−121.3589	Yes
05KJ001	Saskatchewan River at the Pas	389,000	53.8381	−101.2087	Yes
07KC001	Peace River at Peace Point	293,000	59.1181	−112.4369	Yes
10ED002	Liard River near the Mouth	275,000	61.7427	−121.2280	No
08MF005	Fraser River at Hope	217,000	49.3860	−121.4542	Yes
08NE058	Columbia River at International Boundary	156,000	49.0008	−117.6283	Yes
07DA001	Athabasca River below Fort McMurray	133,000	56.7804	−111.4022	No
05GG001	North Saskatchewan River at Prince Albert	131,000	53.2034	−105.7721	Yes
10BE001	Liard River at lower Crossing	104,000	59.4125	−126.0972	No
07FD002	Peace River near Taylor	101,000	56.1394	−120.6724	Yes
10MC002	Peel River above Fort McPherson	70,600	67.2589	−134.8888	No
05AJ001	South Saskatchewan River at Medicine Hat	56,368	50.0421	−110.6775	Yes
07OB001	Hay River near Hay River	51,700	60.7430	−115.8596	No
07GJ001	Smoky River at Watino	50,300	55.7146	−117.6231	No
05CK004	Red Deer River near Bindloss	47,849	50.9027	−110.2995	Yes
08EF001	Skeena River at Usk	42,300	54.6319	−128.4306	No
08CE001	Stikine River at Telegraph Creek	29,000	57.9003	−131.1597	No
10CD001	Muskwa River near Fort Nelson	20,300	58.7881	−122.6616	No
08DB001	Nass River above Shumal Creek	18,400	55.2623	−129.0850	No
08BB005	Taku River near Juneau	16,700	58.5386	−133.7000	No
07AD002	Athabasca River at Hinton	9760	53.4243	−117.5694	No
08CG001	Iskut River below Johnson River	9500	56.7344	−131.6690	No
08NM085	Okanagan River near Oliver	7540	49.1146	−119.5667	Yes

Daily precipitation and daily maximum and minimum air temperature data were obtained from the Pacific Climate Impacts Consortium's Pacific North-Western North America meteorological (PNWNAmet) dataset [28]. PNWNAmet is a temporally consistent high-resolution gridded daily meteorological dataset at 1/16° spatial resolution for northwestern North America interpolated from temporally consistent long-term homogenized daily station data covering 1945 through 2012. Additionally, historical gridded SWE data were obtained from the NASA Global Modeling and Assimilation Office's (GMAO) Modern-Era Retrospective Analysis for Research and Applications, Version 2 (MERRA-2) [29]. MERRA-2 provides atmospheric and surface reanalysis data at 50 km spatial resolution from 1980 to the present. As MERRA-2 uses observation-based precipitation data to drive the land surface water budget, its SWE product has lower bias and correlates better against reference data from the Canadian Meteorological Centre than other reanalysis products [30]. For each of the 25 stations, gridded precipitation, temperature and SWE data were extracted and then averaged over the contributing drainage basins. The 1980–2012 time window for the historical period was chosen based on an overlapping time frame between the MERRA-2 data, which starts in 1980, and the PNWNAmet data, which ends in 2012.

3. Application Methods

3.1. Multiple Linear Regression (MLR)

MLR is an extension of ordinary least square regression that can model the linear relationship between several explanatory (predictor) variables and a response (predictand) variable as shown in Equation (1), where i is the index of each observation, y_i is the response variable, x_i represents the explanatory variables, β_0 is the y-intercept, β_p represents slope coefficients and ϵ is the residual. MLR provides a computationally easy and simple to interpret method of predicting streamflow. The main assumptions are that there is a linear relationship between the predictor and predictand variables and that the predictors are not too highly correlated with each other.

$$y_i = \beta_0 + \beta_1 x_{i1} + \beta_2 x_{i2} + \ldots + \beta_p x_{ip} + \epsilon \quad (1)$$

While hydrologic processes have nonlinear characteristics, a number of previous studies have shown that MLR between selected predictors and predictand variables can explain most of the variance in catchment responses [21–24]. In most of the cases, predictors were selected based on an understanding of the physical processes, relevant literature and initial exploratory data analysis. The MLR is also a relatively simple approach for identifying the relative importance of the different potential predictors and the level of significance in the linear relationship established between the predictor and response variables. In this study, the MLR models relate the AMF or MAMJflow (predictands) at each of the 25 hydrometric stations over western Canada with the corresponding basin average values of selected hydroclimatic drivers (predictors). Seven predictors were selected based on the underlying physical processes, relevant literature and initial exploratory data analysis [8,21,24]. As the SWEmax and April 1st SWE are highly correlated, the MLR uses either the magnitude and timing of SWEmax or April 1st SWE, but not both, to avoid multicollinearity. Average spring temperature and precipitation are calculated over the months from March to June (hence MAMJtemp and MAMJpcp respectively). The monthly rates of change in spring temperature (spring warming rate) and precipitation (the spring rate of increasing/decreasing in precipitation) are also calculated as the slope of each variable between March and June (hence MAMJtemp-slope and MAMJpcp-slope, respectively). The nonparametric Spearman's rank correlation between each predictor and predictand measures the strength and direction of monotonic association between two variables and is estimated using Equation (2), where Di is the difference in ranks between the ith pair of predictor and predictand and n is the number of data pairs (see Table 2) [31]. The correlation coefficient, r_s, takes values from +1 to -1, and the strength of the rank correlations was analyzed in terms of statistical significance tests at the 5% significance level.

$$r_s = 1 - \frac{6\sum_{i=1}^{n} D_i^2}{n(n^2-1)} \quad (2)$$

Table 2. Hydroclimatic predictors used in the MLR models to predict AMF and MAMJflow at each of the hydrometric stations.

Predictor Name	Abbreviation	Units	Data Source
Annual maximum snow water equivalent	SWEmax	mm	MERRA-2
Date of annual maximum snow water equivalent	SWEmax-date	Date number	MERRA-2
April 1st snow water equivalent	SWEapril 1st	mm	MERRA-2
Average spring temperature	MAMJtemp	°C	PNWNAmet
Rate of change in spring temperature	MAMJtemp-slope	°C/month	PNWNAmet
Total spring precipitation	MAMJpcp	mm	PNWNAmet
Rate of change in spring precipitation	MAMJpcp-slope	mm/month	PNWNAmet

MLR models are fitted to the predictor/predictand data at each station over the 1980 to 2012 historical period, with MAMJpcp, MAMJpcp-slope, MAMJtemp, MAMJtemp-slope as common predictors and either April 1st SWE or SWEmax and its timing as additional predictors and either AMF or MAMJflow as predictand. First, a best subsets regression approach, as implemented in the R leaps package version 3.0, was employed to fit MLR models to all possible combinations of the predictor variables and each of the two predictands. Then, the best fitting MLR model from the pool of all possible combinations with one to five predictor variables (when using April 1st SWE) or one to six predictor variables (when using SWEmax and its timing) were chosen based on the sum-of-squares of residuals [32]. This approach involved an exhaustive comparison of models from each predictor size group and ended in the selection of models with the lowest Akaike information criterion (AIC) [32].

These five or six models each selected from a predictor size group were then compared in terms of their predictive power by repeated k-fold cross-validation, with k = 5 [33]. The k-fold cross-validation procedure divides the limited dataset into k nonoverlapping folds. Each of the k folds is given an opportunity to be used as a held-back test set, whilst all other folds collectively are used as a training dataset. A total of k models were fit and evaluated on the k hold-out test sets and the mean performance was reported. The model with the lowest root-mean-square error (RMSE) from k-fold cross-validation was selected as the final model if it was statistically significant at a p-value of 0.05. All selected models were also tested for their fulfillment of the regression assumptions by computing their variance inflation factors (VIF) with the intention of removing highly correlated predictors with VIF above the most common acceptable threshold of 5 [33,34].

3.2. Predictors' Relative Importance

Predictor relative importance refers to the quantification of an individual predictor's contribution to a multiple regression model. To identify which predictors are the most influential in explaining variation in AMF and MAMJflow, the total explained variance, R^2, of each selected model for a station was decomposed into the proportion explained by each individual predictor using the Lindemann, Merenda and Gold (LMG) method as implemented in the R package *relimpo* [35]. In addition to the variance explained by each predictor, the percentage of the total variance (explained by the model, R^2) that is contributed by each predictor variable provides a measure of the relative importance of each variable. Analysis of these metrics included the frequency of dominant predictors among all station models, where a dominant predictor is defined as the predictor in a model with the highest relative importance. The spatial distribution of dominant predictors was also mapped as it may be a good indicator of the effect of watershed features on predictor–predictand relationships.

3.3. Future Projection of Annual Peak and Spring Flows

To evaluate the applicability of the selected MLR models for estimating future magnitudes of AMF and MAMJflow, climatic predictors derived from the Canadian Regional Climate Model (CanRCM4; Scinocca et al., 2016) [36] under RCP4.5 and RCP8.5 future scenarios were used. The CanRCM4 output used in this study is from the CORDEX Experiments for North America (NAM-22), which is at 0.22° or approximately 25 km resolution and driven by the CanESM2 GCM [36]. Recognizing that RCM outputs usually have systematic biases [37], this study employed the "delta change method", to account for the biases in future predictor values. Mean projected changes in each of the predictor variables between the 1976–2005 baseline and the two future periods (2041–2070 and 2071–2100) were computed from the CanRCM4 data and then averaged over the basin area contributing to each of the 25 hydrometric stations. The future predictor values for each hydrometric station were computed by applying the CanRCM4 delta changes on the observed values derived from MERRA-2 and PNWNAmet over the historical period as shown in Equation (3) below:

$$\begin{cases} \Delta_m = CanRCM4_{future} - CanRCM4_{baseline} \text{ or } \Delta_m = CanRCM4_{future}/CanRCM4_{baseline} \\ Scenario_{future} = Observed_{historical} + \Delta_m \text{ or } Scenario_{future} = Observed_{historical} * \Delta_m \end{cases} \quad (3)$$

Changes between the baseline and future periods were calculated in terms of ratios for SWEmax, SWEapril 1st, MAMJpcp, MAMJpcp-slope and MAMJtemp-slope or differences for MAMJtemp and SWEmax-date and applied to the historical observations to obtain the corresponding future projections (Equation (3)). The MLR model for each station was then forced with the adjusted hydroclimatic predictors to compute AMF and MAMJflow corresponding to the different RCP scenarios and future periods.

4. Results and Discussion

4.1. Spearman's Rank Correlation

The Spearman correlations between the selected predictors and predictands at each of the 25 hydrometric stations are presented in Figure 2. The correlation coefficients (as expressed by r_s) indicate that both the AMF and MAMJflow at most of the stations are positively correlated with SWEmax, April 1st SWE and spring precipitation (MAMJpcp) and negatively correlated with mean spring temperature (MAMJtemp), though the correlation strength varies. However, there are some smaller watershed stations in the Pacific coastal region (e.g., 08CG001 and 08DB001) and some regulated stations (e.g., 05AJ001) that show weaker correlations with these predictors. This is mainly because peak flows in those coastal watersheds are less controlled by snowmelt and more by rainfall, and in regulated stations, they are affected by regulations. The correlations with SWEmax timing (SWEmax-date) and mean spring precipitation and temperature slopes (MAMJtemp-slope and MAMJpcp-slope, respectively) are generally weaker, although they are significant at some specific stations. While there are some differences in the correlation matrix between regulated and unregulated stations, the results do not indicate one group as being more correlated than the other.

4.2. MLR Model Performances

The MLR models fitted to each of the 25 stations and four combinations (with either AMF or MAMJflow as predictand and either SWEmax and its timing or April 1st SWE as predictors) over the 1980 to 2012 period resulted in a total of 100 optimized models. None of the selected models were found to have problematic multicollinearity, with VIF less than 5 indicating no significant correlation between predictor variables [31]. This may be explained in part by the careful initial variable selection which included the decision to build models with either SWEmax or April 1st SWE but not both variables. However, 12 models (9 for AMF and 3 for MAMJflow, out of the 100) that are not statistically significant (p-value > 0.05) were removed, resulting in the final selection of 88 MLR models. Some of the stations where the MLR models are not statistically significant are located in the Pacific coastal region (e.g., 08CG001, 08DB001), while others are in the interior of the Mackenzie basin with relatively large drainage areas that are affected by the Peace River regulations (e.g., 10KA001, 10LC014). Nonetheless, MLR models also show good performances in most other regulated basins comparable to those of the nonregulated ones. The list of all selected models corresponding to each hydrometric station along with models' R^2 and p-values is provided in Supplementary Materials.

The MLR performances are presented in Figure 3, which summarizes the mean percentage differences between model-predicted and observed AMF and MAMJflow over the 1980–2012 historical period for those models with p-value < 0.05 along with error bars indicating the 95% confidence interval. The results generally indicate a good model performance with the historical period mean prediction error at most stations being within ±10% of the observed values. However, few stations have prediction errors in some years in the order of ±20%. It is also not surprising that most of the stations with relatively higher prediction error are located in highly regulated watersheds (e.g., 08NM085 and 05AJ001 have water withdrawals, and 07KC001 is affected by reservoir regulation). Other factors including the size and hypsometry of the contributing watershed may also affect the performance of the MLR models. However, since the effects of the different factors are overlapping, it is hard to clearly identify the specific reason for some of the unsatisfactory performances. The error in predicting AMF is relatively larger than that in predicting MAMJflow. This is also expected because MAMJflow is averaged flow over four months where the daily variations are smoothened out while AMF is a peak of daily flows occurring at any time during the spring or summer high flow seasons. For the majority of the stations, the MLR models with either SWEmax or April 1st SWE as predictor show similar performances, except in a few stations where using April 1st SWE resulted in a relatively smaller prediction error.

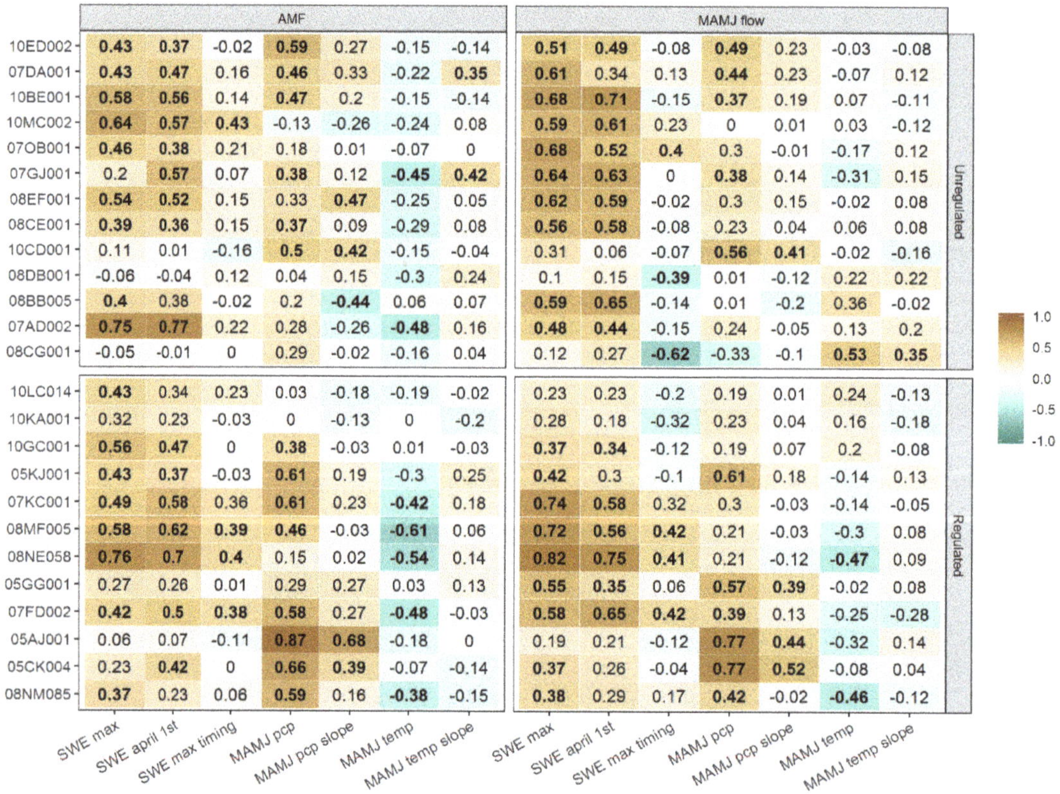

Figure 2. Spearman correlation of AMF (**left panel**) and MAMJflow (**right panel**) with area-averaged predictors corresponding to each of the 25 hydrometric stations. Top panels represent stations not affected by regulation while the bottom panels represent regulated basins. Bold entries correspond to Spearman correlations with $p < 0.05$.

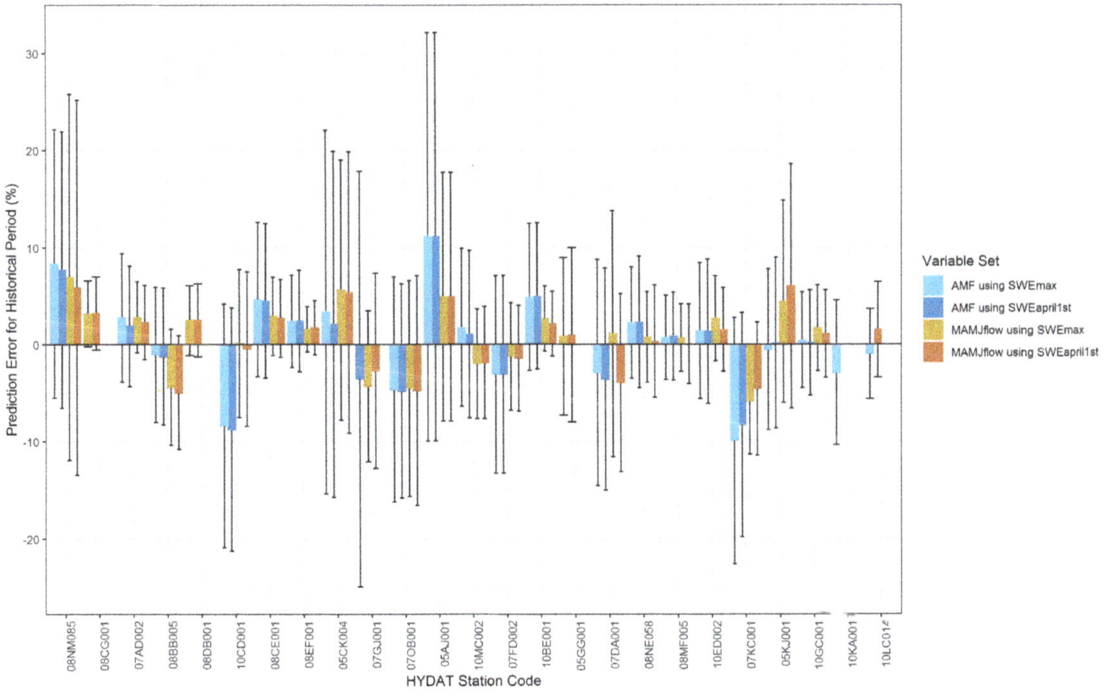

Figure 3. Mean percentage differences between MLR model-predicted and observed annual maximum flow (AMF) and mean spring flow (MAMJflow) over the 1980–2012 period. Error bars indicate the 95% confidence interval for predicted values.

4.3. Predictors' Relative Importance

Figure 4 shows the relative importance of each climatic predictor in the MLR models in explaining the variances of the predictands at each hydrometric station. Top and bottom panels correspond to AMF and MAMJflow while left and right sides display SWEmax and April 1st SWE, respectively. For the majority of stations, April 1st SWE or SWEmax contributed the highest proportion of explained variance, indicating the relatively higher importance of snow accumulation in predicting AMF and MAMJflow. For most stations, MAMJPrc has the second-highest predictor importance. The proportion of explained variance in predicting both AMF and MAMJflow averaged across all stations is in the order of 44% for April 1st SWE and 51.7% for SWEmax. This is followed by MAMJprc with the proportion of explained variance ranging between 29.4% and 37.6% under the different categories. However, for some regulated stations (e.g., 08NM085, 05CK004, 05AJ001, 05KJ001), the predictor importance of MAMJprc is higher than that of SWE. MAMJprc-slope contributed more to the explained variance of AMF while MAMJtemp and MAMJtemp-slope contributed more to the explained variance of MAMJflow at about half of the stations. While MAMJprc-slope contributes to around 10% of the explained variance in predicting AMF, its contribution to predicting MAMJflow is almost nonexistent (<0.3%). The results also show that the contribution of MAMJtemp and MAMJtemp-slope to the explained variance is relatively small and with higher contribution to predicting MAMJflow (5–13%) than to AMF (2–5%).

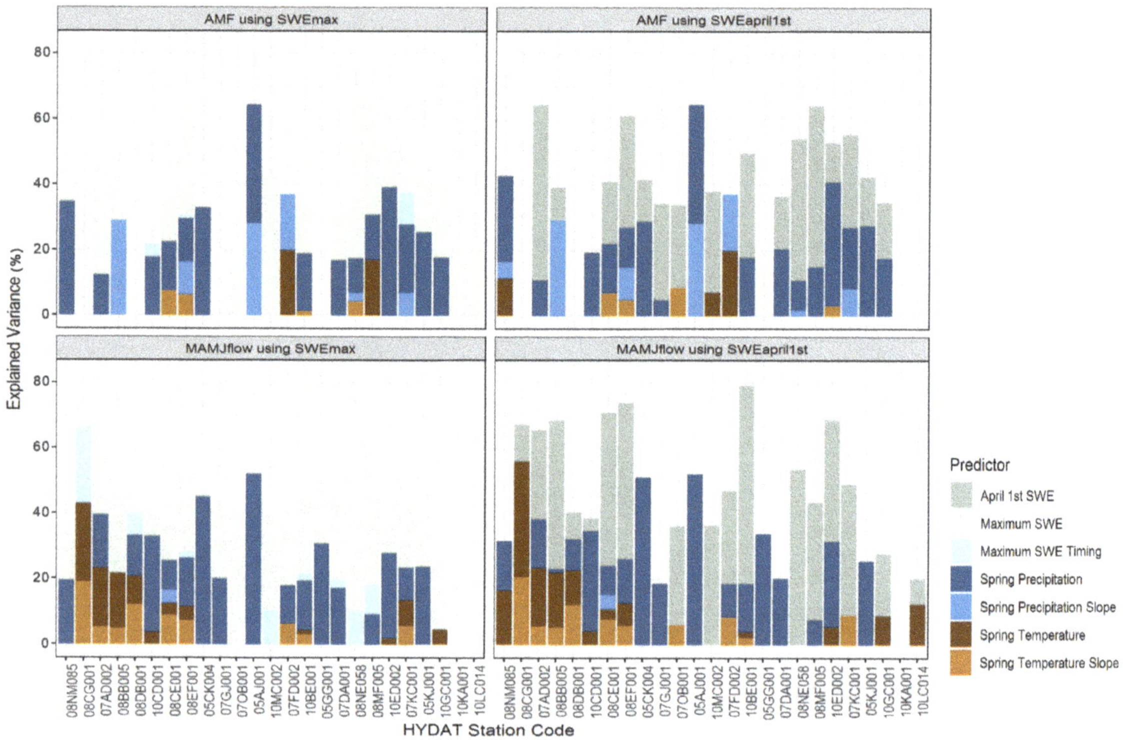

Figure 4. Relative importance of each predictor variable presented in terms of fraction of the explained variance in predicting annual maximum flow (AMF) and mean spring flow (MAMJflow) at each hydrometric station.

Similarly, Figure 5 presents the spatial distribution of predictors' relative importance for the AMF and MAMJflows at each of the 25 stations over western Canada. Once again the figure shows SWEmax as the most important predictor at most stations (21/25 for AMF and 24/25 for MAMJflow), followed by MAMJprc, which is also an important predictor (20/25 for AMF and 19/25 for MAMJflow). MAMJtemp and MAMJtemp-slope are important predictors of MAMJflow for about one-third of the stations, mostly located in the middle part of the study region, while their contribution to predicting AMF is limited to a couple of stations. MAMJprc-slope is important in predicting AMF in the southern and Pacific coast regions but not that important in predicting MAMJflow in most parts of the study region. The contribution of SWEmax timing in predicting both AMF and MAMJflows is limited only to a few stations. The generally smaller importance of spring temperature and its rate of increase in predicting AMF is an indication that spring peak flows in most basins are affected more by accumulated snow and spring precipitation than temperature. The predictability of MAMJflow is also mostly dependent on more variables than that of AMF, a possible reason for the MAMJflow models being more robust with relatively smaller prediction errors. While the importance of SWEmax in predicting flow is higher towards the northern part of the study region, the importance of MAMJprc is higher towards the south, possibly resulting from the effect of temperature on partitioning precipitation into rain and snow. In general, the results presented in Figure 5 show that predictors' relative importance is spatially variable and cannot be generalized for the whole region.

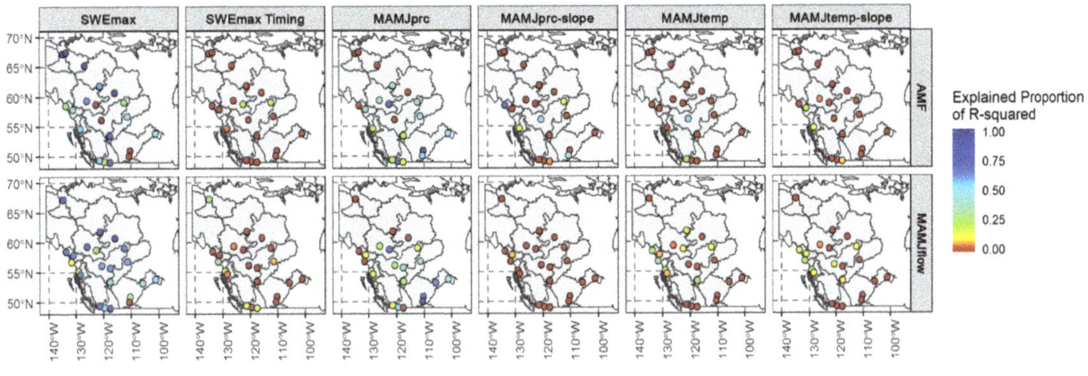

Figure 5. Spatial distribution of predictors' relative importance in predicting annual maximum flow (AMF) and mean spring flow (MAMJflow) at each station over western Canada expressed as the proportion of the total explained variance (R^2) contributed by each predictor.

4.4. Projected Changes in Hydroclimatic Variables

Figure 6 shows mean CanRCM4 projected changes in each of the predictor variables between the 1976–2005 baseline and the two future periods (2041–2070 and 2071–2100) for the two scenarios (RCP4.5 and RCP8.5). The results indicate that both MAMJpcp and MAMJtemp are projected to increase from 10% to 50% for the former and from 2 to 7 °C for the latter depending on the RCP scenarios and the future time windows considered. As expected, the biggest increases for both precipitation and temperature occur under RCP8.5 and far future (2071–2100) period while the smaller increases correspond to RCP4.5 and the near future (2041–2070) period. While there are some regional variations in the projected changes in spring precipitation and temperature with slightly higher increases in the northern than the southern basins, the differences are relatively small.

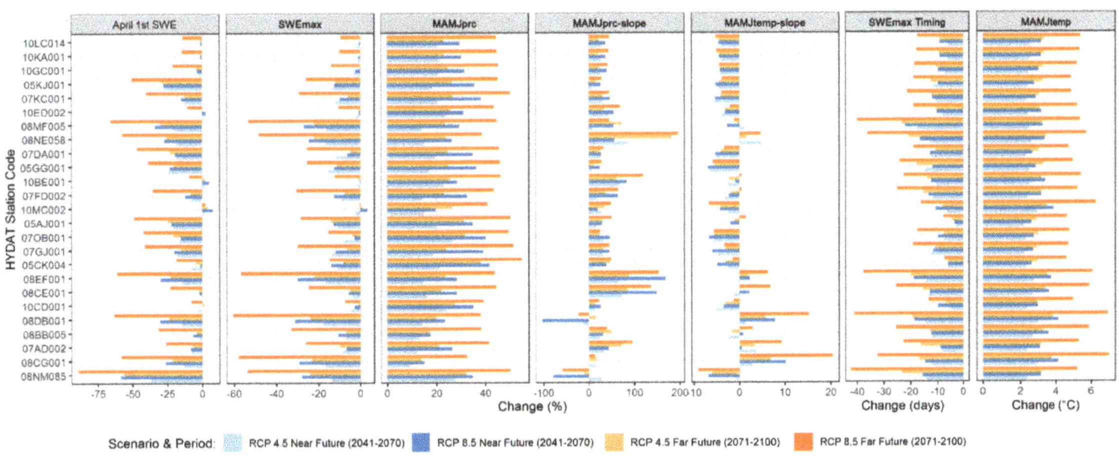

Figure 6. Projected changes in catchment-averaged predictor variables between each of the two future periods (2041–2070 and 2071–2100) and the baseline period (1976–2005) based on the CanRCM4 model projection over the study area corresponding to the RCP4.5 and RCP8.5 scenarios.

SWEmax and April 1st SWE are both projected to decrease in the majority of the basins under both scenarios and time periods with values ranging from −80% to +5%. The greatest decreases are mostly in the Fraser, Colombia, and Pacific coast basins, while the smallest decreases or even slight increases correspond to the most northern basins, as was also found in previous studies [15,38]. At the same time, SWEmax timing is projected to advance by about 5 to 25 days in all river basins. MAMJpcp-slope increases in almost all basins (except two), indicating more increase in precipitation in the later part rather than the earlier part of spring. In contrast, MAMJtemp-slope show decreases for most basins, indicating higher increases in early spring than late spring, except for a few southern stations where the reverse is true. In general, while the overall direction of projected changes in the predictor variables over most parts of the study region is similar, there are some north-to-south variations in the magnitude of those changes.

4.5. Projected Changes in AMF and MAMJflow

Future projections of AMF and MAMJflow were computed using the MLR models developed for the historical period, with delta changes from the CanRCM4 projections under the two RCP scenarios applied on the observed predictor values (Equation (3)). The results indicate overall projected increases in both the AMF and MAMJflow for most of the stations, with a few stations showing no change or some decreases (Figure 7). For RCP8.5 and the far future period, the average increase in AMF (12%, ranging from −69% to +126%) has more spatial variability than that of MAMJflow (28%, ranging from −48% to +81%) (Figure 8). The largest percent increases in both are located in Saskatchewan (05AJ001, 05CK004), although large increases in MAMJflow are also projected in the Athabasca (07AD002) and other northwestern rivers. The largest decreases in both AMF and MAMJflow are located at the Peace River at Taylor (07FD002) and Peace Point (07KC001), respectively. Most northern stations are projected to have relatively smaller percentage changes in AMF compared to the south, although the absolute magnitude of these changes can be larger for those with relatively larger flow magnitudes, such as in the Mackenzie River mainstem stations.

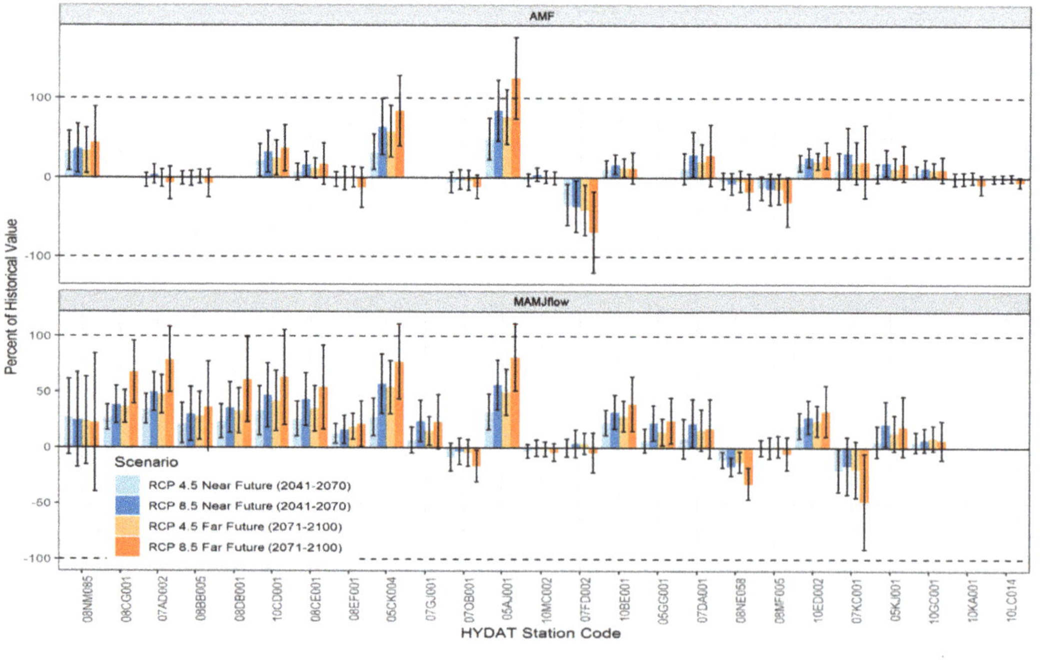

Figure 7. Mean projected changes in annual maximum flow (AMF) and mean spring flow (MAMJflow) for the two RCPs and two future periods computed with the MLR models at each of the hydrometric stations (predictions made with SWEmax; error bars indicate the 95% confidence interval in future values).

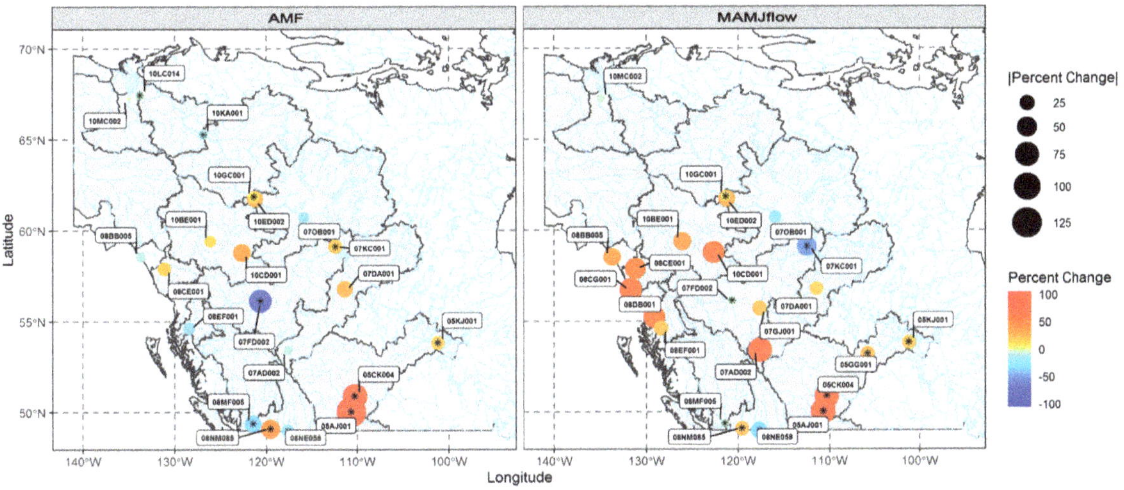

Figure 8. Spatial distribution of projected changes in annual maximum flow (AMF) and mean spring flow (MAMJflow) corresponding to the RCP8.5 and the far future (2071–2100) period (predictions made with SWEmax; asterisks indicate regulated stations).

Projected increases in AMF are mostly located in the interior of western Canada, while increases in MAMJflow are found throughout the region. However, the projected AMF in the coastal basins could be subject to large uncertainties since peak flow events in the region are sometimes influenced by atmospheric rivers, which is not taken into account in the MLR model [39]. No clear pattern was found differentiating the results from regulated and unregulated stations since some regulated stations (e.g., in Peace, Fraser and Colombia rivers) show projected decreases in AMF and MAMJflow while other regulated stations (e.g., South Saskatchewan and Liard rivers) show increases. The wide variation in the direction and magnitude of projected changes in AMF among the river basins could partly be due to changes in the synchrony of mainstem and tributary streamflow during high-flow periods at the mainstem–tributary confluence [40]. Decreasing synchrony may dampen the forced increases in AMFs along mainstem stations, but its relative effects may vary in space and time, as well as in future climate scenarios.

A comparison of the projected changes in flows using the MLR models and future climate projections (applying delta change method) with those of other previous studies using process-based hydrological models driven by statistically downscaled GCMs shows both agreement and disagreement depending on basin characteristics. For example, the directions of MLR-projected changes, such as increases in AMF and MAMJflow for the unregulated Liard (10BE001, 10ED001) and Athabasca (07DA001) basins, are consistent with previous studies in the two basins using process-based hydrologic models [15,41]. Similarly, for the mildly regulated station in the Fraser basin (08MF005), the MLR model projections of decreases in AMF and no change in MAMJflow are comparable with the projections of decreases in AMF and some increases in MAMJflow of Shrestha et al. [27]. However, the MLR model projections for the Peace basin (07KC001, 07FD002) are not consistent with the results of Schnorbus et al. [42], which showed projected increases in both AMF and MAMJflow. Likewise, the MLR projected decreases in AMF and MAMJflow at another regulated station in the Columbia River (08NE058), in contrast to the projected increases for the same basin reported by Werner et al. [43]. The discrepancies could be partly explained by the fact that both the Peace and Columbia rivers are highly regulated by large hydroelectric dams/reservoirs. Schnorbus et al. [42] and Werner et al. [43] also used naturalized flows, while the MLR models are based on the regulated flows. Besides, the lack of representation of physical processes in the MLR model is one of its limitations in extrapolating beyond the range of historical observation affecting flow projections for some basins. Therefore, the overall ability of the MLR model to project AMF and MAMJflow may be limited, especially when applied to regulated basins.

5. Summary and Conclusions

Flows originating from cold and mountainous watersheds are highly dependent on their snow regimes that are determined by temperature and precipitation patterns over the region. Climatic variations generally affect the mean seasonal flows and the magnitude and timing of peak streamflow events with implications on infrastructures and ecosystem services. This study applied the MLR approach for assessing the climatic controls of AMF and MAMJflow in 25 western Canadian river basins using multiple station-specific hydroclimatic predictors. The analysis indicated that the basin-integrated SWEmax or April 1st SWE and mean spring precipitation (MAMJPr) are the most prominent predictors of AMF and MAMJflow throughout the region, together explaining over 80% of the predictand's variance. The proportion of explained variance is in the order of 44% for April 1st SWE and 51.7% for SWEmax, followed by MAMJprc explaining between 29.4% and 37.6% of the predictand's variance. However, predictors' relative importance was spatially variable and could not be generalized for the whole region. While the relative importance of SWEmax in predicting flow is higher towards the north and that of MAMJprc is higher towards the south, the relative importance of MAMJtemp is mostly higher in the middle part of the study region. The best-fitting MLR models were statistically significant at a p-value of 0.05 at the majority (88%) of the stations, and mean prediction errors over the historical period were below $\pm$10% at most stations (although a few stations had prediction errors ranging in the order of $\pm$20%).

The results of this study are highly relevant to other regions of the world where runoff processes are dominated by mountain snowpack. The main lesson is that while winter snow accumulation and spring precipitation are the major drivers of mean spring and annual peak flows in cold mountainous watersheds, their relative contribution to predictability largely depends on the location and other physiographic characteristics of the watersheds. Other studies have also shown that the transformation from snow accumulation to runoff generation in cold regions is dominated by snowmelt and infiltration processes that are highly heterogeneous [44]. In particular, direct human impacts, such as regulation and diversion in the basin, can alter its flow regime and affect the dependency between those predictor and response variables. Therefore, there is a need to carefully consider physiographic characteristics as well as human impacts in using regression models for streamflow simulation.

A warming climate will bring a shift in precipitation from snow towards rain, affecting the snowpack volume and snowmelt timing. Therefore, by adjusting the snowpack together with precipitation and temperature based on CanRCM4 projections, the MLR models were tested to see if they can be applied to predict future changes in AMF and MAMJflow. CanRCM4 outputs show a general increase in mean spring temperature (2 to 7 °C) and precipitation (10% to 55%) and an overall decrease in SWEmax and April 1st SWE for all but two river basins (+5% to −95%) by the end of this century. Application of the MLR models with adjusted hydroclimatic predictors revealed considerable spatial variations, with the projected increase in spring precipitation mostly compensating the opposite effect of increasing spring temperature and decrease in SWE and resulting in AMF changes ranging from −69% to +126% and MAMJflow changes ranging from −48% to +81%. Projected changes are mostly higher for the RCP8.5 and end-of-century scenarios. Projected changes in MAMJflow across the region are more consistent than those of AMF. Other things being equal, change in phase of precipitation from snow towards rain (because of increasing temperature) usually decreases the mean streamflow [1]. However, the current study indicated that the projected increase in precipitation can sometimes compensate for the effect of the decreasing fraction of snowfall and may result in an increase in spring flow.

Comparisons of the MLR model projections with some previous studies using process-based hydrological models driven by statistically downscaled GCMs show good agreement in the direction of change for most of the unregulated rivers, while there are substantial disagreements for the regulated river basins. This is in part due to the lack of physical representation in these models, as well as limitations in extrapolating future flow magnitudes beyond the range of historical observation. Therefore, there is a need to exercise caution in the use of such statistical models for projecting future changes, especially in regulated basins. More research is needed to better understand the extent of the limitation and ways of incorporating relevant information in the modelling process to reduce those limitations. Future research may also look at possible improvements by applying nonlinear methods such as artificial neural networks (ANNs) and other machine learning techniques.

Supplementary Materials: The Supplementary Materials are available online at https://www.mdpi.com/article/10.3390/w13121617/s1.

Author Contributions: Conceptualization, Y.B.D., R.R.S., B.B. and P.C.; methodology, Y.B.D., R.R.S. and C.J.; data analysis, Y.B.D. and C.J.; original draft preparation, Y.B.D. and C.J.; review and editing, Y.B.D., R.R.S., C.J., B.B. and P.C.; visualization, Y.B.D. and C.J.; supervision, Y.B.D. All authors have read and agreed to the published version of the manuscript.

Funding: This study was conducted with internal funding from Environment and Climate Change Canada.

Institutional Review Board Statement: Not applicable.

Informed Consent Statement: Not applicable.

Data Availability Statement: Streamflow data used in the study are available through the Water Survey of Canada (WSC) Historical Hydrometric Data Archive, http://wateroffice.ec.gc.ca/ (accessed in 15 May 2019). Daily precipitation and air temperature data (PNWNAmet) were obtained from the Pacific Climate Impacts Consortium (PCIC), https://data.pacificclimate.org/portal/gridded_observations/map/ (accessed in 15 May 2019). Historical gridded SWE data (MERRA-2) were obtained from the NASA Global Modeling and Assimilation Office (GMAO), https://disc.gsfc.nasa.gov/datasets?project=MERRA-2 (accessed in 15 June 2019). The CanRCM4 data (CORDEX NAM-22) used in this study were obtained from the Canadian Centre for Climate Modelling and Analysis (CCCma) https://climate-modelling.canada.ca/climatemodeldata/canrcm/CanRCM4/index_cordex.shtml (accessed in 15 July 2019).

Acknowledgments: We acknowledge WSC, PCIC, NASA and CCCma for all data used in this study.

Conflicts of Interest: The authors declare no conflict of interest.

References

1. Berghuijs, W.R.; Woods, R.A.; Hrachowitz, M. A precipitation shift from snow towards rain leads to a decrease in streamflow. *Nat. Clim. Chang.* **2014**, *4*, 583–586. [CrossRef]
2. Burn, D.H. Climatic influences on streamflow timing in the headwaters of the Mackenzie River Basin. *J. Hydrol.* **2008**, *352*, 225–238. [CrossRef]
3. Dudley, R.W.; Hodgkins, G.A.; McHale, M.R.; Kolian, M.J.; Renard, B. Trends in snowmelt-related streamflow timing in the conterminous United States. *J. Hydrol.* **2017**, *547*, 208–221. [CrossRef]
4. Sandink, D. Urban flooding and ground-related homes in Canada: An overview. *J. Flood Risk Manag.* **2016**, *9*, 208–223. [CrossRef]
5. Dibike, Y.; Prowse, T.; Bonsal, B.; O'Neil, H. Implications of future climate on water availability in the western Canadian river basins. *Int. J. Climatol.* **2017**, *37*, 3247–3263. [CrossRef]
6. Bawden, A.J.; Burn, D.H.; Prowse, T.D. Recent changes in patterns of western Canadian river flow and association with climatic drivers. *Hydrol. Res.* **2015**, *46*, 551–565. [CrossRef]
7. Bonsal, B.; Shrestha, R.R.; Dibike, Y.; Peters, D.L.; Spence, C.; Mudryk, L.; Yang, D. Western Canadian Freshwater Availability: Current and Future Vulnerabilities. *Environ. Rev.* **2020**, *28*, 528–545. [CrossRef]
8. Kang, D.H.; Gao, H.; Shi, X.; ul Islam, S.; Déry, S.J. Impacts of a rapidly declining mountain snowpack on streamflow timing in Canada's Fraser River basin. *Sci. Rep.* **2016**, *6*, 1–8. [CrossRef] [PubMed]
9. Barnett, T.P.; Adam, J.C.; Lettenmaier, D.P. Potential impacts of a warming climate on water availability in snow-dominated regions. *Nature* **2005**, *438*, 303–309. [CrossRef]
10. Buttle, J.M.; Allen, D.M.; Caissie, D.; Davison, B.; Hayashi, M.; Peters, D.L.; Pomeroy, J.W.; Simonovic, S.; St-Hilaire, A.; Whitfield, P.H. Flood processes in Canada: Regional and special aspects. *Can. Water Resour. J.* **2016**, *41*, 7–30. [CrossRef]

11. Erler, A.R.; Peltier, W.R. Projected changes in precipitation extremes for western Canada based on high-resolution regional climate simulations. *J. Clim.* **2016**, *29*, 8841–8863. [CrossRef]
12. Shrestha, R.R.; Cannon, A.J.; Schnorbus, M.A.; Alford, H. Climatic controls on future hydrologic changes in a subarctic river basin in Canada. *J. Hydrometeorol.* **2019**, *20*, 1757–1778. [CrossRef]
13. Najafi, M.R.; Zwiers, F.; Gillett, N. Attribution of the Observed Spring Snowpack Decline in British Columbia to Anthropogenic Climate Change. *J. Clim.* **2017**, *30*, 4113–4130. [CrossRef]
14. Dibike, Y.; Eum, H.I.; Prowse, T. Modelling the Athabasca watershed snow response to a changing climate. *J. Hydrol. Reg. Stud.* **2018**, *15*, 134–148. [CrossRef]
15. Shrestha, R.R.; Bonsal, B.R.; Bonnyman, J.M.; Cannon, A.J.; Najafi, M.R. Heterogeneous snowpack response and snow drought occurrence across river basins of northwestern North America under 1.0 °C to 4.0 °C global warming. *Clim. Chang.* **2021**, *164*, 1–21. [CrossRef]
16. Stewart, I.T. Changes in snowpack and snowmelt runoff for key mountain regions. *Hydrol. Process.* **2009**, *23*, 78–94. [CrossRef]
17. Rood, S.B.; Pan, J.; Gill, K.M.; Franks, C.G.; Samuelson, G.M.; Shepherd, A. Declining summer flows of Rocky Mountain rivers: Changing seasonal hydrology and probable impacts on floodplain forests. *J. Hydrol.* **2008**, *349*, 397–410. [CrossRef]
18. Milner, A.M.; Khamis, K.; Battin, T.J.; Brittain, J.E.; Barrand, N.E.; Füreder, L.; Cauvy-Fraunié, S.; Gíslason, G.M.; Jacobsen, D.; Hannah, D.M.; et al. Glacier shrinkage driving global changes in downstream systems. *Proc. Natl. Acad. Sci. USA* **2017**, *114*, 9770–9778. [CrossRef] [PubMed]
19. Cunderlik, J.M.; Ouarda, T.B. Trends in the timing and magnitude of floods in Canada. *J. Hydrol.* **2009**, *375*, 471–480. [CrossRef]
20. Dibike, Y.; Eum, H.I.; Coulibaly, P.; Hartmann, J. Projected Changes in the Frequency of Peak Flows along the Athabasca River: Sensitivity of Results to Statistical Methods of Analysis. *Climate* **2019**, *7*, 88. [CrossRef]
21. Curry, C.L.; Zwiers, F.W. Examining controls on peak annual streamflow and floods in the Fraser River Basin of British Columbia. *Hydrol. Earth Syst. Sci.* **2018**, *22*, 2285–2309. [CrossRef]
22. Tsakiri, K.; Marsellos, A.; Kapetanakis, S. Artificial neural network and multiple linear regression for flood prediction in Mohawk River, New York. *Water* **2018**, *10*, 1158. [CrossRef]
23. Moradi, A.M.; Dariane, A.B.; Yang, G.; Block, P. Long-range reservoir inflow forecasts using large-scale climate predictors. *Int. J. Climatol.* **2020**, *40*, 5429–5450. [CrossRef]
24. Shrestha, R.R.; Pesklevits, J.; Yang, D.; Peters, D.L.; Dibike, Y.B. Climatic Controls on Mean and Extreme Streamflow Changes Across the Permafrost Region of Canada. *Water* **2021**, *13*, 626. [CrossRef]
25. Environment and Climate Change Canada. National Hydrometric Network Basin Polygons. Available online: https://open.canada.ca/data/en/dataset/0c121878-ac23-46f5-95df-eb9960753375 (accessed on 15 May 2019).
26. Neteler, M.; Bowman, M.H.; Landa, M.; Metz, M. GRASS GIS: A multi-purpose open source GIS. *Environ. Model Softw.* **2012**, *31*, 124–130. [CrossRef]
27. Shrestha, R.R.; Schnorbus, M.A.; Werner, A.T.; Berland, A.J. Modelling spatial and temporal variability of hydrologic impacts of climate change in the Fraser River basin, British Columbia, Canada. *Hydrol. Process.* **2012**, *26*, 1840–1860. [CrossRef]
28. Werner, A.T.; Schnorbus, M.A.; Shrestha, R.R.; Cannon, A.J.; Zwiers, F.W.; Dayon, G.; Anslow, F. A long-term, temporally consistent, gridded daily meteorological dataset for northwestern North America. *Sci. Data* **2019**, *6*, 1–16. [CrossRef]
29. Gelaro, R.; McCarty, W.; Suárez, M.J.; Todling, R.; Molod, A.; Takacs, L.; Randles, C.A.; Darmenov, A.; Bosilovich, M.G.; Reichle, R.; et al. The modern-era retrospective analysis for research and applications, version 2 (MERRA-2). *J. Clim.* **2017**, *30*, 5419–5454. [CrossRef]
30. Reichle, R.H.; Draper, C.S.; Liu, Q.; Girotto, M.; Mahanama, S.P.; Koster, R.D.; De Lannoy, G.J. Assessment of MERRA-2 land surface hydrology estimates. *J. Clim.* **2017**, *30*, 2937–2960. [CrossRef]
31. Gaál, L.; Szolgay, J.; Kohnová, S.; Hlavčová, K.; Parajka, J.; Viglione, A.; Merz, R.; Blöschl, G. Dependence between flood peaks and volumes: A case study on climate and hydrological controls. *Hydrol. Sci. J.* **2015**, *60*, 968–984. [CrossRef]
32. Hastie, T.; Tibshirani, R.; Friedman, J. *The Elements of Statistical Learning: Data Mining, Inference, and Prediction*, 2nd ed.; Springer Science & Business Media: Berlin, Germany, 2009; pp. 57–58.
33. Laio, F.; Di Baldassarre, G.; Montanari, A. Model selection techniques for the frequency analysis of hydrological extremes. *Water Resour. Res.* **2009**, *45*, W07416. [CrossRef]
34. James, G.; Witten, D.; Hastie, T.; Tibshirani, R. *An Introduction to Statistical Learning with Applications in R*; Springer: New York, NY, USA, 2013; pp. 59–126.
35. Schoups, G.; Van de Giesen, N.C.; Savenije, H.H.G. Model complexity control for hydrologic prediction. *Water Resour. Res.* **2008**, *44*, W00B03. [CrossRef]
36. Grömping, U. Relative importance for linear regression in R: The package relaimpo. *J. Stat. Softw.* **2006**, *17*, 1–27. [CrossRef]
37. Scinocca, J.F.; Kharin, V.V.; Jiao, Y.; Qian, M.W.; Lazare, M.; Solheim, L.; Flato, G.M.; Biner, S.; Desgagne, M.; Dugas, B. Coordinated global and regional climate modeling. *J. Clim.* **2016**, *29*, 17–35. [CrossRef]
38. Shrestha, R.R.; Schnorbus, M.A.; Werner, A.T.; Zwiers, F.W. Evaluating hydroclimatic change signals from statistically and dynamically downscaled GCMs and hydrologic models. *J. Hydrometeorol.* **2014**, *15*, 844–860. [CrossRef]
39. Sharma, A.R.; Déry, S.J. Linking atmospheric rivers to annual and extreme river runoff in British Columbia and southeastern Alaska. *J. Hydrometeorol.* **2020**, *21*, 2457–2472. [CrossRef]

40. Rupp, D.E.; Chegwidden, O.S.; Nijssen, B.; Clark, M.P. Changing river network synchrony modulates projected increases in high flows. *Water Resour. Res.* **2021**. [CrossRef]
41. Eum, H.I.; Dibike, Y.; Prowse, T. Climate-induced alteration of hydrologic indicators in the Athabasca River Basin, Alberta, Canada. *J. Hydrol.* **2017**, *544*, 327–342. [CrossRef]
42. Schnorbus, M.; Werner, A.; Bennett, K. Impacts of climate change in three hydrologic regimes in British Columbia, Canada. *Hydrol. Process.* **2014**, *28*, 1170–1189. [CrossRef]
43. Werner, A.T.; Schnorbus, M.A.; Shrestha, R.R.; Eckstrand, H.D. Spatial and Temporal Change in the Hydro-Climatology of the Canadian Portion of the Columbia River Basin under Multiple Emissions Scenarios. *Atmos. Ocean* **2013**, *51*, 357–379. [CrossRef]
44. Shook, K.; Pomeroy, J.; van der Kamp, G. The transformation of frequency distributions of winter precipitation to spring streamflow probabilities in cold regions; case studies from the Canadian Prairies. *J. Hydrol.* **2015**, *521*, 395–409. [CrossRef]

water

MDPI

Article

Hydrological Extremes and Responses to Climate Change in the Kelantan River Basin, Malaysia, Based on the CMIP6 HighResMIP Experiments

Mou Leong Tan [1,*], Ju Liang [2], Narimah Samat [1], Ngai Weng Chan [1], James M. Haywood [2,3] and Kevin Hodges [4]

[1] GeoInformatic Unit, Geography Section, School of Humanities, Universiti Sains Malaysia, Penang 11800, Malaysia; narimah@usm.my (N.S.); nwchan@usm.my (N.W.C.)
[2] College of Engineering, Mathematics and Physical Sciences, University of Exeter, Exeter EX4 4QE, UK; J.Liang@exeter.ac.uk (J.L.); J.M.Haywood@exeter.ac.uk (J.M.H.)
[3] Met Office, FitzRoy Road, Exeter Devon EX1 3PB, UK
[4] Department of Meteorology, University of Reading, Berkshire RG6 6UR, UK; k.i.hodges@reading.ac.uk
* Correspondence: mouleong@usm.my or mouleong@gmail.com

Abstract: This study introduces a hydro-climatic extremes assessment framework that combines the latest climate simulations from the Coupled Model Intercomparison Project Phase 6 (CMIP6) HighResMIP with the Soil and Water Assessment (SWAT) model, and examines the influence of the different climate model resolutions. Sixty-six hydrological and environmental flow indicators from the Indicators of Hydrologic Alteration (IHA) were computed to assess future extreme flows in the Kelantan River Basin (KRB), Malaysia, which is particularly vulnerable to flooding. Results show that the annual precipitation, streamflow, maximum and minimum temperatures are projected to increase by 6.9%, 9.9%, 0.8 °C and 0.9 °C, respectively, by the 2021–2050 period relative to the 1985–2014 baseline. Monthly precipitation and streamflow are projected to increase especially for the Southwest Monsoon (June–September) and the early phase of the Northeast Monsoon (December) periods. The magnitudes of the 1-, 3-, 7-, 30- and 90-day minima flows are projected to increase by 7.2% to 8.2% and the maxima flows by 10.4% to 28.4%, respectively. Lastly, changes in future hydro-climatic extremes are frequently quite different between the high-resolution and low-resolution models, e.g., the high-resolution models projected an increase of 11.8% in mean monthly flow in November-December-January compared to 3.2% for the low-resolution models.

Keywords: climate change; CMIP6; extreme; SWAT; flood; IHA; global warming; drought; Malaysia; Kelantan

Citation: Tan, M.L.; Liang, J.; Samat, N.; Chan, N.W.; Haywood, J.M.; Hodges, K. Hydrological Extremes and Responses to Climate Change in the Kelantan River Basin, Malaysia, Based on the CMIP6 HighResMIP Experiments. *Water* **2021**, *13*, 1472. https://doi.org/10.3390/w13111472

Academic Editors: Rajesh R. Shrestha and Mohammad Reza Najafi

Received: 10 April 2021
Accepted: 20 May 2021
Published: 24 May 2021

Publisher's Note: MDPI stays neutral with regard to jurisdictional claims in published maps and institutional affiliations.

1. Introduction

Climate change features pervasive global warming driven by anthropogenic emissions of greenhouse gases (GHGs), and is one of the major global threats that strongly affect the environment, ecosystems and human society. Intensification of precipitation and increases in temperature due to global warming have been observed in Asia in the past few decades [1–3]. These changes have a major impact on different hydrological systems and consequently increase the risk of regional water hazards such as flood and drought [4,5]. For example, damage caused by floods exceeds more than USD 10 billion a year in China [6,7]. Severe droughts can significantly reduce agricultural yields and freshwater supplies, resulting in social-economic losses [8]. Therefore, quantification of the climate change impact on precipitation extremes and subsequent extreme flows is important in developing better adaptation, more effective mitigation and greater resilience against water hazards.

The numerical modelling of the climate systems is one of the fundamental pillars in studying the changes in hydro-climatic extremes under different scenarios of anthropogenic

GHG emissions. For such studies, general circulation models (GCMs) are extensively used to project scenarios of potential climate change across the earth system. Previous studies based on GCM experiments produced for the Coupled Model Intercomparison Project phase 5 (CMIP5) have projected a continuous global warming and increase in temperature extremes across the globe in the 21st century [9]. For hydrological research, outputs from the GCMs can provide useful driving data of hydrological models simulating the interaction between climate change and hydrological processes. Betts, et al. [10] applied the HadGEM3A-GA3.0, a high-resolution CMIP5 GCM (~60 km), to examine global freshwater availability under the RCP8.5 scenario. They noticed the water cycle changes are complicated and varied in different geographical regions, with mean river flow increase two times greater than the historical period in South and East Asia. A comparison of GCM resolution on a global hydrological cycle assessment was conducted by Vannière, et al. [11], who found increases in global precipitation from low to high resolutions models. However, uncertainties in the GCM simulations still require careful and more detailed consideration at basin scale.

A new generation of GCM simulations produced for Coupled Model Intercomparison Project Phase 6 (CMIP6) have been developed and released [12]. These experiments are based on state-of-the-art GCMs which are more capable of describing the complex physical processes within the climate system compared to the previous GCM versions [13]. The standard set of CMIP6 simulations are at relatively low resolutions of 100–200 km. However, the release of simulations from the High-Resolution Model Intercomparison Project (HighResMIP, Haarsma et al. [14]) for CMIP6 provides a good opportunity to study hydrological changes at unprecedented resolutions under present and future climates. Hence, application of the high resolution CMIP6 experiments in hydrological modelling at basin scale is likely to be a hotspot of hydrological research in the near future, especially for the impact assessment of hydrological extremes [15]. Previous studies based on regional climate models (RCMs) have found a considerable sensitivity of simulated basin-scale precipitation to changes in RCM model resolution, and such a resolution-dependance can have an impact on hydrological simulations [16,17]. However, considerable uncertainty remains both in RCM and GCM simulations requiring further investigation in order to understand how hydrological changes are influenced by the increase in model resolution. The availability of high-resolution GCMs from HighResMIP allows the comparison of different GCM resolutions and hydrological outputs, especially in the simulation of tropical extreme river flows, and provides a good opportunity to study the effect of model resolution on hydrological assessments. In particular HighResMIP provides simulations at resolutions similar to previous RCM based studies without the methodological uncertainties inherent in using RCMs, i.e., definition of the domain, nesting, nudging, etc.

In tropical Asia, extreme flows are sensitive to extreme climatic events, particularly in small catchments [18]. Numerous studies have investigated the impacts of climate change on extreme flows in different river basins over Southeast Asia [19–21], where the reported changes vary from place to place. Based on the recent Coordinated Regional Climate Downscaling Experiment—Southeast Asia (CORDEX-SEA) simulations, a drier climate condition is projected for the southern part of Southeast Asia by the end of the 21st century, while a wetter condition is mainly found in the northern region [22]. The recent study of Tan, et al. [23] introduced a SouthEast Asia HydrO-meteorological droughT (SEA-HOT) framework that integrates the RCMs simulations from CORDEX-SEA and SWAT for hydro-meteorological drought assessment in current and future climates. Although RCMs can provide useful high-resolution climate inputs for hydrological simulations, there are several known caveats. These include the lack of two-way interactions between the downscaled climate on the one-way nested grids and the external large-scale circulations [24,25]. The performance of RCMs is also strongly dependent on the selection of the driving lateral boundary conditions, e.g., Tangang, et al. [26], which can introduce considerable model uncertainties to climate simulations. Therefore, further studies are required to examine the hydrological simulations of SWAT driven by the latest high-resolution climate simulations, e.g., the CMIP6 HighResMIP experiments, which have a resolution similar to that of CORDEX-SEA.

To assess the hydrologic impacts on ecosystems, The Nature Conservancy (TNC) has developed the Indicators of Hydrologic Alteration (IHA) program, which is a software system that computes various hydrological and environmental indicators for quantifying the frequency, magnitude and duration of flows [27]. These indicators provide useful approaches to study flow behavior so as to improve the understanding of water hazards and their ecological and environmental impacts at a regional level. Several studies have combined the SWAT simulations with IHA to evaluate the impact of climate change on regional extreme flows [28–30]. Kiesel, et al. [31] integrated SWAT and 32 IHA indicators to evaluate future changes in extreme flows in three European catchments. Zhang et al. [32] used a similar approach to evaluate the impacts of cascade dams and climate variability on the streamflow of the Jiulong River watershed in the southeast of China. However, the simulation of SWAT combined with the analyses of IHA in tropical regions has been inadequately investigated in previous studies and requires further research, especially for Southeast Asia.

The Kelantan River Basin (KRB) is a typical tropical river basin in Peninsular Malaysia, which is frequently affected by monsoon floods during the early phase of the Northeast monsoon (NEM) season in almost every year [33]. In fact, previous major flood events in Malaysia were caused by continuous extreme precipitation episodes that were brought by the NEM associated with Madden–Julian oscillation (MJO) and the Borneo vortex [34,35]. According to Chan [36], one of the most destructive floods occurred in 1967, when more than 50% of the state's population were affected. In 2014, the total amount of heavy precipitation from 21 to 23 December reached 1295 mm, and resulted in the highest water level of 22.7 m at the Jambatan Guillermard station [37], which is two times higher than the normal level. Tan et al. [38] evaluated the impact of climate change on water resources in the KRB based on the CMIP5 projections and found that the future annual streamflow will increase by 14.6 to 27.2%. This study also projected an increase in the monthly mean streamflow in November, December and January during the NEM season. However, these previous studies have provided limited information on the projected changes in extreme flows, which is critical for decision making in local water hazard risk management and environmental protection [39] across Kelantan. In addition, the SWAT calibration and validation of KRB have been mostly limited to monthly scale assessments. This study has further evaluated the capability of SWAT for a 30-year daily-scale simulation up to 2014.

Therefore, this study aims to introduce a framework to assess the impact of climate change on extreme flows in tropical regions through a novel approach that integrates SWAT, CMIP6 HighResMIP and the IHA indicators. The specific objectives are: (1) to evaluate the capability of CMIP6 HighResMIP at the basin scale; (2) to assess the SWAT capability in simulating long-term daily streamflow in the KRB; and (3) to quantify the hydrological extremes of the KRB in the mid-21st century period (1985–2014 vs 2021–2050) under high-resolution and low-resolution CMIP6 GCMs. The results of this study will enhance the understanding of how different HighResMIP CMIP6 GCMs resolutions influence the tropical extreme flow simulations. Besides, this study provides a comprehensive and the most up-to-date framework for assessing future hydro-climatic extremes for Southeast Asia as well as other tropical regions both in developing and less developed countries for climate adaptation and environmental protection policy formulations. Moreover, the future hydro-climatic projections can be used as a reference by the local authorities to design flood and drought related policies.

2. Materials and Methods

2.1. Study Area

The KRB is located in the northeast part of Peninsular Malaysia, between latitudes 4°–6° and longitudes 101°–103°, as shown in Figure 1. The Kelantan River originates from Mount Ulu Sepat. The elevation of the river basin ranges from −2 m to 2174 m. The river has a total length of 248 km and drains a basin area of 12,685 km^2, covering more than 85% of Kelantan state. The outlet of this study was selected at the river mouth, so the drainage area is slightly larger than Tan et al. [38] who selected the Jambatan Guillermard streamflow station as the basin's outlet. In 2013, the basin was dominated by forest (70.8%), followed by rubber (13.3%), oil palm (11.2%), agricultural (2.8%), paddy-fields (1.3%) and urban development (0.6%). Based on the flood reports prepared by the Department of Irrigation and Drainage Malaysia (DID), floods have normally occurred in Kuala Krai, Tanah Merah, Machang and Pasir Mas that are located in the middle and downstream parts of KRB. As reported in the DID's 2014/2015 annual flood report, floods in Kelantan for the year 2014 resulted in the evacuation of more than 300,000 people, 14 casualties and about RM156 million total losses [40].

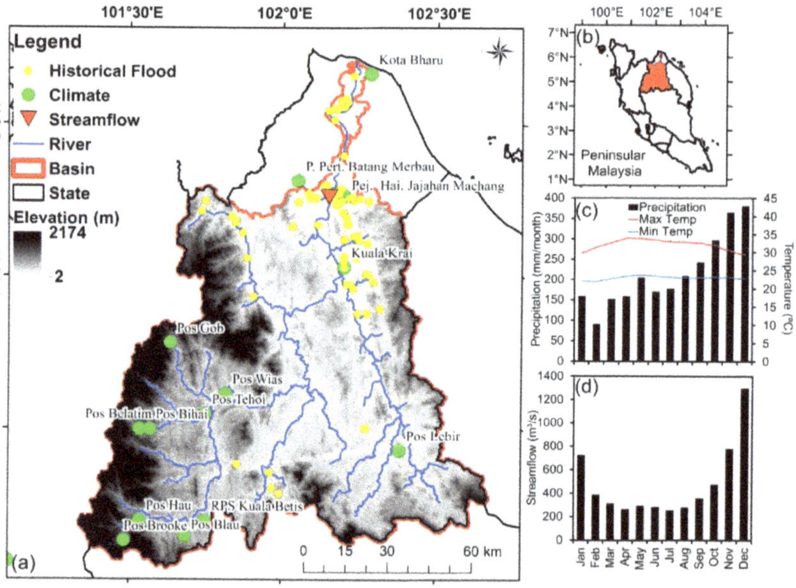

Figure 1. (**a**) Topography, (**b**) location, (**c**) monthly average climate and (**d**) monthly average streamflow of the Kelantan River Basin, Malaysia.

The KRB received an average annual precipitation of 2609.35 mm/year from 1985 to 2014, while average annual streamflow at the Jambatan Guillermard was 475.81 m^3/s. Figure 1 shows the average monthly precipitation, maximum temperature, minimum temperature and streamflow for the basin from 1985 to 2014. The basin receives heavy precipitation during the early phase of the NEM season, mainly in November (365.4 mm/month) and December (380.74 mm/month). Relatively drier conditions can be found in the second phase of the NEM, where only about 91.3 mm/month precipitation is received in February. Average maximum and minimum temperatures of the basin vary from 29.3 °C to 34.1 °C and 22.0 °C to 23.8 °C, respectively. The highest temperature value is normally observed in April, the inter-monsoon period between the NEM and Southwest monsoon (SWM). The highest average monthly streamflow at Jambatan Guillermard is found in December (1296.3 m^3/s), followed by November (781.9 m^3/s) and January (721.3 m^3/s), mainly during the flood periods in this region. As seen in Figure 1, precipitation is the dominant factor influencing the streamflow in the NEM, whereas temperature increases in the SWM raise evaporation and reduce flows. A moderate streamflow rate was found in the driest month of February, largely due to the lag of flows accumulated from the extreme high precipitation between November and January, where a similar situation was reported in other regions [41,42].

2.2. SWAT

The SWAT model is a semi-distributed and continuous hydrological model developed for water resources managers to decide the most appropriate strategy or solution by considering the impact of different management practices on streamflow and non-point source pollution [43]. SWAT has undergone continuous improvement over the past few decades [44,45] and its applicability and credibility have been verified in Southeast Asia [46]. SWAT is typically run at daily or monthly temporal resolution for a continuous period of time. The SWAT simulations can be driven by simulations of average flow climatology, while the representation of extreme flood events usually requires simulations driven at a finer (daily at least) temporal resolution. To simulate the long-term climatology of both extreme droughts and floods, a long-term climate simulation for at least 30-years at daily scale is recommended to be used as the SWAT input [15]. A review of SWAT-based hydro-climatic extremes studies has recently been reported by Tan [15]. About 47% of the studies were conducted in the United States and China. Besides that, only around 10% of the reviewed research has further evaluated both low and high flow conditions. Therefore, more studies in different geographical and climate conditions are needed to evaluate the capability of SWAT in capturing extreme flows.

The SWAT Calibration and Uncertainty Programs (SWAT-CUP), a tool designed specifically to calibrate and validate SWAT, is used in the SWAT model assessment [47,48]. The Sequential Uncertainty Fitting Version 2 (SUFI2) calibration program in the SWAT-CUP was selected to evaluate the model performance at daily and monthly scales. During the calibration, SWAT was run with 500 simulations per iteration under different parameter combinations. The Coefficient of determination (R^2) and Nash-Sutcliffe Efficiency (NSE) [49] are the most commonly used statistical approaches to rate the performance of SWAT for each simulation. The R^2 and NSE values range from 0 to 1 and $-\infty$ to 1, respectively, with 1 being the optimal value for both metrics. Detailed descriptions and formulas of the two statistical approaches are available in previous SWAT studies [50]. Moriasi et al. [51] recommended that the performance of SWAT can be considered as satisfactory if NSE and R^2 values are greater than 0.5 and 0.6, respectively.

2.3. CMIP6 HighResMIP Models

As an integral protocol of CMIP6, HighResMIP [14] provides high-resolution GCM ensemble simulations together with their coarse resolution versions. The HighResMIP experiments allow the assessment of the impact of model resolution on climate simulations and aim to improve the understanding of model biases and uncertainties [14]. This protocol acts as an important input to the Intergovernmental Panel on Climate Change (IPCC) sixth assessment report (AR6). The high resolution (<50-km) GCM simulations from HighResMIP also help to solve the issue of coarse model resolution in the CMIP5 GCMs [52] and are comparable to resolutions used in many RCM based studies [53,54]. In this paper, ten HighResMIP experiments for the present-day (1980–2014) and future (2015–2050) climates are chosen and the data are collected via the data platform of the Earth System Grid Federation. The future period simulations are based on the high-emission 8.5 scenario of the Shared Socioeconomic Pathways (SSPs-8.5), which is part of the new scenario framework for the latest IPCC climate change assessment. Descriptions of these models including the model developers and horizontal model resolutions are given in Table 1.

High resolution models have been demonstrated to offer greater fidelity than coarse resolution models in representing both the observed mean and extreme precipitation events over Peninsula Malaysia [55]. Nevertheless, even the high resolution models of HighResMIP inevitably show biases in their representation of extreme precipitation. Prior to applying the selected HighResMIP experiments to the simulation of SWAT and the subsequent assessment of hydrological impacts, the model biases relative to climate observations in the simulated climatic variables (i.e., daily precipitation, maximum and minimum temperatures) are corrected using a quantile mapping (QM) approach similar to Boé et al. [56] and Kim et al. [57]. QM first calculates the correction parameter (defined as the observed mean value divided by the simulated value) at different quantile levels. Corrected variables are then obtained by the product of the simulated value and the calculated correction parameter for each quantile level with an interval of 0.5 percentile rank. To correct the model bias in simulating the annual cycle of climatic variables, QM is applied to data for each month, respectively. To compare the simulations between low and high resolutions, the ensemble of the low resolution (LR) simulations from HadGEM3-LM, CNRM and FGOALS-L is analyzed and compared to the ensemble of the high resolution (Ens_HR) simulations from the other seven models.

Table 1. Summary of the GCMs used in the CMIP6 HighResMIP experiments.

No.	Modeling Organizations	Model Name	Vertical Resolution (Layers)	Horizontal Resolution (Longitude × Latitude)	Label
1 2 3	The UK Met Office Hadley Centre for Climate Change	HadGEM3-GC31	85	1.875° × 1.25° 0.83° × 0.56° 0.35° × 0.23°	HadGEM3-LM HadGEM3-MM HadGEM3-HM
4 5	French National Centre for Meteorological Research	CNRM-CM6-1	91	1.406° × 1.406° 0.5° × 0.5°	CNRM CNRM-HR
6	27 institutes in Europe (Haarsma et al., 2020)	EC-Earth3P	91	0.703° × 0.703°	EC-Earth
7 8	Meteorological Research Institute (Japan)	MRI-AGCM3-2	60	0.563° × 0.563° 0.188° × 0.188°	MRI-H MRI-S
9	Institute of Atmospheric Physics/Chinese Academy of Sciences	FGOALS-f3	32	1.25° × 1°	FGOALS-L
10	Geophysical Fluid Dynamics Laboratory/ NOAA (U.S.)	GFDL-CM4C192	33	0.625° × 0.5°	GFDL

2.4. IHA Indicators

IHA is a user-friendly tool developed by The Nature Conservancy to measure flow characteristics using 32 IHA (Table 2) and 34 Environmental Flow Component (EFC) (Table 3) indicators [27]. For example, the IHA tool can calculate the magnitude and duration of annual minima and maxima flows for specific periods, e.g., 1-day, 3-day, 7-day, 30-day, 90-day. These indicators provide useful information for policy makers, water managers, hydrologists and researchers to understand the impact of human activity, including land use and anthropogenic climate warming, on rivers and groundwater. Comparative analysis can be conducted to describe and quantify the changes of these extreme elements associated with climate change. The zero-flow day indicator is excluded in this study due to its limited suitability for tropical regions. The IHA version 7.1 is used to calculate the extreme flows based on the SWAT outputs. The computation of IHA is based on the daily streamflow data generated from the SWAT simulations.

Table 2. List of 32 IHA parameters adopted in this study.

Hydrologic Parameters	Symbol
1. Magnitude of monthly water condition (12 parameters) Mean value for each calendar month	January–December
2. Magnitude and duration of annual extreme water conditions (11 parameters)	
Annual minima, 1-day mean	1-day min
Annual minima, 3-day means	3-day min
Annual minima, 7-day means	7-day min

Table 2. *Cont.*

Hydrologic Parameters	Symbol
Annual minima, 30-day means	30-day min
Annual minima, 90-day means	90-day mom
Annual maxima, 1-day mean	1-day max
Annual maxima, 3-day means	3-day max
Annual maxima, 7-day means	7-day max
Annual maxima, 30-day means	30-day max
Annual maxima, 90-day means	90-day max
Base flow index: 7-day minimum flow/mean flow for year	Base flow
3. Timing of annual extreme water conditions (2 parameters)	
Julian date of each annual 1-day maximum	Date min
Julian date of each annual 1-day minimum	Date max
4. Frequency and duration of high and low pulses (4 parameters)	
Number of low pulses within each water year	Lo pulse count
Mean or median duration of low pulses (days)	Lo pulse dura
Number of high pulses within each water year	Hi pulse count
Mean or median duration of high pulses (days)	Hi Pulse dura
5. Rate and frequency of water condition changes (3 parameters)	
Rise rates. Mean of all positive differences between consecutive daily values	Rise rate
Fall rates: Mean of all negative differences between consecutive daily values	Fall rate
Number of hydrologic reversals	Reversals

Table 3. List of 34 Environmental Flow Component (EFC) parameters that adopted in this study.

Environmental Flow Components Parameters	Symbol
1. Monthly low flows (12 parameters)	
Mean values of low flows during each calendar month	January low–December low
2. Extreme low flows (4 parameters)	
Peak flow (minimum flow during event)	EL peak
Duration of extreme low flows (days)	EL duration
Timing of extreme low flows	EL time
Frequency of extreme low flows	EL freq
3. High flow pulses (6 parameters)	
Peak flow (maximum flow during event)	HF peak
Duration of high flow pulse event (days)	HF duration
Timing of high flow pulse event (Julian date of peak flow)	HF time
Frequency of high flow pulse event	HF freq
Rise rate of high flow pulse event	HF rise
Fall rate of high flow pulse event	HF fall
4. Small floods (6 parameters)	
Peak flow of small flood event (maximum flow during event)	SF peak
Duration of small flood event (days)	SF duration
Timing of slow flood event (Julian date of peak flow)	SF time
Frequency of small flood event	SF freq
Rise rate of small flood event	SF Rise
Fall rate of small flood event	SF Fall

Table 3. *Cont.*

Environmental Flow Components Parameters	Symbol
5. Large floods (6 parameters)	
Peak flow of large flood event (maximum flow during event)	LF peak
Duration of large flood event (days)	LF duration
Timing of large flood event (Julian date of peak flow)	LF time
Frequency large flood event	LF freq
Rise rate of large flood event	LF Rise
Fall rate of large flood event	LF Fall

2.5. Model Setup and Input Data

The general framework of this study is shown in Figure 2 and includes the following steps: (1) SWAT input collection, (2) HighResMIP data download and bias correction; (3) SWAT calibration and validation; (4) Incorporation of the bias corrected HighResMIP into calibrated SWAT, (5) calculation of flow extremes based on the IHA indicators, and (6) comparison of the changes between future (2021–2050) and historical (1985–2014) flow extreme.

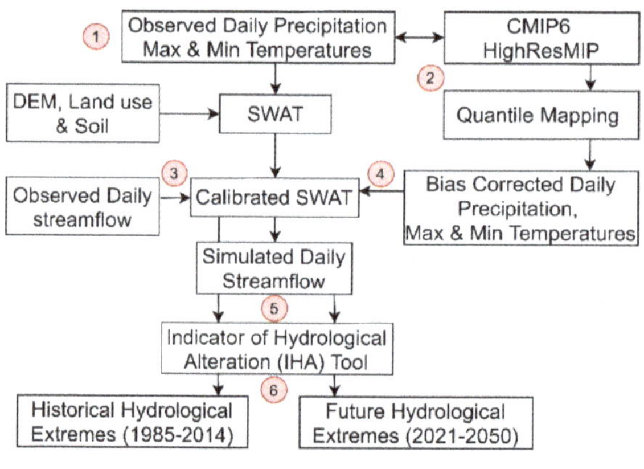

Figure 2. Flow chart of this study.

The simulation of SWAT requires three geophysical inputs, including land use, soil and elevation. The latter information was extracted from satellite-based digital elevation model data, the Shuttle Radar Topography Mission (SRTM). The land use data of 2013 is provided by the Department of Agricultural Malaysia (DOA), while the soil data is from the FAO-UNESCO soil map. The climate driving data of SWAT includes daily precipitation, daily maximum and minimum temperatures. Daily climate data from 1980 to 2014 was collected from the Malaysia Meteorological Department (MMD). To initialize the sub-basin delineation, digital river network data collected from Department of Drainage and Irrigation (DID) was used to integrate into the SRTM DEM for improving the river network formation. Lastly, the streamflow data from the same department is used to calibrate and validate the SWAT model. The minimum threshold value of 10,000 ha was used in the sub-basin delineations. Five slope classes of 0%–10%, 10%–20%, 20%–30%, 30%–40% and >40% were used in the slope definition during the model development. The next step is the formation of hydrologic response units, the smallest spatial unit of SWAT which lumps together all similar land uses, soils and slopes within each sub-basin to perform all model calculations.

SWAT has exhibited a reliable performance in simulating the climatology of monthly streamflow in Kelantan and typical cases of historical drought events [23,38]. This will apply a new configuration of SWAT, calibrated by referring to parameter ranges and sensitivity analysis from previous studies [23,38]. The calibration and validation periods are chosen as 1985–1999 and 2000–2014, respectively. As the SWAT model setup at the monthly scale has been well configured and described in previous studies [23,38], this paper focuses on the daily scale simulation, which is important for allowing the diagnoses of extreme flows.

Bias corrected HighResMIP climate projections as described in Section 2.3 are used to drive the calibrated SWAT in order to simulate the future daily streamflow from 1985 to 2050. Thus, the projected hydro-climatic changes driving the simulation of SWAT are based on multi-GCM ensemble experiments, which help to consider the climate projection uncertainties associated with the different GCM configurations and improve reliability in the assessment of hydro-climatic impacts [58,59]. The simulated streamflow will then be used to calculate the 66 extreme indicators as listed in Tables 2 and 3. The assessment of climate change impact on flow extremes was conducted by comparing the relative differences of extreme indicators between 1985–2014 and 2021–2050. Moreover, the statistical significance of the mean difference between the future and the baseline historical periods at a 95% confidence level (p-value < 0.05) will be determined using the two-tailed Student's t-test.

3. Results

3.1. SWAT Calibration and Validation

Table 4 indicates that the baseflow alpha-factor (ALPHA_BF), initial SCS runoff Curve Number for moisture condition II value (CN2) and effective hydraulic conductivity in main channel alluvium (CH_K2) are among the most sensitive parameters for daily streamflow calibration, which is similar to the previous monthly scale calibration in KRB [23,38]. ALPHA_BF shows the baseflow response to changes in recharge, CN2 represents a function of the soil permeability, land use and soil water condition, while CH_K2 adjusts the relationship of water exchange from groundwater to river [60,61].

Table 4. Final SWAT performance rating as recommended by Moriasi [51].

No	Name	First Iteration	Last Iteration	Fitted
1	v__ALPHA_BF.gw	0.00	1.00	0.00
2	v__CH_K2.rte	0.00	500.00	350.00
3	r__CN2.mgt	−0.50	0.50	−0.45
4	v__GW_DELAY.gw	0.00	500.00	0.00
5	r__SOL_AWC().sol	−0.50	0.50	−1.00
6	v__GW_REVAP.gw	0.02	0.20	0.10
7	v__RCHRG_DP.gw	0.00	1.00	0.00
8	v__GWQMN.gw	0.00	5000.00	1500.00
9	r__CH_N2.rte	−0.50	0.50	−1.00
10	v__REVAPMN.gw	0.00	500.00	128.00
11	v__SURLAG.bsn	0.05	24.00	2.00
12	v__ESCO.bsn	0.00	1.00	0.05

v_ indicates that the original parameter value is replaced with the given value; R indicates that the parameter value is multiplied with 1 + the given value.

The comparison between the observed and simulated streamflow in the case of Jambatan Guillermard during the calibration (1985–1999) and validation (2000–2014) periods is shown in Figure 3. In general, the simulated monthly and daily variations of streamflow are captured reasonably well with respect to the observed streamflow. However, an overestimation of baseflow is found for the period of 2010–2014, which is possibly due to increasing groundwater demand in recent years [62], as there is no accessible groundwater extraction data for the study area. Six groundwater parameters were considered in the SWAT calibration by allowing more water transfer from baseflow to surface or atmosphere so that the impact of inadequate groundwater data is minimized.

The model performance is rated as "very good" in terms of the monthly scale simulation according to the Jambatan Guillemard Bridge for both the calibration (R^2 = 0.84 and NSE = 0.72) and validation (R2 = 0.84 and NSE = 0.63) periods. The performance of

SWAT is rated as satisfactory at the daily scale. For the last iteration of calibration, 78% (p-value = 0.78) of the simulations were bracketed by the 95 Percent Prediction Uncertainty (95PPU) with r-value of 0.89, indicating that the calibration is acceptable. Thus, the calibrated configuration of SWAT will then be used for simulating the hydro-climatic extremes in the KRB and the results will be discussed in Sections 3.3–3.5.

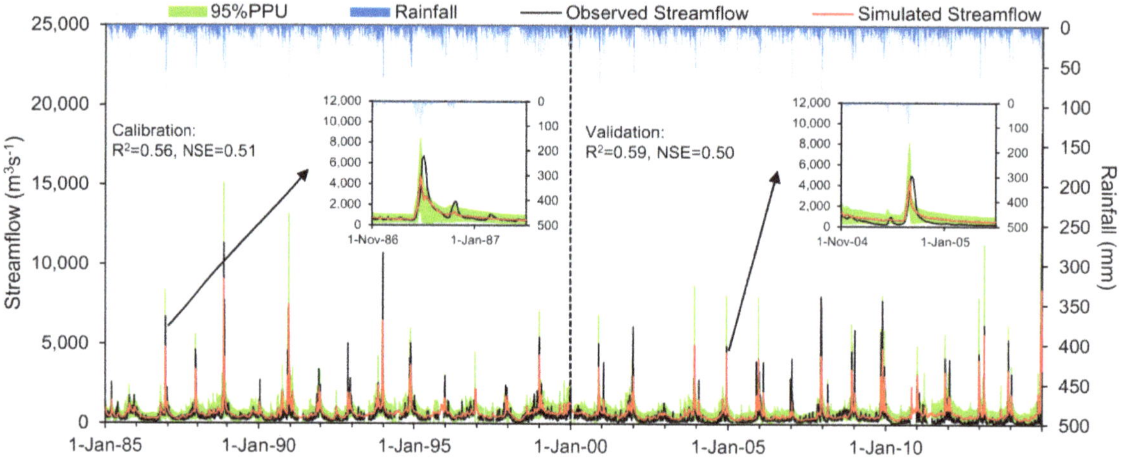

Figure 3. Observed and simulated daily streamflow at Jambatan Guillermard from 1985 to 2014.

3.2. Bias Correction of CMIP6 HighResMIP Models

The biases of the HighResMIP experiments were corrected using the QM, as discussed in Section 2.3. Figure 4 shows the climatology of monthly precipitation and maximum and minimum temperatures from 1980 to 2014 over KRB that are measured from observed data, original and bias-corrected HighResMIP models. The original HighResMIP models tend to underestimate the monthly precipitation in November and December, whereas an inconsistent pattern was found for other months. The HighResMIP models simulate the peak monthly precipitation in November, one month earlier than observed. The FGOALS-f3-L simulation exhibits a generally poorer performance than the other simulations as it significantly underestimates the precipitation amount over KRB, especially during the SWM season (June–September). We also note that the original HighResMIP models generally show a better performance in simulating the precipitation amount compared to the original regional climate model simulations from CORDEX-SEA that dramatically overestimate the observed precipitation by up to five times for certain months in the same basin [23]. As shown in Figure 4d, the QM approach reasonably corrects the biases of the models in representing the peak monthly precipitation in December and precipitation amount for all the simulations. It is also noted that the high resolution (HR) simulations exhibit less underestimation of precipitation during the NEM compared to low resolution (LR).

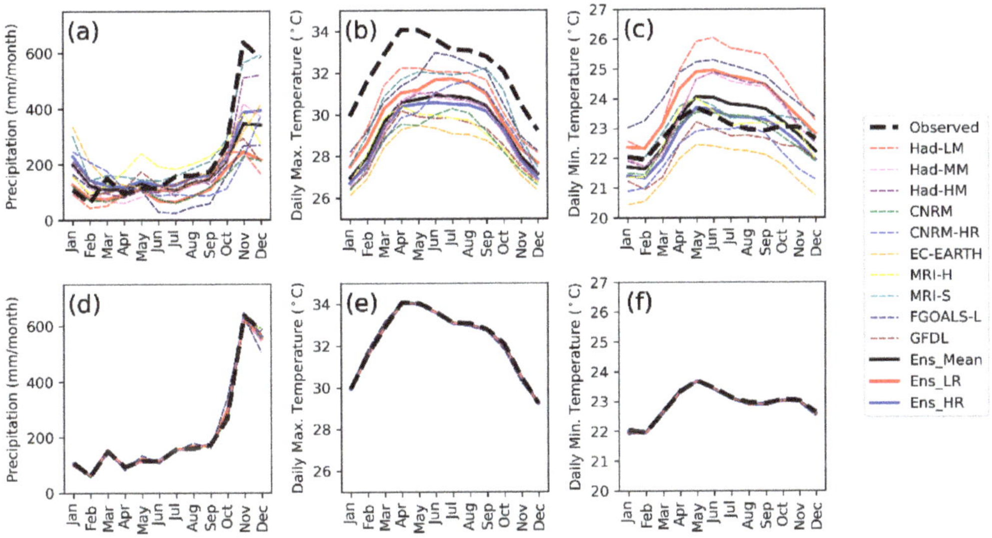

Figure 4. The annual cycles of monthly precipitation amount (**a**,**d**), monthly mean daily maximum (**b**,**e**) and minimum temperature (**c**,**f**) for the ten CMIP6 HighResMIP experiments, compared to the MMD observations during the period 1980~2014: (**a**–**c**) original; (**d**–**f**) bias-corrected for the Kelantan River Basin, Malaysia.

In general, most CMIP6 HighResMIP models reasonably capture the observed warm period in April and September in terms of the monthly mean of daily maximum and minimum temperatures (Figure 4b,c,e,f), mainly during the inter-monsoon periods. However, all the HighResMIP models underestimate the monthly maximum temperature, whereas most of the models show an overestimation in the monthly minimum temperature, except the CNRM-CM6-1-HR, EC-EARTH3P and GFDL-CM4C192 models. Minimum temperatures outperformed maximum temperatures in the climatology simulations since the ensemble mean is much closer to the observed data. Similar to precipitation, maximum and minimum temperature biases reduce significantly after applying the QM bias correction approach, as shown in Figure 4e,f, respectively. Compared to LR, the HR simulations show less overestimation of daily minimum temperature. However, no apparent improvement in simulating maximum temperature is found in HR.

3.3. Climate Change

The projected annual and monthly changes in precipitation, daily maximum and minimum temperatures over the KRB for the period 2021–2050 relative to 1985–2014 are shown in Figure 5. Annual precipitation is projected to increase significantly by 6.9%. For almost every month, the precipitation amount is projected to increase from 0.94 (October) to 15.1% (December), except for April with a reduction of 2.4%. In Figure 5a, statistically significant changes in the monthly average precipitation are seen in June, July, August and December, indicating a general increase in precipitation during the SWM and the early phase of the NEM.

The annual mean of daily maximum and minimum temperatures for the period 2021–2050 relative to 1985–2014 are projected to increase by 0.8 and 0.9 °C, respectively (Figure 5b). The equivalent warming trends are slightly higher than the historical long-term warming trends (0.1 and 0.3 °C/decade for maximum and minimum temperatures respectively) from 1985 to 2018 in Malaysia [1]. For each month, maximum and minimum temperatures are projected to increase by 0.7–1.0 °C, with higher magnitudes of warming in May and November (Figure 5b). There is also a significant difference between the future and historical temperatures in their mean values for both the annual and monthly scales (Figure 5b), supporting the literature regarding a warming situation in the next 30 years.

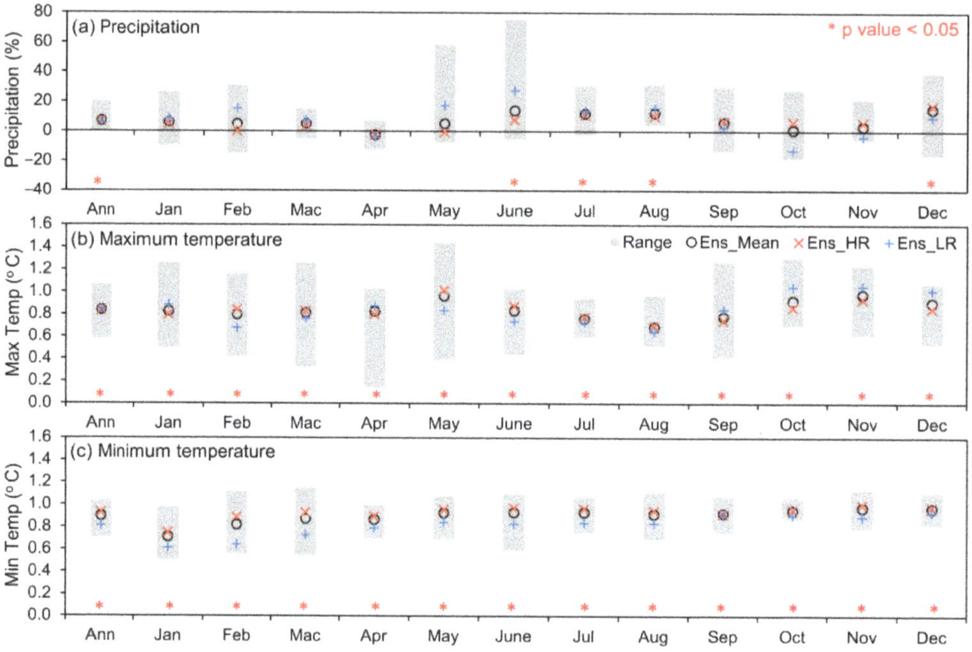

Figure 5. Projected changes of (**a**) precipitation, (**b**) maximum and (**c**) minimum temperatures between 1985–2014 and 2021–2050 in the Kelantan River Basin, Malaysia.

3.4. Hydrologic Extreme Changes

A comparison of the changes in hydrological extremes at the Jambatan Guillemard station between the 2021–2050 and 1985–2014 periods is shown in Figure 6. Annual and monthly streamflow are projected to increase by 9.9% and 3.5% to 16.8% in the future, respectively, as simulated from an ensemble mean of the ten HighResMIP models. A higher rate of increase of more than 10% can be found in June to August and December. Additionally, a significant difference of mean streamflow at 95% confidence level can be found at the time-scales of annual, December to January, March, and June to October, which is similar to the monthly precipitation changes. As December and January are the major flood periods in the KRB, more intense flooding impacts may occur in the future.

Next we analyze the indicators quantifying the annual streamflow extremes in terms of their magnitude for different durations, e.g., the second IHA group listed in Table 1. Figure 6b shows that the 1-, 3-, 7-, 30- and 90-day minima and maxima flows increase significantly from 7.2% to 8.2% and 10.4% to 28.4%, respectively, in the 2021–2050 period. There is a high deviation for the case of extremely high flows at 1-, 3- and 7- day maxima flow, showing that the magnitude of floods might increase in the near future. As for the baseflow index, a slight decrease of 0.9% is seen for the future period. This indicates that the future amount of water available for freshwater supply tends to remain the same as the current value. However, stress on water supply may occur if there is an increased water demand following population growth in the future.

The third IHA group indicates the streamflow extreme events in terms of the occurrence timing. A water year is defined using the Julian dates format where "day 1" refers to 1st January and "day 365" refers to 31st December. As shown in Figure 6c, the occurrence timing of the future annual minimum and maximum events that are projected by the ten models will increase by 1.0% to 25.3% (1.87 to 46.27) and 0.4% to 16.2% (0.73 to 29.67), respectively. This indicates that the future streamflow extreme events are projected to be delayed for a few days to a few weeks. The fourth IHA group shows the number and duration of extreme pulses. The results show that the number of low pulses showed no significant changes; meanwhile, the number of high pulses is projected to increase significantly by 22.7%. The duration of future low and high pulses tends to change slightly compared to the historical period, based on an ensemble mean of 3.3% and 5.9%, respectively. The last IHA group contains the rate and frequency of water condition changes. There is a significant increase in the rise rate and fall rate at the Jambatan Guillermard station by 31.9% and 25.5%, respectively, showing that a rapid increase or decrease in streamflow might happen in the future. The number of hydrological reversals will increase by 11.6% as projected by the model ensemble, which shows an increment in the number of daily streamflow increases after decreasing, and decreases after increasing.

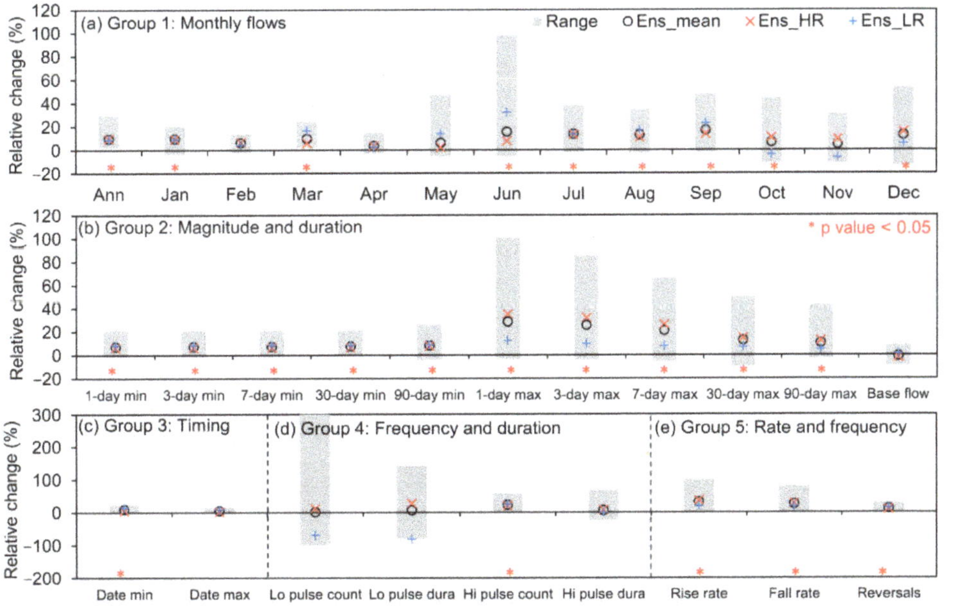

Figure 6. Projected changes in hydrological extremes as indicated by 32 IHA parameters of (**a**) monthly flows, (**b**) magnitude and duration, (**c**) timing, (**d**) frequency and duration and (**e**) rate and frequency between 1985–2014 and 2021–2050 at the Jambatan Guillermard station.

3.5. Environmental Flow Changes

The EFC indicators that divide into monthly low flows, extreme low flows, high flow pulses, small floods and large floods, were also considered in this study because this information is very important to sustain riverine ecological integrity. The first EFC group shows the monthly low flow sustained by groundwater, so any changes in these parameters can be related to groundwater availability. Basically, a slight decrease in the monthly low flow can be found in April, May and October to December by 0.2% to 1.1%, based on the ensemble mean projection as shown in Figure 7. By contrast, monthly low flows from July to September are projected to increase significantly at the 95% confidence level by 4.3 to 9.4%.

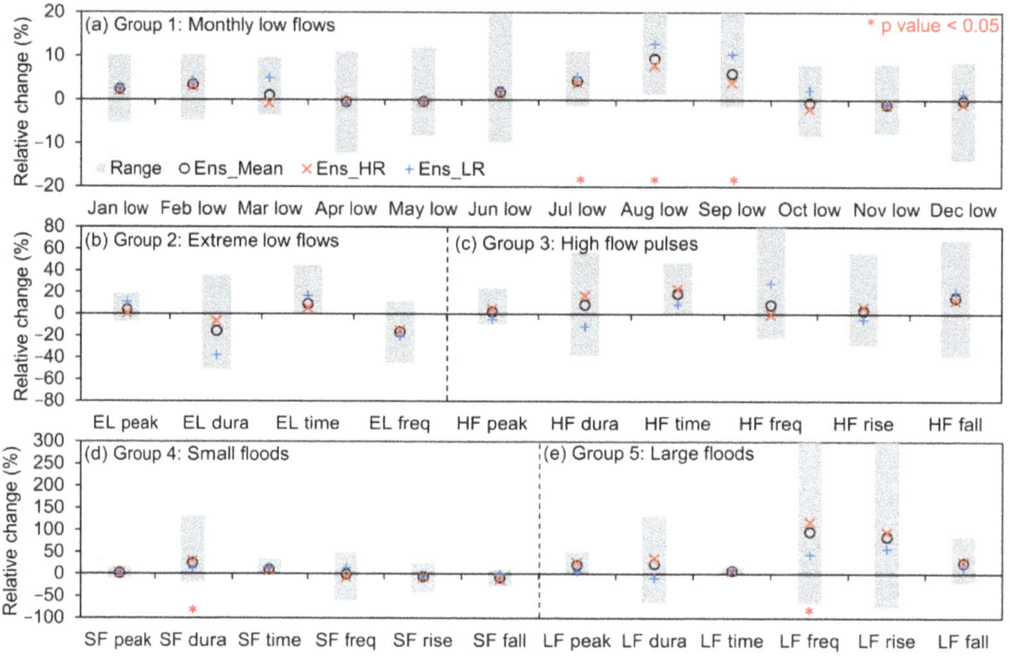

Figure 7. Projected changes in hydrological extremes as indicated by 32 EFC parameters of (**a**) monthly low flows, (**b**) extreme low flows, (c) high flow pulses, (**d**) small floods and (**e**) large floods between 1985–2014 and 2021–2050 at the Jambatan Guillermard station.

4. Discussion

Climate change is expected to have a more significant impact on hydrological extreme flows than environmental flow components, since more IHA indicators (Figure 6) were found to be significant in the 2021–2050 period than EFC indicators (Figure 7). The findings show that extreme high flows are more sensitive to climate than normal and low flows in the KRB. The situation can be highlighted by the fact of continuous heavy precipitation in the first week of January 2021, that resulted in serious flooding in five states of Peninsular Malaysia, including Kelantan. According to the Kelantan e-flood portal, https://ebanjir.kelantan.gov.my/ (accessed on 20 March 2021), water levels at the Jambatan Guillermard station exceeded the danger level of 16 m from 6th January to 9th January 2021, up to the peak level of 18.89 m. The water level is only slightly lower than the Kelantan Big Yellow Flood of 2014 that reached a peak level of 22.74 m. The alteration in extreme flows has a great influence on biodiversity and the ecological system within and surrounding the river [63].

The statistical analysis has shown that the SWAT model is acceptable for daily streamflow simulations in the KRB. In Southeast Asia, a review of SWAT application in hydro-climatic extreme studies reported that the SWAT model tended to have a better performance on monthly time scales compared to daily time scales [46]. A possible explanation for this might be insufficient rain gauges in the tropics to capture the daily-scale extreme precipitation at a better spatial coverage [64]. For example, the only rain gauge in the southeastern part of the KRB as shown in Figure 1 may miss captured extreme events in the nearby region, and therefore the extreme flows were underestimated by the SWAT model. In fact, the reported NSE and R^2 values at daily time scales in previous SWAT extreme related studies commonly ranged from 0.50 to 0.79 [15]. Besides, the SWAT model tends to underestimate the peak flows which are reported in the river basins of Spain [65], Brazil [66] and Hawaii [67]. Based on Krysanova and Arnold [68], the flood-plain deposition algorithm within SWAT needs to be improved in order to increase the accuracy of SWAT in replicating peak flows. This suggestion has been taken into consideration during the SWAT+ development [69]; however, how effective the improvement is still needs further investigation.

Climate projections are always regarded as one of the major uncertainties in hydro-climatic impact modelling [15,70]. Given that the HR models have been shown to have a better fidelity in representing precipitation extreme events when compared to observations than their LR model counterparts [55], one might expect the HR models to provide a more accurate assessment of future precipitation extremes. Of the models assessed in this study, changes in future precipitation extremes are frequently quite different between the HR and LR models. The mean precipitation change is significantly higher during the November-December-January period when most flooding occurs at 9.6% for the HR models compared to 5.0% for the LR models (Figure 5). The associated mean monthly flow changes are significantly higher during the Nov-Dec-Jan period at 11.8% for the HR models compared to 3.2% for the LR models. The changes in the magnitude and duration of the 1-day to 90-day maximum flows are universally greater in the HR models than in the LR models with values for 1-day maxima increases by as much as 35.1% for HR compared to 12.8% for LR models (Figure 6). Large flood event marker changes in future climate scenarios are significantly more prevalent in the HR models than the LR models. For example, the large flood frequency increases by 119% for the HR models compared to 44.4% for the LR models (Figure 7). Similarly, Troin [71] also reported that RCM is superior than GCMs for hydro-climatic assessment, particularly in mountainous regions. Therefore, numerous high-resolution GCMs along with different downscaling techniques should be considered in future hydro-climatic modelling [72].

This intercomparison of simulations at different model resolutions suggests that the simulation results based on climate simulations at relatively low model resolutions should be interpreted with cautions. The linkage between the model performance in simulating regional precipitation and the spatial resolution of the model is complex. For example, Liang et al. [55] found that the high resolution version of HadGEM3-GC3.1 has a stronger ability to simulate the Borneo Vortices during the NEM season and their associated precipitation in Malaysia compared to the low resolution versions. This partly explains that the high-resolution simulations of MRI-S and Had-HM exhibit a better performance in capturing the high precipitation period (November and December) of the study area with respect to the low-resolution experiments, as shown in Figure 4a. We note that the application of all available climate models, or only the models with good performance in hydrological impact assessment, is still a debatable issue regarding what is the optimal choice [70], and offer a caveat that the number of models that provide LR diagnostics is limited to just three in our analyses, compared to seven for the HR models. Further model evaluations should be added as more HighResMIP model output becomes available.

Bias correction of climate model output is important for improving the quality of driving data of hydrological simulation for hydro-climatic impact modelling studies [23,73,74], and is mainly based on statistical approaches. For instance, Tan et al. [38] used a linear scaling approach to correct biases in the CMIP5 GCMs before applying them into SWAT. In this study, a statistical method based on QM is used to correct biases in the HighResMIP experiments. This method has also been used for bias corrections in the dynamical downscaling simulations over Malaysia [75] and Southeast Asia [76] from the CORDEX-SEA experiments. This study of Shrestha, et al. [77] suggested that there is no significant difference between the simple (linear scaling) and complex (QM) bias correction schemes for monthly streamflow studies. However, Luo, et al. [78] compared seven bias correction schemes to downscale precipitation and temperature in the Kaidu River Basin, China, and reported that the effect of different bias correction schemes is larger in precipitation than temperature. More studies on how the use of different approaches influence bias corrections in GCMs and how RCMs influence the daily streamflow simulations are required for the study area of the paper and the surrounding western Maritime Continent.

5. Conclusions

Extreme hydro-climatic events exert substantial impact on the environment and human society. This study incorporates the latest high-resolution GCM simulations from HighResMIP and SWAT to project possible future changes in hydrological extremes over the KRB. The SUFI-2 algorithm was applied in the SWAT sensitivity analysis, calibration and validation for improving the credibility of SWAT for simulating the long-term climatology of daily streamflow. In this process, ALPHA_BF, CN2 and CH_K2 are found to be the most sensitive parameters in the SWAT calibration, which is consistent with previous studies [23].

The original outputs of the HighResMIP experiments tend to underestimate monthly precipitation in November and December. Besides, the models are found to simulate an earlier peak (by about one month) of monthly precipitation compared to observations. Most of the HighResMIP experiments underestimate the monthly maximum temperature, while an overestimation is found for the monthly minimum temperature compared to observations. Based on future climate simulations with the model biases corrected by QM, the annual precipitation, maximum and minimum temperatures are projected to increase significantly by 6.9%, 0.8 °C and 0.9 °C, respectively, for the 2021–2050 period relative to the 1985–2014 baseline period. Monthly precipitation in the basin is expected to increase for almost every month by 0.9 to 15.1%, except for April with a decreasing signal of 2.4%. Monthly maximum and temperatures are projected to increase by 0.7–1.0 °C.

The future simulation of the annual mean streamflow for the period 2021–2050 shows a significant increase by 9.9% relative to the 1985–2014 baseline period. Meanwhile, monthly streamflow is projected to increase for all months by 3.5 to 16.8%, with significant changes mainly found in the SWM and the early NEM periods. The magnitude of 1-, 3-, 7-, 30- and 90-day minima and maxima flows are projected to increase significantly by up to 28.4%. In contrast, the baseflow index is projected to change slightly by about 0.9%. The occurrence timing of the extreme flows is expected to be delayed by a few days to a few weeks in the future. The duration of future low and high pulses shows only minor changes as compared to the baseline period. By contrast, the rise rate and fall rate exhibit a rapid increase or decrease of streamflow that may occur in the future.

For the changes in the monthly distribution of low flows for the period 2021–2050 relative to 1985–2014, slight decreases are seen in April, May, October, November and December, whereas a significant increase is found from July to September by 4.3% to 9.4%. The duration and frequency of extreme low flows are projected to decrease by 15.7% and 16.5% respectively. For the high-flow pulses, future projections show an increase by 1.8% to 18.6% compared to the baseline period. Generally, indicators for both small and large floods are projected to increase in the future. However, only changes in small flood duration and large floods' frequency are statistically significant.

This study constructs a framework for comprehensively assessing hydro-climatic extremes by integrating hydrological modelling and state-of-the-art high-resolution climate simulations. Further studies are needed to understand the limited ability of SWAT in capturing both the peak and low flows in the KRB. Besides, future work is needed to compare SWAT with its latest version, SWAT+ [69], and investigate the potential improvement of SWAT+ in simulating peak flows in the KRB. As more simulations of CMIP6 GCMs with different model resolutions will be released to the public, a comprehensive investigation on how the horizontal and vertical resolutions of GCMs influence the SWAT simulation will be investigated in the near future. Finally, this study indicates that high- and low-resolution model resolutions resulted in quite different changes in future hydro-climatic extremes, so a more reliable climate projection quantification framework and ensembles techniques should be developed to minimize uncertainties in hydro-climatic extreme simulations.

Author Contributions: Conceptualization, M.L.T., J.L. and J.M.H.; methodology, M.L.T. and J.L.; software, M.L.T. and J.L.; validation, M.L.T. and J.L.; formal analysis, M.L.T., J.L., J.M.H. and K.H.; data curation, M.L.T. and J.L.; writing—original draft preparation, M.L.T.; writing—review and editing, J.L., N.S., N.W.C., J.M.H. and K.H.; project administration, M.L.T. and J.M.H.; funding acquisition, M.L.T. and J.M.H. All authors have read and agreed to the published version of the manuscript.

Funding: This research was funded by Ministry of Higher Education Malaysia, grant number 203.PHUMANITI.6780001" and Newton Fund of NERC, grant number NE/S002707/1 under the IMpacts of PRecipitation from Extreme StormS, Malaysia (IMPRESS-MALAYSIA) project supported by the Newton-Ungku Omar Fund. Part of the initial data collection was funded under the FRGS project 203.PHUMANITI.6711693.

Institutional Review Board Statement: Not applicable.

Informed Consent Statement: Not applicable.

Data Availability Statement: The data presented in this study are available on request from the corresponding author.

Acknowledgments: The author also knowledge the Department of Agriculture Malaysia, Department of Irrigation and Drainage Malaysia, Malaysian Meteorological Department and Malaysia Geospatial Data Infrastructure (MyGDI) for supplying the geospatial and hydro-climatic data.

Conflicts of Interest: The authors declare no conflict of interest.

References

1. Tan, M.L.; Juneng, L.; Tangang, F.T.; Chung, J.X.; Radin Firdaus, R.B. Changes in Temperature Extremes and Their Relationship with ENSO in Malaysia from 1985 to 2018. *Int. J. Climatol.* **2021**, *41*, E2564–E2580. [CrossRef]
2. Thoeun, H.C. Observed and projected changes in temperature and rainfall in Cambodia. *Weather Clim. Extrem.* **2015**, *7*, 61–71. [CrossRef]
3. Tong, S.; Li, X.; Zhang, J.; Bao, Y.; Bao, Y.; Na, L.; Si, A. Spatial and temporal variability in extreme temperature and precipitation events in Inner Mongolia (China) during 1960–2017. *Sci. Total Environ.* **2019**, *649*, 75–89. [CrossRef] [PubMed]
4. François, B.; Schlef, K.E.; Wi, S.; Brown, C.M. Design considerations for riverine floods in a changing climate—A review. *J. Hydrol.* **2019**, *574*, 557–573. [CrossRef]
5. Kundzewicz, Z.W.; Kanae, S.; Seneviratne, S.I.; Handmer, J.; Nicholls, N.; Peduzzi, P.; Mechler, R.; Bouwer, L.M.; Arnell, N.; Mach, K.; et al. Flood risk and climate change: Global and regional perspectives. *Hydrol. Sci. J.* **2014**, *59*, 1–28. [CrossRef]
6. Kundzewicz, Z.W.; Su, B.; Wang, Y.; Xia, J.; Huang, J.; Jiang, T. Flood risk and its reduction in China. *Adv. Water Resour.* **2019**, *130*, 37–45. [CrossRef]
7. Kron, W.; Eichner, J.; Kundzewicz, Z.W. Reduction of flood risk in Europe—Reflections from a reinsurance perspective. *J. Hydrol.* **2019**, *576*, 197–209. [CrossRef]
8. Balti, H.; Ben Abbes, A.; Mellouli, N.; Farah, I.R.; Sang, Y.; Lamolle, M. A review of drought monitoring with big data: Issues, methods, challenges and research directions. *Ecol. Inform.* **2020**, *60*, 101136. [CrossRef]
9. Collins, M.; Knutti, R.; Arblaser, J.; Dufresne, J.-L.; Fichefet, T.; Friedlingstein, P.; Gao, X.; Gutowski, W.; Johns, T.; Krinner, G.; et al. Long-term Climate Change: Projections, Commitments and Irreversibility. In *Climate Change 2013-The Physical Science Basis: Contribution of Working Group I to the Fifth Assessment Report of the Intergovernmental Panel on Climate Change*; Cambridge University Press: Cambridge, UK, 2013; pp. 1029–1136.
10. Betts, R.A.; Alfieri, L.; Bradshaw, C.; Caesar, J.; Feyen, L.; Friedlingstein, P.; Gohar, L.; Koutroulis, A.; Lewis, K.; Morfopoulos, C.; et al. Changes in climate extremes, fresh water availability and vulnerability to food insecurity projected at 1.5 °C and 2 °C global warming with a higher-resolution global climate model. *Philos. Trans. R. Soc. A Math. Phys. Eng. Sci.* **2018**, *376*, 20160452. [CrossRef]
11. Vannière, B.; Demory, M.-E.; Vidale, P.L.; Schiemann, R.; Roberts, M.J.; Roberts, C.D.; Matsueda, M.; Terray, L.; Koenigk, T.; Senan, R. Multi-model evaluation of the sensitivity of the global energy budget and hydrological cycle to resolution. *Clim. Dyn.* **2019**, *52*, 6817–6846. [CrossRef]
12. Eyring, V.; Bony, S.; Meehl, G.A.; Senior, C.A.; Stevens, B.; Stouffer, R.J.; Taylor, K.E. Overview of the Coupled Model Intercomparison Project Phase 6 (CMIP6) experimental design and organization. *Geosci. Model Dev.* **2016**, *9*, 1937–1958. [CrossRef]
13. Kim, Y.-H.; Min, S.-K.; Zhang, X.; Sillmann, J.; Sandstad, M. Evaluation of the CMIP6 multi-model ensemble for climate extreme indices. *Weather Clim. Extrem.* **2020**, *29*, 100269. [CrossRef]
14. Haarsma, R.J.; Roberts, M.J.; Vidale, P.L.; Senior, C.A.; Bellucci, A.; Bao, Q.; Chang, P.; Corti, S.; Fučkar, N.S.; Guemas, V.; et al. High Resolution Model Intercomparison Project (HighResMIP v1.0) for CMIP6. *Geosci. Model Dev.* **2016**, *9*, 4185–4208. [CrossRef]
15. Tan, M.L.; Gassman, P.; Yang, X.; Haywood, J. A Review of SWAT Applications, Performance and Future Needs for Simulation of Hydro-Climatic Extremes. *Adv. Water Resour.* **2020**, *143*, 103662. [CrossRef]

16. Mendoza, P.A.; Mizukami, N.; Ikeda, K.; Clark, M.P.; Gutmann, E.D.; Arnold, J.R.; Brekke, L.D.; Rajagopalan, B. Effects of different regional climate model resolution and forcing scales on projected hydrologic changes. *J. Hydrol.* **2016**, *541*, 1003–1019. [CrossRef]
17. Pastén-Zapata, E.; Jones, J.M.; Moggridge, H.; Widmann, M. Evaluation of the performance of Euro-CORDEX Regional Climate Models for assessing hydrological climate change impacts in Great Britain: A comparison of different spatial resolutions and quantile mapping bias correction methods. *J. Hydrol.* **2020**, *584*, 124653. [CrossRef]
18. Ghausi, S.A.; Ghosh, S. Diametrically Opposite Scaling of Extreme Precipitation and Streamflow to Temperature in South and Central Asia. *Geophys. Res. Lett.* **2020**, *47*, e2020GL089386. [CrossRef]
19. Okwala, T.; Shrestha, S.; Ghimire, S.; Mohanasundaram, S.; Datta, A. Assessment of climate change impacts on water balance and hydrological extremes in Bang Pakong-Prachin Buri river basin, Thailand. *Environ. Res.* **2020**, *186*, 109544. [CrossRef] [PubMed]
20. Hoang, L.P.; van Vliet, M.T.H.; Kummu, M.; Lauri, H.; Koponen, J.; Supit, I.; Leemans, R.; Kabat, P.; Ludwig, F. The Mekong's future flows under multiple drivers: How climate change, hydropower developments and irrigation expansions drive hydrological changes. *Sci. Total Environ.* **2019**, *649*, 601–609. [CrossRef]
21. Raghavan, S.V.; Tue, V.M.; Shie-Yui, L. Impact of climate change on future stream flow in the Dakbla river basin. *J. Hydroinform.* **2013**, *16*, 231–244. [CrossRef]
22. Supari; Tangang, F.; Juneng, L.; Cruz, F.; Chung, J.X.; Ngai, S.T.; Salimun, E.; Mohd, M.S.F.; Santisirisomboon, J.; Singhruck, P.; et al. Multi-model projections of precipitation extremes in Southeast Asia based on CORDEX-Southeast Asia simulations. *Environ. Res.* **2020**, *184*, 109350. [CrossRef]
23. Tan, M.L.; Juneng, L.; Tangang, F.T.; Samat, N.; Chan, N.W.; Yusop, Z.; Ngai, S.T. SouthEast Asia HydrO-meteorological droughT (SEA-HOT) framework: A case study in the Kelantan River Basin, Malaysia. *Atmos. Res.* **2020**, *246*, 105155. [CrossRef]
24. Harris, L.M.; Durran, D.R. An Idealized Comparison of One-Way and Two-Way Grid Nesting. *Mon. Weather Rev.* **2010**, *138*, 2174–2187. [CrossRef]
25. Bowden, J.H.; Otte, T.L.; Nolte, C.G.; Otte, M.J. Examining Interior Grid Nudging Techniques Using Two-Way Nesting in the WRF Model for Regional Climate Modeling. *J. Clim.* **2012**, *25*, 2805–2823. [CrossRef]
26. Tangang, F.; Chung, J.X.; Juneng, L.; Supari; Salimun, E.; Ngai, S.T.; Jamaluddin, A.F.; Mohd, M.S.F.; Cruz, F.; Narisma, G.; et al. Projected future changes in rainfall in Southeast Asia based on CORDEX–SEA multi-model simulations. *Clim. Dyn.* **2020**, *55*, 1247–1267. [CrossRef]
27. Richter, B.D.; Baumgartner, J.V.; Powell, J.; Braun, D.P. A Method for Assessing Hydrologic Alteration within Ecosystems. *Conserv. Biol.* **1996**, *10*, 1163–1174. [CrossRef]
28. Chen, Q.; Chen, H.; Wang, J.; Zhao, Y.; Chen, J.; Xu, C. Impacts of Climate Change and Land-Use Change on Hydrological Extremes in the Jinsha River Basin. *Water* **2019**, *11*, 1398. [CrossRef]
29. López-Ballesteros, A.; Senent-Aparicio, J.; Martínez, C.; Pérez-Sánchez, J. Assessment of future hydrologic alteration due to climate change in the Aracthos River basin (NW Greece). *Sci. Total Environ.* **2020**, *733*, 139299. [CrossRef]
30. Vu, T.T.; Kiesel, J.; Guse, B.; Fohrer, N. Analysis of the occurrence, robustness and characteristics of abrupt changes in streamflow time series under future climate change. *Clim. Risk Manag.* **2019**, *26*, 100198. [CrossRef]
31. Kiesel, J.; Gericke, A.; Rathjens, H.; Wetzig, A.; Kakouei, K.; Jähnig, S.C.; Fohrer, N. Climate change impacts on ecologically relevant hydrological indicators in three catchments in three European ecoregions. *Ecol. Eng.* **2019**, *127*, 404–416. [CrossRef]
32. Zhang, Z.; Liu, J.; Huang, J. Hydrologic impacts of cascade dams in a small headwater watershed under climate variability. *J. Hydrol.* **2020**, *590*, 125426. [CrossRef]
33. Tan, M.L.; Ramli, H.P.; Tam, T.H. Effect of DEM Resolution, Source, Resampling Technique and Area Threshold on SWAT Outputs. *Water Resour. Manag.* **2018**, *32*, 4591–4606. [CrossRef]
34. Tangang, F.T.; Juneng, L.; Salimun, E.; Vinayachandran, P.N.; Seng, Y.K.; Reason, C.J.C.; Behera, S.K.; Yasunari, T. On the roles of the northeast cold surge, the Borneo vortex, the Madden-Julian Oscillation, and the Indian Ocean Dipole during the extreme 2006/2007 flood in southern Peninsular Malaysia. *Geophys. Res. Lett.* **2008**, *35*. [CrossRef]
35. Hai, O.S.; Samah, A.A.; Chenoli, S.N.; Subramaniam, K.; Ahmad Mazuki, M.Y. Extreme Rainstorms that Caused Devastating Flooding across the East Coast of Peninsular Malaysia during November and December 2014. *Weather Forecast.* **2017**, *32*, 849–872. [CrossRef]
36. Chan, N.W. Flood disaster management in Malaysia: An evaluation of the effectiveness of government resettlement schemes. *Disaster Prev. Manag. Int. J.* **1995**, *4*, 22–29. [CrossRef]
37. Baharuddin, K.A.; Abdull Wahab, S.F.; Nik Ab Rahman, N.H.; Nik Mohamad, N.A.; Tuan Kamauzaman, T.H.; Md Noh, A.Y.; Abdul Majod, M.R. The Record-Setting Flood of 2014 in Kelantan: Challenges and Recommendations from an Emergency Medicine Perspective and Why the Medical Campus Stood Dry. *Malays. J. Med. Sci.* **2015**, *22*, 1–7. [PubMed]
38. Tan, M.L.; Ibrahim, A.L.; Yusop, Z.; Chua, V.P.; Chan, N.W. Climate change impacts under CMIP5 RCP scenarios on water resources of the Kelantan River Basin, Malaysia. *Atmos. Res.* **2017**, *189*, 1–10. [CrossRef]
39. Stefanidis, K.; Panagopoulos, Y.; Mimikou, M. Response of a multi-stressed Mediterranean river to future climate and socio-economic scenarios. *Sci. Total Environ.* **2018**, *627*, 756–769. [CrossRef] [PubMed]
40. Department of Irrigation and Drainage Malaysia. *Summary of the 2014/2015 Floods*; 2015. Available online: https://info.water.gov.my/index.php/databank/view_contribution/18/3967 (accessed on 20 May 2021).
41. Sazib, N.; Bolten, J.; Mladenova, I. Exploring Spatiotemporal Relations between Soil Moisture, Precipitation, and Streamflow for a Large Set of Watersheds Using Google Earth Engine. *Water* **2020**, *12*, 1371. [CrossRef]

42. Mohseni, O.; Stefan, H.G. A monthly streamflow model. *Water Resour. Res.* **1998**, *34*, 1287–1298. [CrossRef]
43. Arnold, J.G.; Srinivasan, R.; Muttiah, R.S.; Williams, J.R. Large area hydrologic modeling and assessment part I: Model development. *JAWRA J. Am. Water Resour. Assoc.* **1998**, *34*, 73–89. [CrossRef]
44. Arnold, J.G.; Moriasi, D.N.; Gassman, P.W.; Abbaspour, K.C.; White, M.J.; Srinivasan, R.; Santhi, C.; Harmel, R.D.; Van Griensven, A.; Van Liew, M.W.; et al. SWAT: Model use, calibration, and validation. *Trans. ASABE* **2012**, *55*, 1491–1508. [CrossRef]
45. Gassman, P.W.; Reyes, M.R.; Green, C.H.; Arnold, J.G. The Soil and Water Assessment Tool: Historical Development, Applications, and Future Research Directions. *Trans. ASABE* **2007**, *50*, 1211–1250. [CrossRef]
46. Tan, M.L.; Gassman, P.W.; Srinivasan, R.; Arnold, J.G.; Yang, X. A Review of SWAT Studies in Southeast Asia: Applications, Challenges and Future Directions. *Water* **2019**, *11*, 914. [CrossRef]
47. Abbaspour, K.C.; Vaghefi, S.A.; Srinivasan, R. A Guideline for Successful Calibration and Uncertainty Analysis for Soil and Water Assessment: A Review of Papers from the 2016 International SWAT Conference. *Water* **2018**, *10*, 18. [CrossRef]
48. Abbaspour, K.C.; Vejdani, M.; Haghighat, S. SWAT-CUP calibration and uncertainty programs for SWAT. In *Modsim International Congress on Modelling & Simulation Land Water & Environmental Management Integrated Systems for Sustainability*; Modelling and Simulation Society of Australia and New Zealand: Wageningen, New Zealand, 2007; Volume 364, pp. 1596–1602.
49. Nash, J.E.; Sutcliffe, J.V. River flow forecasting through conceptual models part I—A discussion of principles. *J. Hydrol.* **1970**, *10*, 282–290. [CrossRef]
50. Zhang, H.; Wang, B.; Liu, D.L.; Zhang, M.; Leslie, L.M.; Yu, Q. Using an improved SWAT model to simulate hydrological responses to land use change: A case study of a catchment in tropical Australia. *J. Hydrol.* **2020**, *585*, 124822. [CrossRef]
51. Moriasi, D.N.; Gitau, M.W.; Pai, N.; Daggupati, P. Hydrologic and water quality models: Performance measures and evaluation criteria. *Trans. ASABE* **2015**, *58*, 1763–1785. [CrossRef]
52. Forsythe, N.; Archer, D.R.; Pritchard, D.; Fowler, H. Chapter 7—A Hydrological Perspective on Interpretation of Available Climate Projections for the Upper Indus Basin. In *Indus River Basin*; Khan, S.I., Adams, T.E., Eds.; Elsevier: Amsterdam, The Netherlands, 2019; pp. 159–179.
53. Musie, M.; Sen, S.; Srivastava, P. Application of CORDEX-AFRICA and NEX-GDDP datasets for hydrologic projections under climate change in Lake Ziway sub-basin, Ethiopia. *J. Hydrol. Reg. Stud.* **2020**, *31*, 100721. [CrossRef]
54. Tessema, N.; Kebede, A.; Yadeta, D. Modelling the effects of climate change on streamflow using climate and hydrological models: The case of the Kesem sub-basin of the Awash River basin, Ethiopia. *Int. J. River Basin Manag.* **2020**, 1–12. [CrossRef]
55. Liang, J.; Catto, J.L.; Hawcroft, M.; Hodges, K.I.; Tan, M.L.; Haywood, J.M. Climatology of Borneo Vortices in the HadGEM3-GC3.1 General Circulation Model. *J. Clim.* **2021**, *34*, 3401–3419. [CrossRef]
56. Boé, J.; Terray, L.; Habets, F.; Martin, E. Statistical and dynamical downscaling of the Seine basin climate for hydro-meteorological studies. *Int. J. Climatol.* **2007**, *27*, 1643–1655. [CrossRef]
57. Kim, M.-K.; Kim, S.; Kim, J.; Heo, J.; Park, J.-S.; Kwon, W.-T.; Suh, M.-S. Statistical downscaling for daily precipitation in Korea using combined PRISM, RCM, and quantile mapping: Part 1, methodology and evaluation in historical simulation. *Asia-Pac. J. Atmos. Sci.* **2016**, *52*, 79–89. [CrossRef]
58. Reshmidevi, T.V.; Nagesh Kumar, D.; Mehrotra, R.; Sharma, A. Estimation of the climate change impact on a catchment water balance using an ensemble of GCMs. *J. Hydrol.* **2018**, *556*, 1192–1204. [CrossRef]
59. Pesce, M.; Critto, A.; Torresan, S.; Giubilato, E.; Pizzol, L.; Marcomini, A. Assessing uncertainty of hydrological and ecological parameters originating from the application of an ensemble of ten global-regional climate model projections in a coastal ecosystem of the lagoon of Venice, Italy. *Ecol. Eng.* **2019**, *133*, 121–136. [CrossRef]
60. Neitsch, S.L.; Arnold, J.G.; Kiniry, J.R.; Grassland, J.R.W. *Soil and Water Assessment Tool Theoretical Documentation Version 2009*; Agricultural Research Service Blackland Research Center: Temple, TX, USA, 2011.
61. Arnold, J.G.; Kiniry, J.R.; Srinivasan, R.; Williams, J.R.; Haney, E.B.; Neitsch, S.L. *Soil and Water Assessment Tool Input/Output File Documentation: Version 2012 (Texas Water Resources Institute TR-439)*; USDA-ARS, Grassland, Soil and Water Research Laboratory, and Texas AgriLife Research, Blackland Research and Extension Center: Temple, TX, USA, 2012.
62. Hussin, N.H.; Yusoff, I.; Raksmey, M. Comparison of Applications to Evaluate Groundwater Recharge at Lower Kelantan River Basin, Malaysia. *Geosciences* **2020**, *10*, 289. [CrossRef]
63. Poff, N.L.; Zimmerman, J.K.H. Ecological responses to altered flow regimes: A literature review to inform the science and management of environmental flows. *Freshw. Biol.* **2010**, *55*, 194–205. [CrossRef]
64. Bador, M.; Alexander, L.V.; Contractor, S.; Roca, R. Diverse estimates of annual maxima daily precipitation in 22 state-of-the-art quasi-global land observation datasets. *Environ. Res. Lett.* **2020**, *15*, 035005. [CrossRef]
65. Jimeno-Sáez, P.; Senent-Aparicio, J.; Pérez-Sánchez, J.; Pulido-Velazquez, D. A Comparison of SWAT and ANN Models for Daily Runoff Simulation in Different Climatic Zones of Peninsular Spain. *Water* **2018**, *10*, 192. [CrossRef]
66. Pereira, D.D.R.; Martinez, M.A.; Pruski, F.F.; da Silva, D.D. Hydrological simulation in a basin of typical tropical climate and soil using the SWAT model part I: Calibration and validation tests. *J. Hydrol. Reg. Stud.* **2016**, *7*, 14–37. [CrossRef]
67. Leta, O.T.; El-Kadi, A.I.; Dulai, H.; Ghazal, K.A. Assessment of climate change impacts on water balance components of Heeia watershed in Hawaii. *J. Hydrol. Reg. Stud.* **2016**, *8*, 182–197. [CrossRef]
68. Krysanova, V.; Arnold, J.G. Advances in ecohydrological modelling with SWAT-a review. *Hydrol. Sci. J.* **2008**, *53*, 939–947. [CrossRef]

69. Bieger, K.; Arnold, J.G.; Rathjens, H.; White, M.J.; Bosch, D.D.; Allen, P.M.; Volk, M.; Srinivasan, R. Introduction to SWAT+, A Completely Restructured Version of the Soil and Water Assessment Tool. *JAWRA J. Am. Water Resour. Assoc.* **2017**, *53*, 115–130. [CrossRef]
70. Kundzewicz, Z.W.; Krysanova, V.; Benestad, R.E.; Hov, Ø.; Piniewski, M.; Otto, I.M. Uncertainty in climate change impacts on water resources. *Environ. Sci. Policy* **2018**, *79*, 1–8. [CrossRef]
71. Troin, M.; Caya, D.; Velázquez, J.A.; Brissette, F. Hydrological response to dynamical downscaling of climate model outputs: A case study for western and eastern snowmelt-dominated Canada catchments. *J. Hydrol. Reg. Stud.* **2015**, *4*, 595–610. [CrossRef]
72. Duan, W.; He, B.; Takara, K.; Luo, P.; Nover, D.; Hu, M. Impacts of climate change on the hydro-climatology of the upper Ishikari river basin, Japan. *Environ. Earth Sci.* **2017**, *76*, 490. [CrossRef]
73. Tan, M.L.; Ficklin, D.; Ibrahim, A.L.; Yusop, Z. Impacts and uncertainties of climate change on streamflow of the Johor River Basin, Malaysia using a CMIP5 General Circulation Model ensemble. *J. Water Clim. Chang.* **2014**, *5*, 676–695. [CrossRef]
74. Wang, D.; Hejazi, M.; Cai, X.; Valocchi, A.J. Climate change impact on meteorological, agricultural, and hydrological drought in central Illinois. *Water Resour. Res.* **2011**, *47*. [CrossRef]
75. Ngai, S.T.; Juneng, L.; Tangang, F.; Chung, J.X.; Salimun, E.; Tan, M.L.; Amalia, S. Future projections of Malaysia daily precipitation characteristics using bias correction technique. *Atmos. Res.* **2020**, 104926. [CrossRef]
76. Ngai, S.T.; Tangang, F.; Juneng, L. Bias correction of global and regional simulated daily precipitation and surface mean temperature over Southeast Asia using quantile mapping method. *Glob. Planet. Chang.* **2017**, *149*, 79–90. [CrossRef]
77. Shrestha, M.; Acharya, S.C.; Shrestha, P.K. Bias correction of climate models for hydrological modelling—Are simple methods still useful? *Meteorol. Appl.* **2017**, *24*, 531–539. [CrossRef]
78. Luo, M.; Liu, T.; Meng, F.; Duan, Y.; Frankl, A.; Bao, A.; De Maeyer, P. Comparing Bias Correction Methods Used in Downscaling Precipitation and Temperature from Regional Climate Models: A Case Study from the Kaidu River Basin in Western China. *Water* **2018**, *10*, 1046. [CrossRef]

water

MDPI

Article

Concurrent Changes in Extreme Hydroclimate Events in the Colorado River Basin

Katrina E. Bennett [1,*], Carl Talsma [1] and Riccardo Boero [2]

1 Los Alamos National Laboratory, Earth and Environmental Sciences, Los Alamos, NM 87545, USA; talsmac83@lanl.gov

2 Los Alamos National Laboratory, Analytics, Intelligence and Technology, Los Alamos, NM 87545, USA; riccardo@lanl.gov

* Correspondence: kbennett@lanl.gov

Abstract: Extreme events resulting in catastrophic damage have more than doubled in the last five years, costing hundreds of lives and thousands of homes, and heavily undermining regional economic stability. At present, most of these hydroclimatic extreme events are documented by the media as individual events; however, in scientific terms, many are better understood as concurrent events—concurrent extremes of both temperature and precipitation (e.g., drought, floods). This paper considers concurrent changes in hydroclimate extremes, including heatwaves, drought, flooding, and low flows, in six historical-to-future (1970–1999, 2070–2099) Earth System Model (ESM) climate scenarios for the Colorado River basin. Results indicate that temperature-driven *Impacts* (heatwaves, drought) have the strongest responses while percipitation-driven *Impacts* have weaker responses. All *Impacts* exhibit an increase in magnitude from synoptic to annual time scales, with heatwaves increasing in strength about three times at the annual time scale versus the synoptic, while low flows only increase slightly. Critical watersheds in the Colorado were identified, highlighting the Blue River basin, Uncompahgre, East Taylor, Salt/Verde watersheds, locations of important water infrastructures, water resources, and hydrological research. Our results indicate that concurrent extreme hydroclimate events are projected to increase in the future and intensify within critical regions of the Colorado River basin. Considering extreme hydroclimate events concurrently is an important step towards linking economic and social effects of these events and their associated instabilities on a regional scale.

Keywords: extreme events; hydrology; concurrent; climate change; Colorado River basin; heatwaves; drought; flooding; low flows

Citation: Bennett, K.E.; Talsma, C.; Boero, R. Concurrent Changes in Extreme Hydroclimate Events in the Colorado River Basin. *Water* **2021**, *13*, 978. https://doi.org/10.3390/w13070978

Academic Editor: Athanasios Loukas

Received: 11 February 2021
Accepted: 30 March 2021
Published: 1 April 2021

Publisher's Note: MDPI stays neutral with regard to jurisdictional claims in published maps and institutional affiliations.

1. Introduction

Extreme events resulting in catastrophic damage (i.e., loss of life and costs exceeding a billion dollars in response expenses, property damage, and economic disruption) have more than doubled in the last five years, significantly undermining regional economic stability [1]. The increase in both the frequency and intensity of these extreme events has been directly linked to climate change (i.e., 6 events in a 2011 publication, versus 28 events in more recent work just four years later) [2–7]. These extreme events occur locally, regionally, and globally, with major consequences for every aspect of human society and economy [3,7].

The growing body of research on this topic reflects the fact that it has become one of the most urgent issues of the last decade. Studies with the keywords "extreme events" in their titles doubled between 2010 and 2018 (Google Scholar and Scopus, 26 June 2020, Figure 1), and, in a recent study on the "Twenty three unsolved problems in hydrology," "extreme events" was listed as a major outstanding research topic, with particular emphasis placed on the causes of flood/drought periods and recently documented changes to these periods [8]. Extreme events were also the focus of a 2012 *Intergovernmental Panel on Climate Change*

Special Report [7], an updated 2018 *Special Report* [3], a "Key Finding" issued by the US National Climate Assessment (NCA 2014), numerous databases and data sets (e.g., NOAA's Billion-Dollar Weather and Climate Disasters https://www.ncdc.noaa.gov/billions/time-series, accessed on 20 October 2020), and a multitude of recent research papers [9–11]. We urgently need to understand how extremes events are changing and we need to better characterize and track such changes in critical regions, such as major watersheds.

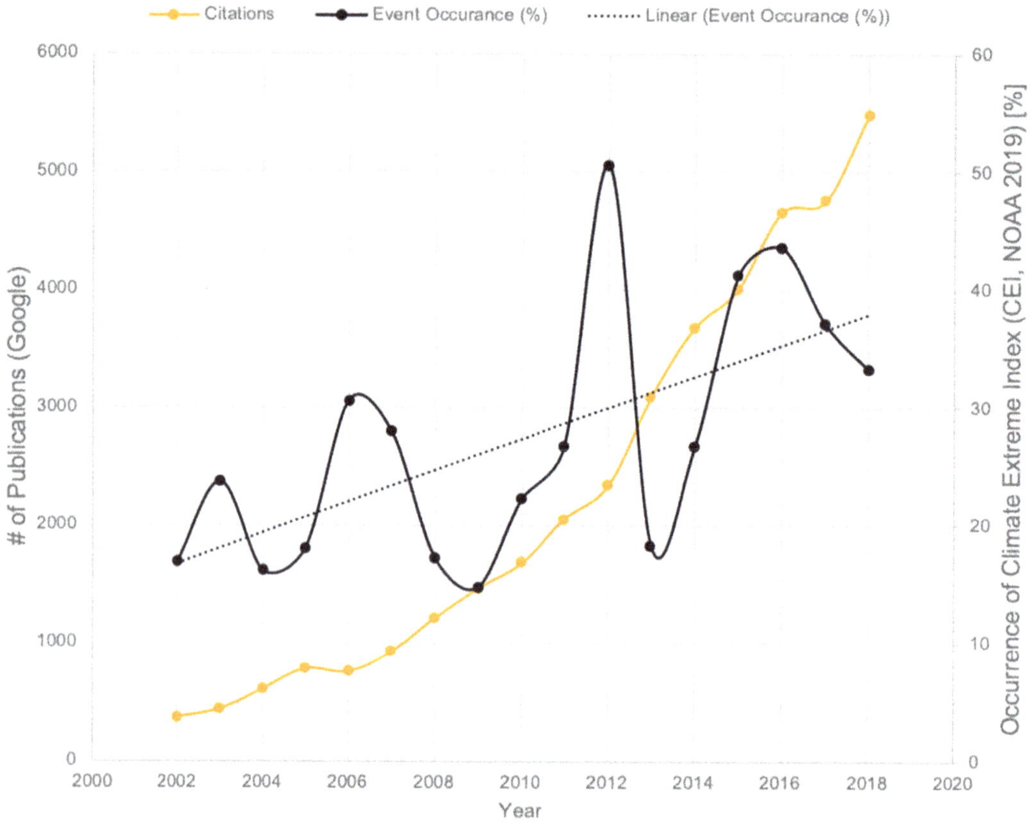

Figure 1. Peer reviewed literature documenting extremes events has tracked increases in extremes over the past ~10 years. Data source: NOAA CEI database, 2019.

In the United States, watersheds play a fundamental role in regulating water resources for commercial and domestic purposes. For example, the Colorado River basin (CRB) provides water for over 40 million people [12] and directly facilitates 1.4 trillion dollars in agricultural and commercial applications (roughly 1/13th of the entire US economy, based on the 2014 Gross Domestic Product). However, in the Southwest Climate region alone (UT, CO, AZ, NM), flooding, drought, freezing events, wildfire, severe storms, and winter storms have cost approximately 40 billion dollars between 1980–2020, with a more than 5-fold increase in extreme events from the 1980s to the 2010s [1].

Unfortunately, despite the vital importance of understanding extreme hydroclimate events in watersheds such as the CRB, several fundamental steps have been overlooked. Even the most basic task of comprehensively defining, classifying, and categorizing extreme events has proven to be problematic, complicating comparisons across research fields, which may have different standards for describing the occurrence, duration, and severity of such events [13,14]. Additionally, many of the current tools used to consider extreme events use simple algorithms (e.g., NOAA's Climate Extreme Index, or CEI) based on simple indicators (minimum temperature changes, precipitation days, etc.), which allow policy makers and the non-scientific community to understand basic changes in extreme event trends. However, these resources cannot effectively track, for example, coupled extreme events (e.g., drought/wildfire, winter storm/flood), and may even overestimate the occurrences of extremes that impact society and underestimate the impacts of concurrent extremes. Consequently, much of the current work on extreme events has focused largely on the analysis of individual or univariate indicators in documenting changing extremes [15–18]. However, more complex indicators and statistical approaches are needed to improve the scientific community's ability to identify and characterize coupled, joint, concurrent, or multivariate climate extremes [19–26] and to document the clear effects of these concurrent events on the economy and society [27,28].

Despite the strong focus on individual indicators, research has emerged that characterizes future changes in extreme events concurrently [22,26,29–38]. Review studies on concurrent hydroclimate extremes provide an assessment of all studies and methodological approaches to date of publication [39,40]. A Special Issue on concurrent or compound extremes was also published [41]. The argument for consideration of concurrent extremes over univariable extreme event analysis is given by several authors, offering definitions, frameworks, paradigms shifts, and generally making the case that examining extreme events in this manner is a means to improve projections of future changes, and that rethinking the traditional univariate approach will allow for physical sciences to be more clearly linked to socioeconomic impacts of extremes [42–46].

This paper describes modeled scenarios of concurrent extreme events in the CRB. We present the results for six different Earth System Models (ESMs), and four different *Impacts* (heat waves, drought, low flows, and flooding), calculated as coupled, joint, multivariate, or concurrent indicators for two time periods, historical (1970–1999) and future (2070–2099). For the purposes of this paper, extreme *Impacts* are considered as the 95th percentile exceedance of concurrent indicators (from here on in, joint, coupled, multivariate indicators are referred to as concurrent). We consider the changes in *Impacts* across synoptic, monthly, seasonal, and annual time scales; and we also identify the most critical watersheds within the CRB where concurrent changes are compounded, based on the accumulated changes in *Impacts* between historical and future models. We conclude the paper with a summary of the work and next steps.

2. Materials and Methods

2.1. Study Site

Our study area is the CRB, covering an area of 640,000 km^2 in the semi-arid to arid Southwestern United States and Northern Mexico. The basin stretches from 30° N to 44° N and from 106° W to 116° W, extending from the alpine regions of the Southern Rocky Mountains to the Gulf of California, and covers elevations from sea level to more than 4000 m, providing water resources to a vast number of adjacent infrastructures (e.g., agricultural, water routing structures, and cities, Figure 2). The large range of spatially and temporally variable CRB landscapes and ecosystems span multiple climatic zones, with observed annual average temperatures ranging from 4 °C to 24 °C (average 11 °C) and annual average precipitation total ranging from 79 mm to 1699 mm (average 363 mm) [47]. Most precipitation in the basin falls as snow at high elevations, so roughly 85% of the CRB flow originates between its upper headwaters and Lees Ferry at Glen Canyon, Arizona [48].

Figure 2. The domain of the Colorado River Basin with adjacent areas that receive Colorado River water. Adapted from USGS (accessed 11th January 2021)

2.2. Climate Simulations

Our hydrological model was forced using downscaled projections of daily temperature, precipitation, and wind speed from the Multivariate Adaptive Constructed Analogue (MACA) database [49]. We extracted the following six climate data projections from MACA based on Earth System Model (ESM) simulations from the Coupled Model Intercomparison Project, phase 5 (CMIP5) [50] because they represent a range of future climate responses for the CRB and also include dynamic vegetation components, listed in the brackets below: HadGEM2-ES (TRIFFID) [51,52], MIROC-ESM (SIEB-DGVM) [53,54], MPI-ESM-LR (JSBACH) [55,56], IPSL-CM5A-LR (ORCHIDEE) [57,58], and GFDL-ESM2M, and GFDL-ESM2G (LM3V), [59,60]. Each ESM differs by a multitude of factors, including but not limited to the country where the model was developed, initial conditions, physics representations, and tuning mechanisms [61]. For this work, we used the representative concentration pathway (RCP) 8.5 emissions scenario, which tracks closely with changing emissions levels over time [62] and anticipates strongly increasing emissions by 2100 [63].

2.3. Hydrological Simulations

To generate simulations of extreme hydrological indicators, we used the Variable Infiltration Capacity (VIC) model version 4.2 [64,65] at a 1/16th degree (6 km) spatial resolution. For each grid cell in the simulation domain, VIC simulates vertical energy and water dynamics at an hourly time step for land cover tiles situated above a 3-layer soil column. Heterogeneity in VIC infiltration is represented by a sub-grid scale statistical distribution (the variable infiltration capacity curve). Surface runoff is generated via saturation excess, while sub-surface runoff is characterized by a non-linear baseflow curve [66].

We ran VIC using publicly available gridded historical climate data (daily precipitation, minimum and maximum temperature, and wind speed) for the CRB (Livneh et al., 2015) and calibrated VIC for 2006–2010 using the United States Geological Survey (USGS) naturalized gauged monthly streamflow data [67] and an automatic calibration tool [68] to correct streamflow peaks, volume, and low flow biases. See Bennett, et al. [69] for a complete description of VIC parameterizations and calibration details.

2.4. Extreme Indicators and Impacts

We calculated two time periods for our analysis: historical (1970–1999) and future (2070–2099) for synoptic (5 days, with 73 5-day intervals in each year), monthly (12-months), four-seasonal (December-January-February, DJF; March-April-May, MAM; June-June-August, JJA; and September-October-November, SON), and annual intervals. Leap-year days were removed from the time series, so all modeled years are 365 days. We calculated extreme climatic and hydrologic indicators (e.g., maximum temperature, freezing days, minimum streamflow) for each time period and interval, as described in Table 1. Some indicators were VIC model input (e.g., temperature) while others were generated from VIC model output (e.g., evapotranspiration). Our defined *Impacts* were: heatwaves, drought, flooding, and low flows. For each *Impact*, we choose two to three indicators that showed the least correlation with one another, yet remained relevant to the assigned *Impact* (Table 2).

Table 1. Extreme indicators, description and units, and abbreviations used in the text.

Indicators	Description and Units	Abbreviation
Maximum temperature	Maximum temperature achieved over the time period (°C)	*tx*
Maximum precipitation	Maximum daily precipitation over the time period (mm)	*precx*
Low precipitation days	Number of days when accumulated precipitation is <0.01 mm. (count)	*dryd*
Maximum streamflow	Maximum daily streamflow over the time period (mm)	*qx*
Minimum streamflow	Minimum daily streamflow over the time period(mm)	*qn*
Maximum soil moisture	Maximum daily soil moisture from the third soil moisture layer over the time period (mm)	*soilmx*
Minimum soil moisture	Minimum daily soil moisture from the third soil moisture layer over the time period (mm)	*soilmn*
Maximum evapotranspiration (ET)	Maximum daily evapotranspiration over the time period (mm)	*evapx*

Table 2. Extreme impacts and the relevant indicators used to construct the impact.

Impacts	Indicators
Heat Waves	Maximum temperature, maximum ET
Drought	Maximum temperature, low precipitation days, minimum soil moisture
Low Flows	Minimum streamflow, minimum soil moisture, maximum ET
Flooding	Maximum precipitation, maximum streamflow, maximum soil moisture

2.4.1. Peaks Over Threshold Extreme Exceedance

We defined the occurrence of extreme indicator values by calculating the historical 95th percentile (or 5th percentile for low extreme indicators) at each grid cell location for each indicator and using the 95th percentile as the threshold. Following a Peaks Over Threshold (POT) approach, any instance where an indicator value exceeded the historical 95th percentile threshold for a given grid cell was considered extreme. The exceedance of the threshold for each indicator at each grid cell and timestep was expressed as a binary value and is referred to as "exceedance" from here on.

2.4.2. Distance Number

Distance Number (D_n) is a metric that indicates the distance between data pairs normalized by the variance of the observed or historical values. By normalizing the difference by the historical standard deviation, we can obtain a value analogous to a Z-score, describing the change that occurs between historical and future data [70–72]. We define the Distance Number as:

$$Dn_{xy} = \frac{1}{n}\sum\nolimits_{i=1}^{n} \frac{\left(F_{i,xy} - H_{i.xy}\right)}{S_{xy}}, \tag{1}$$

where F is the future indicator exceedance, H is the historical indicator exceedance, S is the temporal standard deviation of the historical exceedance for each cell, i is the timestep, xy is the spatial cell location, and n is the number of timesteps. We calculated the distance number using equation (1) and estimate the change in exceedance of extreme *Impacts* by averaging D_n for each indicator relevant to a given *Impact*. For example, the concurrent D_n value for flooding is the average D_n value of the D_n of maximum precipitation, maximum streamflow, and maximum soil moisture, respectively. By normalizing the exceedance values by the historical standard deviation, the concurrent D_n value allows for a comparison of change across large spatial distances and four different time scales, and also allows for a comparison between different *Impacts* and ESMs. D_n values were examined for individual grid cell responses, and by the Natural Resource Counsel (NRCS), United States Department of Agriculture (USDA) hydrological unit basins. We used the cataloguing unit level eight (HUC8) watersheds, of which there are 134 watersheds in the CRB.

We estimated critical D_n values based on the mean of the D_n for each grid cells in all 134 HUC8 watershed in the CRB. The HUC8 watersheds where the mean D_n exceeded the 95th percentile of the mean D_n were identified as critical for all *Impacts* and ESMs. For all HUC8s, we summed instances for *Impacts* and ESMs; watersheds with values above 4 are identified as notably critical (described as having criticality and compounded concurrent extremes in Results and Discussion sections).

3. Results

The six ESMs examined in this study exhibit very similar temperature and precipitation data across the historical time period. Each ESM exhibits a mean temperature across the entire CRB of between 11.7 °C and 11.9 °C, and an average annual precipitation between 363.5 mm/yr and 369.9 mm/yr (*std. dev.* = 6.4 mm/yr, Table 3). The ESMs were chosen to represent the range of changes in precipitation and temperature projected by the larger CMIP5 model suite (not shown). Among the six ESMs, the projected change in temperature for the future scenario (2070–2099) differenced from the historical scenario (1970–1999) ranges from an increase of 4.1 °C (GFDL-ESM2M) to 7.0 °C (MIROC-ESM), with a multi-model average of 5.5 °C (*std. dev.* = 1.2 °C). Projections of future precipitation show large variance among ESMs, with ESMs projecting both increases and decreases in annual precipitation. HadGEM2-ES365 shows the largest decrease in precipitation (−52.8 mm/yr) while GFDL-ESM2G shows the largest increase (37.2 mm/yr). The mean of the six ESMs shows a small increase of only 2.1 mm/yr in precipitation and a standard deviation between ESMs of 32.6 mm/yr.

Table 3. Future and historical average annual temperature and precipitation values from the six GCMs considered. We also include the multi-model average and standard deviation for each field (bold italics). Temp. = temperature, Prec. = precipitation.

	Historical		Future		Change	
	Temp. (°C)	Precip. (mm)	Temp. (°C)	Precip. (mm)	Temp. (°C)	Precip. (mm)
GFDL-ESM2G	11.7	365.9	16.3	403	4.5	37.2
GFDL-ESM2M	11.7	366.5	15.8	377.9	4.1	11.3
HadGEM2-ES365	11.8	366.8	18	360.4	6.2	−6.4
IPSL-CM5A-LR	11.9	351.7	18.2	298.9	6.3	−52.8
MIROC-ESM	11.8	369.9	18.8	400.7	7	30.8
MPI-ESM-LR	11.9	363.5	17	356.3	5.1	−7.3
Average	**11.8**	**364.1**	**17.3**	**366.2**	**5.5**	**2.1**
Standard Deviation	**0.1**	**6.4**	**1.2**	**38.3**	**1.2**	**32.6**

3.1. Individual Indicators

Figure 3 shows the average across all six ESMs and four time scales for the individual indicators contributing to each *Impact*. Maximum temperature shows the largest D_n values of any indicator, exhibiting an average of 2.07 across each ESM and time scale. Additionally, maximum temperature (*tx*) shows little spatial variance (*std. dev.* = 0.07), which is not surprising given that temperature often exhibits coherent and strong responses under climate change [73]. Maximum precipitation (*precx*) shows little spatial variance (*std. dev.* = 0.09) with modestly positive D_n values (0.22) across the entire basin, indicating an increase in maximum precipitation for the CRB as a whole. D_n values are slightly higher in the upper CRB than in the lower basin for maximum precipitation. Similarly, maximum evaporation (*evapx*) and dry days (*dryd*) show slightly positive values (0.16, and 0.14, respectively) across nearly the entire CRB with little spatial variability (*std. dev.* = 0.08 to 0.13). Maximum evaporation shows slightly more elevated D_n values in the highest elevations regions of the CRB, as well as in a narrow band in the lower CRB known as the Arizona Transition Zone, where topography transitions from the high elevation Colorado Plateau to the basin and range region of the US southwest [74,75]. This region is also characterized as experiencing the highest average annual rainfall in the state of Arizona [76].

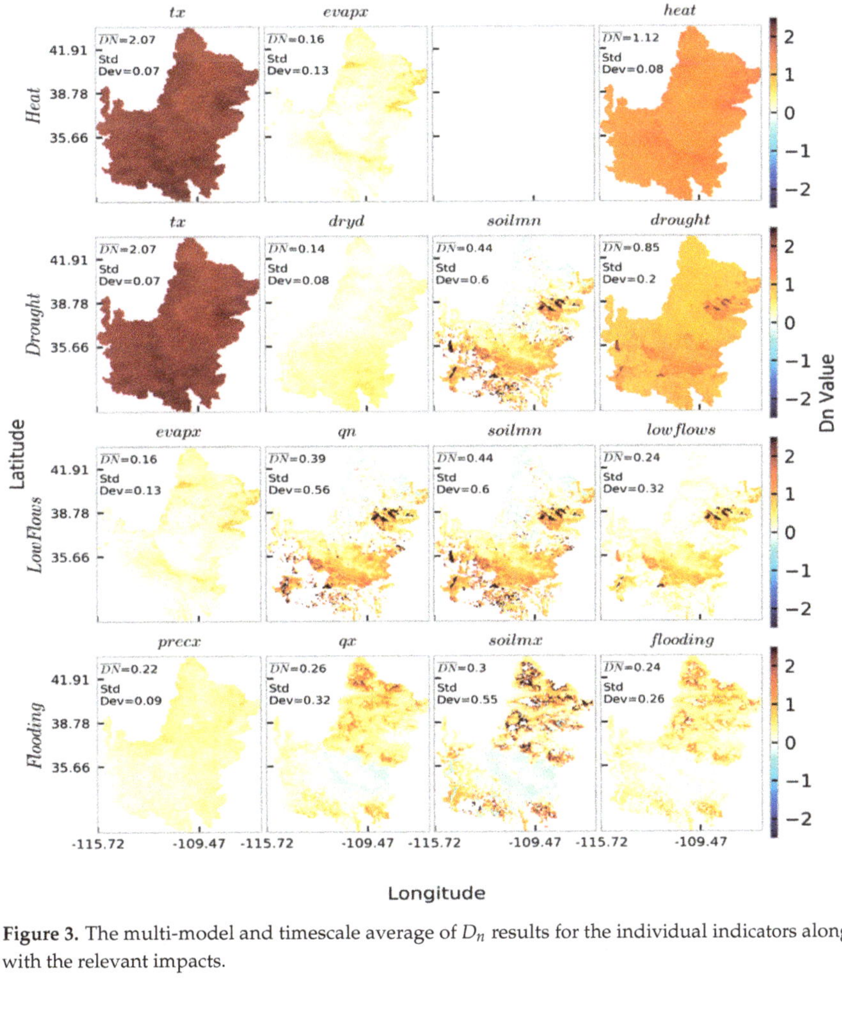

Figure 3. The multi-model and timescale average of D_n results for the individual indicators along with the relevant impacts.

Indicators related to minimum (*std. dev.*= 0.56) or maximum streamflow (*std. dev.*= 0.32), or minimum (*std. dev.*= 0.60) or maximum soil moisture (*std. dev.*= 0.55) exhibit far more spatial variance than the other indicators (*qn*, *qx*, *soilmn*, soilmx, respectively). Maximum soil moisture and maximum streamflow have larger and more positive D_n values in the upper CRB compared to the lower CRB, while minimum soil moisture and maximum soil moisture show larger and more positive D_n values in the lower CRB compared to the upper CRB.

3.2. Impacts

D_n results as difference between the future scenario and historical scenario for each *Impact* and time scale are shown in Figures 4–8 and Table 4. Overall, we observe an increase in each of the four *Impacts* when averaged over all ESMs and across the entire CRB (Table 3, Figure 4). Generally, the D_n values increase as the time scale lengthens, with the largest values occurring on the annual time scale (Figure 4). However, the change in D_n values depends on the *Impact*. Temperature-driven *Impact*s (heatwaves, drought) exhibit an increase in magnitude from synoptic to annual time scales, with heatwaves resulting in approximately three times the D_n value at the annual time scale versus the synoptic (0.60 compared to 2.01, Table 3, Figure 4). On the other hand, precipitation-driven *Impacts* (low flows, flooding) exhibit only a slight increase when moving from synoptic to annual time scales (e.g., 0.21 compared to 0.32 for low flows, Table 3, Figure 4). This may be because precipitation-driven extremes have a shorter memory in the climate system than temperature-driven extremes, which often occur over longer periods of time and have longer memories in terms of both reoccurrence and frequency [77].

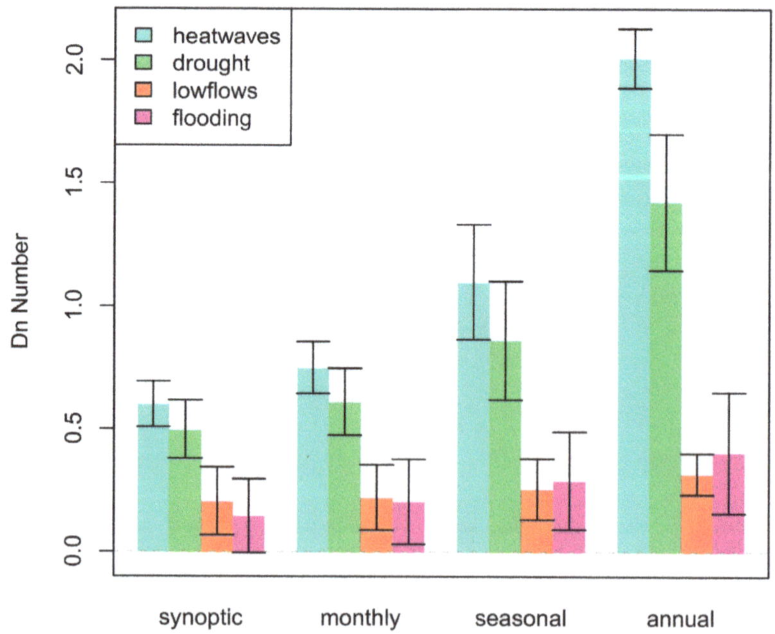

Figure 4. The multi-model average D_n value across the CRB for each impact and timescale. The whiskers represent +/− one standard deviation between the different ESMs.

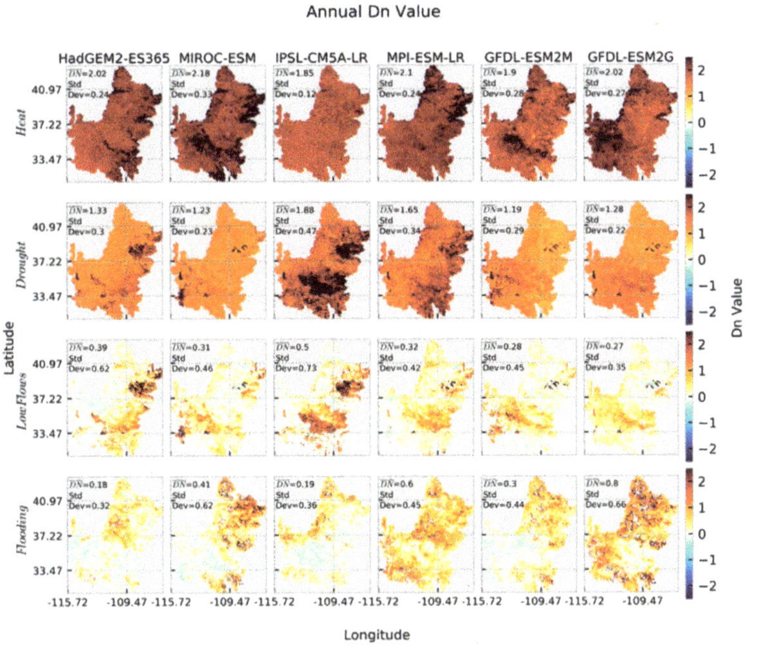

Figure 5. The D_n value results across the CRB at an annual timescale for each ESM and Impact. The average D_n value and spatial standard deviation is also shown for each panel.

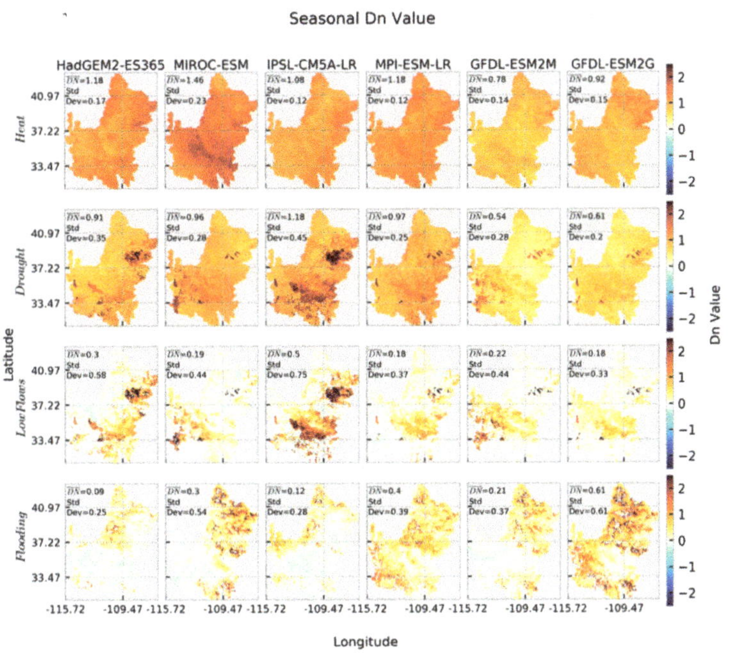

Figure 6. The D_n value results across the CRB at a seasonal timescale for each ESM and Impact. The average D_n value and spatial standard deviation is also shown for each panel.

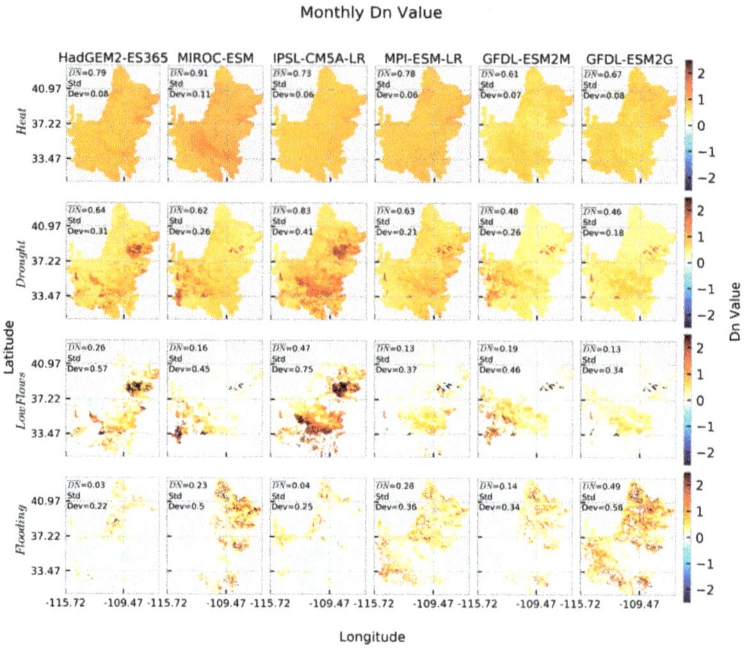

Figure 7. The D_n value results across the CRB at a monthly timescale for each ESM and Impact. The average D_n value and spatial standard deviation is also shown for each panel.

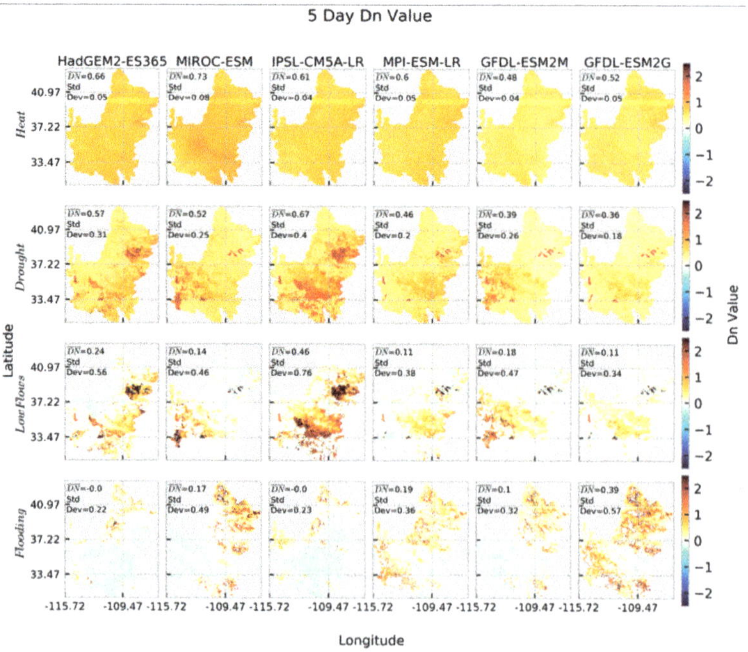

Figure 8. The D_n value results across the CRB at a synoptic timescale for each ESM and Impact. The average D_n value and spatial standard deviation is also shown for each panel.

Table 4. Multi-Model Average D_n values for each impact and timescale. The table includes D_n values for the entire CRB, Upper CRB, and Lower CRB as well as the standard deviation between GCMs in parenthesis.

		Synoptic	**Monthly**	**Seasonal**	**Annual**
CRB	Heatwaves	0.60 (0.09)	0.75 (0.11)	1.10 (0.23)	2.01 (0.12)
	Drought	0.50 (0.12)	0.61 (0.14)	0.86 (0.24)	1.42 (0.28)
	Low Flows	0.21 (0.14)	0.22 (0.13)	0.26 (0.12)	0.32 (0.08)
	Flooding	0.14 (0.15)	0.20 (0.17)	0.29 (0.20)	0.40 (0.25)
Upper CRB	Heatwaves	0.60 (0.08)	0.75 (0.10)	1.11 (0.19)	2.02 (0.13)
	Drought	0.44 (0.13)	0.55 (0.15)	0.79 (0.25)	1.35 (0.29)
	Low Flows	0.14 (0.15)	0.15 (0.15)	0.19 (0.15)	0.27 (0.14)
	Flooding	0.27 (0.20)	0.34 (0.22)	0.44 (0.24)	0.60 (0.29)
Lower CRB	Heatwaves	0.60 (0.10)	0.75 (0.12)	1.09 (0.27)	2.00 (0.14)
	Drought	0.55 (0.12)	0.66 (0.14)	0.92 (0.25)	1.49 (0.28)
	Low Flows	0.27 (0.15)	0.28 (0.14)	0.31 (0.13)	0.36 (0.08)
	Flooding	0.04 (0.12)	0.09 (0.15)	0.16 (0.18)	0.24 (0.23)

Figures 5–8 show the spatial detail of D_n value results across the CRB for each of the *Impacts*, ESMs, and time scales. The D_n values show strong and spatially coherent increases in both heatwaves and drought across the entire basin for each of the four time scales. As temperature increases is occurring across the CRB, maximum temperature presumably drives many of the spatially consistent increases for both drought and heatwaves [73]. MIROC-ESM, the ESM with the warmest projected temperature across the basin, also showed the largest *Impacts* and D_n values of the six ESMs (2.18, Figure 5 and repeated in Figures 6–8).

The D_n values of drought often exhibit more spatial variability than heatwaves. Of the six ESMs, IPSL-CM5A-LR is the driest of the ESMs (1.88, Figure 5 but repeated in Figures 6–8), generally exhibiting the highest D_n values and showing especially severe drought in the Little Colorado River basin and in areas along the southern half of the CRB and the Colorado headwaters (e.g., Figure 5, column 3). Of the indicators, minimum soil moisture generally exhibited the largest D_n values in these areas as well, and was the most spatially variable indicator contributing to drought (see Figure 3). Both of the GFDL ESMs exhibited the lowest D_n values for drought, likely due to the more modest temperature increases exhibited by these ESMs (1.19 and 1.28, Figure 5, Table 3).

In contrast, low flows and flooding exhibited both positive and negative values throughout the CRB, although the entirety of the CRB, when averaged, exhibited positive D_n values for both *Impacts* at all time scales (Figures 5–8). Flooding is the stronger indicator at annual time scales in the entire CRB, but low flows have higher D_n values at the synoptic scales, and both indicators have a similar magnitude of response at monthly and seasonal scales (Table 4). Agreement across ESMs illustrates higher coherency (lower variances) in low flows at longer time scales, while lower coherency (higher variance) is shown at longer time scales for flooding. Figures 5–8 show the spatial variations in responses across ESMs, with the driest ESM having the strongest responses for low flows (0.50, IPSL-CM5A-LR), and the wettest ESM (0.80, GFDL-ESM-2G) having the strongest responses for flooding.

The high-elevation, snow-dominated upper CRB and the low-elevation, arid lower CRB respond differently to climate change effects, and this dichotomy carries through the D_n results for *Impacts* and time scales for the Upper and Lower CRB (Table 4). Heatwaves show very little difference between Upper and Lower CRB (0.00 to 0.03) and generally have very little spatial variance, as discussed above. Both drought (−0.11 to −0.14) and low flows (−0.10 to −0.13) exhibited lower D_n values in the Upper CRB than in the Lower CRB. Flooding exhibits the greatest difference between Upper and Lower basins (0.23 to 0.37), with larger D_n values in the Upper CRB. Overall, the lowest D_n values are flooding in the Lower CRB at synoptic time scales (0.04), while low flows is lowest in the Upper CRB at synoptic time scales (0.14).

To account for the regions of the CRB where cumulative concurrent changes are occurring, we summed the instances when the *Impacts* or ESMs for HUC8 watersheds were greater than the 95th percentile of mean D_n value (Figure 9). The watersheds with the highest count of these instances are the Blue River basin (10), followed by Little Colorado River Headwaters (7). After these two HUC8 watersheds, the East-Taylor, Uncompahgre, San Miguel, and Big Chino-Williamson Valley basins each have six instances of critical concurrent extreme *Impacts*. Gunnison, Dolores, Verde, Lower Salt, and all the other watersheds identified in Figure 9 not already discussed have 4 concurrent extreme *Impacts*. The Uncompahgre had the most instances of criticality for low flows (4), followed by the adjacent San Miguel basin (3). San Miguel also had the greatest number of instances of drought of any watershed (3). Across the *Impacts* heatwaves has the most instances of criticality, with Blue River having the greatest (5), followed by the Little Colorado River Headwaters, East-Taylor, and the Big Chino-Williamson Valley watersheds (3 for each HUC8).

Figure 9. Critical basins in the Colorado River basin.

4. Discussion

The projected changes in temperature in the CRB over the historical to future time periods are drastic, but not unexpected. The average increase in temperature reflects estimates of regional change over the same period [78]. While there is some uncertainty between the ESMs in the magnitude of temperature increase, it is relatively small compared to the uncertainty in the ESM's estimates of future precipitation, a result that is also found across many studies of climate change [79]. Unlike estimated changes in temperature, estimated changes in precipitation vary greatly between the ESMs, and there is no agreement on the direction of change, with an almost equal average annual precipitation *decrease* modeled by IPSL-CM5A-LR and an annual precipitation *increase* modeled by GFDL-ESM2G. However, it should be noted that a wet ESM projection for precipitation may not lead to a less arid environment in the CRB, as potential increases in evaporation due to air temperature increases could skew the region towards greater aridity despite precipitation increases [80].

Despite the wide range of climate projections from a range of ESMs, the D_n values show (with a few minor exceptions) a remarkably similar spatial pattern for the extremes that include temperature indicators, such as heatwaves and drought. These *Impacts* also have the strongest signal of extremes across the CRB [81]. Therefore, heatwaves and drought will likely affect the CRB on a uniform basis and most strongly, with some pockets of higher and lower responses related to antecedent moisture conditions owing to drier or wetter soils in the case of droughts [82].

On the other hand, the spatial pattern for low flows and flooding is more variable across the CRB, owing to their spatially varying drivers and indicators such as precipitation, streamflow, and soil moisture. For both *Impacts*, soil moisture (minimums or maximums for low flows or flooding, respectively) had the strongest response among the indicators, illustrating the importance of soil moisture on these extremes. The effect of antecedent soil moisture on flooding is well documented in studies of historical events, forecasting, and climate change [83–85]. Although studies on low flows appear less common, Castillo, et al. [86] suggest that antecedent soil moisture is an important control on runoff, particularly during medium- and low-intensity storms, which are common in semi-arid locations such as the Lower CRB. These *Impacts* were thus dependent upon the intensity of those indicators in time within particular regions of the CRB, with notable differences between the Upper and the Lower CRB, indicating the need to examine these *Impacts* within specific regions or watersheds.

In general, the time scale of responses for concurrent extreme events was strongest for D_n values at the annual time scale. *Impacts* such as heatwaves and drought more than doubled from synoptic to annual time scales. This may be because temperature-driven extremes, which often occur over longer periods of time have longer memories in climate systems in terms of both reoccurrence and frequency [77]. Consequently, researchers looking to examine these temperature-driven extremes, the annual time scale is perhaps the most appropriate scale at which to examine these *Impacts*. Although results for low flows and flooding were also strongest at the annual time scale, these *Impacts* had much less change in their response across time scales across the CRB, suggesting that the time scale of analysis for these may be aligned with research drivers and/or science questions at hand. For example, a forecasting research method should examine flooding *Impacts* at a synoptic time scale, since forecasting is most concerned with synoptic scale changes in extremes. Additionally, important was the fact that flooding and low flows, also had the greatest disagreement across ESMs at annual time scales.

The Upper CRB and Lower CRB responded differently to concurrent extreme events for both spatial patterns and time scales. Within the Upper CRB, the coherency of low flows did not shift much across time scales, while, for the Lower CRB, coherency decreased (spatial variability increased). Flooding response was higher and low flows was lower in the Upper CRB, a pattern that was reversed for the Lower CRB, indicating the importance of changing floods for the Upper basin, while low flows are a larger issue for the Lower CRB. This result is not unexpected owing to the moisture deficits regimes of the Lower CRB, and the moisture excess regime of the Upper Colorado [48].

Our findings concur with the previous, limited research on extreme concurrent events, although to our knowledge, studies have not been undertaken specifically for the CRB. For example, univariate conditions for temperature and droughts showed no change, but concurrent droughts and heatwaves were shown to be occurring across the US [22];

a similar study showed that wind speeds were not increasing univariately across the Midwest, hot, dry, and windy events were increasing [87]. Analysis of concurrent extreme events was undertaken in a study by Hao, AghaKouchak and Phillips [26] where they used 13 CMIP5 models to examine concurrent extremes as scenarios, showing that concurrent warm/dry and warm/wet global extreme events have increased gradually since the ~1960s and substantially since the ~1990s, with warm/wet extreme events increasing in high latitudes and tropical regions, and warm/dry extremes increasing in many areas, including central Africa, eastern Australia, northern China, parts of Russia, and the Middle East [26]. Recent extremes in California were examined by Diffenbaugh, Swain and Touma [42], who highlights the role of record high annual mean temperatures in combination with record low annual mean precipitation in 2013, which led to increased evapotranspiration, more intense drought, and intensifying wildfire occurrences. Likewise, studies that consider changes such as forest disturbances under a warming climate are also looking at concurrent extreme events [88], although studies may not necessarily self-identify as examples of concurrent extreme event analysis.

Identifying critical regions within the CRB is important to determine the vulnerability of specific watersheds to concurrent extreme events. The watersheds identified in this work had instances of multiple ESMs and/or multiple *Impacts* projected to occur under the annual interval, a time scale most noted for its implications for water management of the system χ. Examining the entire CRB and being able to rank the most critical watersheds allows us to focus more closely on those systems in terms of resources and research. The critical basins identified in this work are indeed systems that have been well studied and/or have important water resources located within them. For example, Livneh, et al. [89] examined the Uncompahgre River basin in his investigation of forest disturbance and dust-on-snow implications, a study on concurrent events, although not specifically identified as such. The East-Taylor basin was selected as a 'representative community watershed' for intensive study by The Watershed Function project, funded by the USDOE Biological and Environmental Research Subsurface Biogeochemistry Program [90]. Additionally, the Salt/Verde complex has experienced damaging extreme events leading to effects on humans and economy [91–93]. The Blue River watershed is an important system as it includes the largest reservoir in the Upper CRB, and has been identified as threatened by climate change such as increasing temperatures [94].

This paper examines future changes in concurrent extreme events for different time scales in the CRB, focusing largely on a basic statistical metric to combine extremes. Our study is limited in that is leaves out important work to understand which extreme indicators to combine together to represent *Impacts*, and also more advanced statistics metrics, is fundamental to testing the applications and results illustrated herein. The next phase of this work includes a more advanced statistical method using copulas to characterize and describe changing concurrent extremes across the CRB. Under the same project, we have developed an economic model to consider the effects of modeled flooding in the CRB on the US economy under future climate change.

5. Conclusions

We analyzed extreme concurrent *Impacts* of heatwaves, drought, low flows, and flooding for the CRB under multiple future (2070–2099) climate change scenarios. Temperature and precipitation are projected to change dramatically in the CRB: all ESMs predict temperature increases, but different ESMs predict either significant increases or significant decreases in precipitation. Despite this range in response, extreme concurrent events are projected to increase across the CRB in all time scales and for all *Impacts*.

Temperature-driven concurrent extremes (heatwaves and drought) are strongest and most spatially coherent across the CRB, while precipitation-driven concurrent extremes (flooding and low flows) are less strong and more spatially variable across the CRB. Annual time scales for analysis of concurrent extremes show the strongest responses for all variables. Temperature-driven concurrent extreme *Impacts* shift the most across the time scales of analysis, with a more than doubling of response from synoptic to annual time scales. Precipitation-driven concurrent extremes do not show as much of a change across time scales, with a slight increase in response moving from synoptic, monthly, seasonal, to annual time scales. However, there is generally greater agreement moving from the annual to synoptic time scales; with the exception of low flows that has greater ESM coherence at annual time scales.

The Upper and Lower CRB act similarly temperature-driven concurrent extremes, again, but different in their response to changes in precipitation-driven concurrent extremes; with flooding having the strongest response in the Upper CRB, and low flows having a stronger response in the Lower CRB. We identified critical watersheds in the CRB that are projected to experience compounded concurrent extreme events, watersheds that include important water management structures in the Upper CRB, such as the Blue River basin, and where vital water research resources are located, such as the Uncompahgre and the East-Taylor basins. In the Lower CRB, the Salt/Verde basin provide important sources of in-basin surface water flow in an otherwise arid environment, and water for agriculture and hydropower at the Roosevelt Dam [95]. Our work verifies that concurrent extreme events are likely to increase in frequency and magnitude under future climate change [7,96,97].

As the science of extreme events evolves, examining concurrent events will likely be an important step in understanding the changing nature of extremes, capturing and analyzing the largest and most damaging of these events, tying instances of extremes in climate and land surface model simulations to observations, and linking the physical responses to other indicators of instability, such as economy and society.

Author Contributions: Conceptualization, K.E.B.; methodology, K.E.B., C.T., and R.B.; validation, C.T.; formal analysis, K.E.B. and C.T.; writing—original draft preparation, K.E.B. and C.T.; writing—review and editing, K.E.B., C.T., and R.B.; visualization, C.T.; project administration, K.E.B.; funding acquisition, K.E.B. All authors have read and agreed to the published version of the manuscript.

Funding: This work was funded under the Los Alamos National Laboratory Lab Directed Research and Development (LDRD) Early Career Research program (20180621ECR).

Institutional Review Board Statement: Not applicable.

Informed Consent Statement: Not applicable.

Data Availability Statement: Downscaled CMIP5 climate model projections may be downloaded via the MACA web portal: https://climate.northwestknowledge.net/MACA/ (accessed on 20 October 2020). VIC model may be downloaded via GitHub: https://github.com/UW-Hydro/VIC (accessed on 20 October 2020). Historical VIC forcing data may be obtained from ftp://gdo-dcp.ucllnl.org/pub/dcp/archive/OBS/livneh2014.1_16deg/ (accessed on 20 October 2020). Naturalized streamflow data for the Colorado River basin may be obtained from USBR: https://www.usbr.gov/lc/region/g4000/NaturalFlow/current.html (accessed on 20 October 2020) (U.S. Bureau of Reclamation, 2018). Other model parameter files and model outputs may be obtained by contacting the authors.

Acknowledgments: We would like to thank Jorge Urrego-Blanco for his help with statistical measures to formulate concurrent extreme events that became a part of this paper. Thank you to the help of an anonymous reviewer for their suggestions and improvements to the text of this paper. Additionally, thank you to Richard Middleton, who has brought KB to Los Alamos and provided initial support under LDRD DR (20150397DR), which allowed KB to develop the VIC modeling work that our analysis is based on (completed under Los Alamos Director's Postdoctoral Fellowship, 20160654PRD).

Conflicts of Interest: The authors declare no conflict of interest. The funders had no role in the design of the study; in the collection, analyses, or interpretation of data; in the writing of the manuscript, or in the decision to publish the results.

References

1. NOAA. NOAA National Centers for Environmental Information (NCEI) U.S. Billion-Dollar Weather and Climate Disasters. 2021. Available online: https://www.ncdc.noaa.gov/billions/ (accessed on 30 March 2021).
2. Stott, P. How climate change affects extreme weather events. *Science* **2016**, *352*, 1517–1518. [CrossRef]

3. Chen, Y.; Moufouma-Okia, W.; Masson-Delmotte, V.; Zhai, P.; Pirani, A. Recent Progress and Emerging Topics on Weather and Climate Extremes Since the Fifth Assessment Report of the Intergovernmental Panel on Climate Change. *Annu. Rev. Environ. Resour.* **2018**, *43*, 35–59. [CrossRef]
4. Trenberth, K.E.; Fasullo, J.T.; Shepherd, T.G. Attribution of climate extreme events. *Nat. Clim. Chang.* **2015**, *5*, 725–730. [CrossRef]
5. National Academies of Sciences Engineering and Medicine. *Attribution of Extreme Weather Events in the Context of Climate Change*; National Academies Press: Washington, DC, USA, 2016. [CrossRef]
6. Swain, D.L.; Singh, D.; Touma, D.; Diffenbaugh, N.S. Attributing extreme events to climate change: A new frontier in a warming world. *One Earth* **2020**, *2*, 522–527. [CrossRef]
7. Field, C.B.; Barros, V.; Stocker, T.F.; Qin, D.; Dokken, D.J.; Ebi, K.L.; Mastrandrea, M.D.; Mach, K.J.; Plattner, G.-K.; Allen, S.K.; et al. *Managing the Risks of Extreme Events and Disasters to Advance Climate Change Adaptation: Special Report of the Intergovernmental Panel on Climate Change*; Cambridge University Press: Cambridge, UK; New York, NY, USA, 2012; p. 582.
8. Blöschl, G.; Bierkens, M.F.; Chambel, A.; Cudennec, C.; Destouni, G.; Fiori, A.; Kirchner, J.W.; McDonnell, J.J.; Savenije, H.H.; Sivapalan, M. Twenty-three unsolved problems in hydrology (UPH)—A community perspective. *Hydrol. Sci. J.* **2019**, *64*, 1141–1158. [CrossRef]
9. Fares, A.; Habibi, H.; Awal, R. Extreme events and climate change: A multidisciplinary approach. In *Climate Change and Extreme Events*; Elsevier: Amsterdam, The Netherlands, 2021; pp. 1–7.
10. Driouech, F.; ElRhaz, K.; Moufouma-Okia, W.; Arjdal, K.; Balhane, S. Assessing future changes of climate extreme events in the CORDEX-MENA region using regional climate model ALADIN-climate. *Earth Syst. Environ.* **2020**, *4*, 477–492. [CrossRef]
11. Bao, J.; Sherwood, S.C.; Alexander, L.V.; Evans, J.P. Future increases in extreme precipitation exceed observed scaling rates. *Nat. Clim. Chang.* **2017**, *7*, 128–132. [CrossRef]
12. Maupin, M.A.; Ivahnenko, T.; Bruce, B. *Estimates of Water Use and Trends in the Colorado River Basin, Southwestern United States, 1985–2010*; Scientific Investigations Report 2018-5049; United States Geological Survey: Reston, VA, USA, 2018; p. 75.
13. Broska, L.H.; Poganietz, W.-R.; Vögele, S. Extreme events defined—A conceptual discussion applying a complex systems approach. *Futures* **2020**, *115*, 102490. [CrossRef]
14. McPhillips, L.E.; Chang, H.; Chester, M.V.; Depietri, Y.; Friedman, E.; Grimm, N.B.; Kominoski, J.S.; McPhearson, T.; Méndez-Lázaro, P.; Rosi, E.J.; et al. Defining Extreme Events: A Cross-Disciplinary Review. *Earth Future* **2018**, *6*, 441–455. [CrossRef]
15. Sillmann, J.; Kharin, V.; Zhang, X.; Zwiers, F.; Bronaugh, D. Climate extremes indices in the CMIP5 multi-model ensemble. Part 1: Model evaluation in the present climate. *J. Geophys. Res. Atmos.* **2013**. [CrossRef]
16. Sillmann, J.; Kharin, V.; Zwiers, F.; Zhang, X.; Bronaugh, D. Climate extreme indices in the CMIP5 multi-model ensemble. Part 2: Future climate projections. *J. Geophys. Res. Atmos.* **2013**, *118*, 2473–2493. [CrossRef]
17. Bennett, K.E.; Cannon, A.J.; Hinzman, L. Historical trends and extremes in boreal Alaska river basins. *J. Hydrol.* **2015**, *527*, 590–607. [CrossRef]
18. Bennett, K.E.; Walsh, J. Spatial and temporal changes in indices of extreme precipitation and temperature for Alaska. *Int. J. Climatol.* **2015**, *35*, 1434–1452. [CrossRef]
19. Tencer, B.; Weaver, A.; Zwiers, F. Joint occurrence of daily temperature and precipitation extreme events over Canada. *J. Appl. Meteorol. Climatol.* **2014**, *53*, 2148–2162. [CrossRef]
20. Sadegh, M.; Moftakhari, H.; Gupta, H.V.; Ragno, E.; Mazdiyasni, O.; Sanders, B.; Matthew, R.; AghaKouchak, A. Multihazard scenarios for analysis of compound extreme events. *Geophys. Res. Lett.* **2018**, *45*, 5470–5480. [CrossRef]
21. Tian, F.; Klingaman, N.P.; Dong, B. The driving processes of concurrent hot and dry extreme events in China. *J. Clim.* **2020**. [CrossRef]
22. Mazdiyasni, O.; AghaKouchak, A. Substantial increase in concurrent droughts and heatwaves in the United States. *Proc. Natl. Acad. Sci. USA* **2015**, *112*, 11484–11489. [CrossRef]
23. Dowdy, A.J.; Catto, J.L. Extreme weather caused by concurrent cyclone, front and thunderstorm occurrences. *Sci. Rep.* **2017**, *7*, 40359. [CrossRef]
24. Zhou, P.; Liu, Z. Likelihood of concurrent climate extremes and variations over China. *Environ. Res. Lett.* **2018**, *13*, 094023. [CrossRef]
25. Zhou, S.; Williams, A.P.; Berg, A.M.; Cook, B.I.; Zhang, Y.; Hagemann, S.; Lorenz, R.; Seneviratne, S.I.; Gentine, P. Land-atmosphere feedbacks exacerbate concurrent soil drought and atmospheric aridity. *Proc. Natl. Acad. Sci. USA* **2019**, *116*, 18848–18853. [CrossRef]
26. Hao, Z.; AghaKouchak, A.; Phillips, T.J. Changes in concurrent monthly precipitation and temperature extremes. *Environ. Res. Lett.* **2013**, *8*, 034014. [CrossRef]
27. Toreti, A.; Cronie, O.; Zampieri, M. Concurrent climate extremes in the key wheat producing regions of the world. *Sci. Rep.* **2019**, *9*. [CrossRef]
28. Kornhuber, K.; Coumou, D.; Vogel, E.; Lesk, C.; Donges, J.F.; Lehmann, J.; Horton, R.M. Amplified Rossby waves enhance risk of concurrent heatwaves in major breadbasket regions. *Nat. Clim. Chang.* **2020**, *10*, 48–53. [CrossRef]
29. Beniston, M. Trends in joint quantiles of temperature and precipitation in Europe since 1901 and projected for 2100. *Geophys. Res. Lett.* **2009**, *36*. [CrossRef]
30. Martin, J.-P.; Germain, D. Large-scale teleconnection patterns and synoptic climatology of major snow-avalanche winters in the Presidential Range (New Hampshire, USA). *Int. J. Climatol.* **2017**, *37*, 109–123. [CrossRef]

31. Wazneh, H.; Arain, M.A.; Coulibaly, P.; Gachon, P. Evaluating the Dependence between Temperature and Precipitation to Better Estimate the Risks of Concurrent Extreme Weather Events. *Adv. Meteorol.* **2020**, *2020*, 8763631. [CrossRef]
32. Luo, L.; Apps, D.; Arcand, S.; Xu, H.; Pan, M.; Hoerling, M. Contribution of temperature and precipitation anomalies to the California drought during 2012–2015. *Geophys. Res. Lett.* **2017**, *44*, 3184–3192. [CrossRef]
33. Miao, C.; Sun, Q.; Duan, Q.; Wang, Y. Joint analysis of changes in temperature and precipitation on the Loess Plateau during the period 1961–2011. *Clim. Dyn.* **2016**, *47*, 3221–3234. [CrossRef]
34. Fischer, E.M.; Knutti, R. Robust projections of combined humidity and temperature extremes. *Nat. Clim. Chang.* **2013**, *3*, 126–130. [CrossRef]
35. Estrella, N.; Menzel, A. Recent and future climate extremes arising from changes to the bivariate distribution of temperature and precipitation in Bavaria, Germany. *Int. J. Climatol.* **2013**, *33*, 1687–1695. [CrossRef]
36. AghaKouchak, A.; Cheng, L.; Mazdiyasni, O.; Farahmand, A. Global warming and changes in risk of concurrent climate extremes: Insights from the 2014 California drought. *Geophys. Res. Lett.* **2014**, *41*, 8847–8852. [CrossRef]
37. Sedlmeier, K.; Feldmann, H.; Schädler, G. Compound summer temperature and precipitation extremes over central Europe. *Theor. Appl. Climatol.* **2018**, *131*, 1493–1501. [CrossRef]
38. Sedlmeier, K.; Mieruch, S.; Schädler, G.; Kottmeier, C. Compound extremes in a changing climate—A Markov chain approach. *Nonlinear Process. Geophys.* **2016**, *23*, 375–390. [CrossRef]
39. Hao, Z.; Singh, V.P. Drought characterization from a multivariate perspective: A review. *J. Hydrol.* **2015**, *527*, 668–678. [CrossRef]
40. Hao, Z.; Singh, V.P.; Hao, F. Compound extremes in hydroclimatology: A review. *Water* **2018**, *10*, 718. [CrossRef]
41. Beevers, L.; White, C.J.; Pregnolato, M. Editorial to the Special Issue: Impacts of Compound Hydrological Hazards or Extremes. *Geosciences* **2020**, *10*, 496. [CrossRef]
42. Diffenbaugh, N.S.; Swain, D.L.; Touma, D. Anthropogenic warming has increased drought risk in California. *Proc. Natl. Acad. Sci. USA* **2015**, *112*, 3931–3936. [CrossRef]
43. Raymond, C.; Horton, R.M.; Zscheischler, J.; Martius, O.; AghaKouchak, A.; Balch, J.; Bowen, S.G.; Camargo, S.J.; Hess, J.; Kornhuber, K.; et al. Understanding and managing connected extreme events. *Nat. Clim. Chang.* **2020**, *10*, 611–621. [CrossRef]
44. Zscheischler, J.; Westra, S.; van den Hurk, B.J.J.M.; Seneviratne, S.I.; Ward, P.J.; Pitman, A.; AghaKouchak, A.; Bresch, D.N.; Leonard, M.; Wahl, T.; et al. Future climate risk from compound events. *Nat. Clim. Chang.* **2018**, *8*, 469–477. [CrossRef]
45. Leonard, M.; Westra, S.; Phatak, A.; Lambert, M.; van den Hurk, B.; McInnes, K.; Risbey, J.; Schuster, S.; Jakob, D.; Stafford-Smith, M. A compound event framework for understanding extreme impacts. *WIREs Clim. Chang.* **2014**, *5*, 113–128. [CrossRef]
46. Zscheischler, J.; Seneviratne, S.I. Dependence of drivers affects risks associated with compound events. *Sci. Adv.* **2017**, *3*, e1700263. [CrossRef]
47. Livneh, B.; Bohn, T.J.; Pierce, D.W.; Munoz-Arriola, F.; Nijssen, B.; Vose, R.; Cayan, D.R.; Brekke, L. A spatially comprehensive, hydrometeorological data set for Mexico, the U.S., and Southern Canada 1950–2013. *Sci. Data* **2015**, *2*, 150042. [CrossRef]
48. Christensen, N.S.; Lettenmaier, D.P. A multimodel ensemble approach to assessment of climate change impacts on the hydrology and water resources of the Colorado River Basin. *Hydrol. Earth Syst. Sci.* **2007**, *11*, 1417–1434. [CrossRef]
49. Abatzoglou, J.T.; Brown, T.J. A comparison of statistical downscaling methods suited for wildfire applications. *Int. J. Climatol.* **2012**, *32*, 772–780. [CrossRef]
50. Taylor, K.E.; Stouffer, R.J.; Meehl, G.A. An overview of CMIP5 and the experiment design. *Bull. Am. Meteorol. Soc.* **2012**, *93*, 485–498. [CrossRef]
51. Collins, W.J.; Bellouin, N.; Doutriaux-Boucher, M.; Gedney, N.; Halloran, P.; Hinton, T.; Hughes, J.; Jones, C.D.; Joshi, M.; Liddicoat, S.; et al. Development and evaluation of an Earth-System model—HadGEM2. *Geosci. Model Dev.* **2011**, *4*, 1051–1075. [CrossRef]
52. Cox, P.M. *Description of the TRIFFID Dynamic Global Vegetation Model*; Technical Note 24; Hadley Centre, United Kingdom Meteorological Office: Bracknell, UK, 2001.
53. Sato, H.; Itoh, A.; Kohyama, T. SEIB-DGVM: A new Dynamic Global Vegetation Model using a spatially explicit individual-based approach. *Ecol. Model.* **2007**, *200*, 279–307. [CrossRef]
54. Watanabe, S.; Hajima, T.; Sudo, K.; Nagashima, T.; Takemura, T.; Okajima, H.; Nozawa, T.; Kawase, H.; Abe, M.; Yokohata, T. MIROC-ESM 2010: Model description and basic results of CMIP 5-20 c 3 m experiments. *Geosci. Model Dev.* **2011**, *4*, 845–872. [CrossRef]
55. Giorgetta, M.A.; Jungclaus, J.; Reick, C.H.; Legutke, S.; Bader, J.; Böttinger, M.; Brovkin, V.; Crueger, T.; Esch, M.; Fieg, K. Climate and carbon cycle changes from 1850 to 2100 in MPI-ESM simulations for the Coupled Model Intercomparison Project phase 5. *J. Adv. Model. Earth Syst.* **2013**, *5*, 572–597. [CrossRef]
56. Reick, C.; Raddatz, T.; Brovkin, V.; Gayler, V. Representation of natural and anthropogenic land cover change in MPI-ESM. *J. Adv. Model. Earth Syst.* **2013**, *5*, 459–482. [CrossRef]
57. Krinner, G.; Viovy, N.; de Noblet-Ducoudré, N.; Ogée, J.; Polcher, J.; Friedlingstein, P.; Ciais, P.; Sitch, S.; Prentice, I.C. A dynamic global vegetation model for studies of the coupled atmosphere-biosphere system. *Glob. Biogeochem. Cycles* **2005**, *19*. [CrossRef]
58. Dufresne, J.-L.; Foujols, M.-A.; Denvil, S.; Caubel, A.; Marti, O.; Aumont, O.; Balkanski, Y.; Bekki, S.; Bellenger, H.; Benshila, R. Climate change projections using the IPSL-CM5 Earth System Model: From CMIP3 to CMIP5. *Clim. Dyn.* **2013**, *40*, 2123–2165. [CrossRef]

59. Shevliakova, E.; Pacala, S.W.; Malyshev, S.; Hurtt, G.C.; Milly, P.; Caspersen, J.P.; Sentman, L.T.; Fisk, J.P.; Wirth, C.; Crevoisier, C. Carbon cycling under 300 years of land use change: Importance of the secondary vegetation sink. *Glob. Biogeochem. Cycles* **2009**, *23*. [CrossRef]
60. Delworth, T.L.; Broccoli, A.J.; Rosati, A.; Stouffer, R.J.; Balaji, V.; Beesley, J.A.; Cooke, W.F.; Dixon, K.W.; Dunne, J.; Dunne, K. GFDL's CM2 global coupled climate models. Part I: Formulation and simulation characteristics. *J. Clim.* **2006**, *19*, 643–674. [CrossRef]
61. Fajardo, J.; Corcoran, D.; Roehrdanz, P.R.; Hannah, L.; Marquet, P.A. GCM compareR: A web application to assess differences and assist in the selection of general circulation models for climate change research. *Methods Ecol. Evol.* **2020**, *11*, 656–663. [CrossRef]
62. Le Quéré, C.; Moriarty, R.; Andrew, R.M.; Canadell, J.G.; Sitch, S.; Korsbakken, J.I.; Friedlingstein, P.; Peters, G.P.; Andres, R.J.; Boden, T. Global carbon budget 2015. *Earth Syst. Sci. Data* **2015**, *7*, 349–396. [CrossRef]
63. Van Vuuren, D.P.; Edmonds, J.; Kainuma, M.; Riahi, K.; Thomson, A.; Hibbard, K.; Hurtt, G.C.; Kram, T.; Krey, V.; Lamarque, J.-F. The representative concentration pathways: An overview. *Clim. Chang.* **2011**, *109*, 5–31. [CrossRef]
64. Bohn, T.J.; Vivoni, E.R. Process-based characterization of evapotranspiration sources over the North American monsoon region. *Water Resour. Res.* **2016**, *52*, 358–384. [CrossRef]
65. Liang, X.; Lettenmaier, D.P.; Wood, E.F.; Burges, S.J. A simple hydrologically based model of land surface water and energy fluxes for general circulation models. *J. Geophys. Res.* **1994**, *99*, 14415–14428. [CrossRef]
66. Franchini, M.; Pacciani, M. Comparative analysis of several conceptual rainfall-runoff models. *J. Hydrol.* **1991**, *122*, 161–219. [CrossRef]
67. USBR. Colorado River Basin Natural Flow and Salt Data. Available online: https://www.usbr.gov/lc/region/g4000/NaturalFlow/supportNF.html (accessed on 24 December 2020).
68. Yapo, P.O.; Gupta, H.V.; Sorooshian, S. Multi-objective global optimization for hydrologic models. *J. Hydrol.* **1998**, *204*, 83–97. [CrossRef]
69. Bennett, K.E.; Bohn, T.J.; Solander, K.; McDowell, N.G.; Vivoni, E.; Middleton, R. Climate-driven disturbances in the San Juan River sub-basin of the Colorado River. *Hydrol. Earth Syst. Sci.* **2018**, *22*, 709–725. [CrossRef]
70. Urrego-Blanco, J.R.; Hunke, E.C.; Urban, N.M.; Jeffery, N.; Turner, A.K.; Langenbrunner, J.R.; Booker, J.M. Validation of sea ice models using an uncertainty-based distance metric for multiple model variables. *J. Geophys. Res. Ocean.* **2017**, *122*, 2923–2944. [CrossRef]
71. Langenbrunner, J.R.; Hemez, F.M.; Booker, J.M.; Ross, T.J. Model choice considerations and information integration using analytical hierarchy process. *Procedia Soc. Behav. Sci.* **2010**, *2*, 7700–7701. [CrossRef]
72. Booker, J.M. Interpretations of Langenbrunner's Dn metric for V&V. In *Los Alamos National Laboratory Technical Report*; LA-UR-06-3720; Los Alamos National Laboratory: Los Alamos, NM, USA, 2006.
73. IPCC. *Climate Change 2013: The Physical Science Basis. Contribution of Working Group I to the Fifth Assessment Report of the Intergovernmental Panel on Climate Change (IPCC)*; Stocker, T., Qin, D., Plattner, G., Tignor, M., Allen, S., Boschung, J., Nauels, A., Xia, Y., Bex, B., Midgley, B., Eds.; Cambridge University Press: Cambridge, UK; New York, NY, USA, 2013; p. 1535.
74. Omernik, J.M.; Griffith, G.E. Ecoregions of the conterminous United States: Evolution of a hierarchical spatial framework. *Environ. Manag.* **2014**, *54*, 1249–1266. [CrossRef]
75. Hendricks, J.; Plescia, J. A review of the regional geophysics of the Arizona transition zone. *J. Geophys. Res. Solid Earth* **1991**, *96*, 12351–12373. [CrossRef]
76. Sheppard, P.R.; Comrie, A.C.; Packin, G.D.; Angersbach, K.; Hughes, M.K. The climate of the US Southwest. *Clim. Res.* **2002**, *21*, 219–238. [CrossRef]
77. Dirmeyer, P.A.; Schlosser, C.A.; Brubaker, K.L. Precipitation, recycling, and land memory: An integrated analysis. *J. Hydrometeorol.* **2009**, *10*, 278–288. [CrossRef]
78. Garfin, G.; Franco, G.; Blanco, H.; Comrie, A.; Gonzalez, P.; Piechota, T.; Smyth, R.; Waskom, R. Southwest: The Third National Climate Assessment. In *Climate Change Impacts in the United States: The Third National Climate Assessment*; US Global Change Research Program: Washington, DC, USA, 2014; p. 25.
79. Knutti, R.; Sedláček, J. Robustness and uncertainties in the new CMIP5 climate model projections. *Nat. Clim. Chang.* **2013**, *3*, 369–373. [CrossRef]
80. Bennett, K.E.; Miller, G.; Talsma, C.; Jonko, A.; Bruggeman, A.; Atchley, A.; Lavadie-Bulnes, A.; Kwicklis, E.; Middleton, R. Future water resource shifts in the high desert Southwest of Northern New Mexico, USA. *J. Hydrol. Reg. Stud.* **2020**, *28*, 100678. [CrossRef]
81. Frich, P.; Alexander, L.V.; Della-Marta, P.; Gleason, B.; Haylock, M.; Tank, A.K.; Peterson, T. Observed coherent changes in climatic extremes during the second half of the twentieth century. *Clim. Res.* **2002**, *19*, 193–212. [CrossRef]
82. Livneh, B.; Hoerling, M.P. The physics of drought in the US central Great Plains. *J. Clim.* **2016**, *29*, 6783–6804. [CrossRef]
83. Schoener, G.; Stone, M.C. Impact of antecedent soil moisture on runoff from a semiarid catchment. *J. Hydrol.* **2019**, *569*, 627–636. [CrossRef]
84. Woldemeskel, F.; Sharma, A. Should flood regimes change in a warming climate? The role of antecedent moisture conditions. *Geophys. Res. Lett.* **2016**, *43*, 7556–7563. [CrossRef]
85. Oubeidillah, A.; Tootle, G.; Piechota, T. Incorporating Antecedent Soil Moisture into Streamflow Forecasting. *Hydrology* **2019**, *6*, 50. [CrossRef]

86. Castillo, V.; Gomez-Plaza, A.; Martınez-Mena, M. The role of antecedent soil water content in the runoff response of semiarid catchments: A simulation approach. *J. Hydrol.* **2003**, *284*, 114–130. [CrossRef]
87. Tavakol, A.; Rahmani, V.; Harrington, J., Jr. Probability of compound climate extremes in a changing climate: A copula-based study of hot, dry, and windy events in the central United States. *Environ. Res. Lett.* **2020**, *15*, 104058. [CrossRef]
88. Buma, B.; Livneh, B. Potential Effects of Forest Disturbances and Management on Water Resources in a Warmer Climate. *For. Sci.* **2015**, *61*, 895–903. [CrossRef]
89. Vano, J.A.; Udall, B.; Cayan, D.R.; Overpeck, J.T.; Brekke, L.D.; Das, T.; Hartmann, H.C.; Hidalgo, H.G.; Hoerling, M.; McCabe, G.J. Understanding uncertainties in future Colorado River streamflow. *Bull. Am. Meteorol. Soc.* **2014**, *95*, 59–78. [CrossRef]
90. Livneh, B.; Deems, J.S.; Buma, B.; Barsugli, J.J.; Schneider, D.; Molotch, N.P.; Wolter, K.; Wessman, C.A. Catchment response to bark beetle outbreak and dust-on-snow in the Colorado Rocky Mountains. *J. Hydrol.* **2015**, *523*, 196–210. [CrossRef]
91. Hubbard, S.S.; Williams, K.H.; Agarwal, D.; Banfield, J.; Beller, H.; Bouskill, N.; Brodie, E.; Carroll, R.; Dafflon, B.; Dwivedi, D.; et al. The East River, Colorado, Watershed: A Mountainous Community Testbed for Improving Predictive Understanding of Multiscale Hydrological-Biogeochemical Dynamics. *Vadose Zone J.* **2018**, *17*, 180061. [CrossRef]
92. Rivera, E.R.; Dominguez, F.; Castro, C.L. Atmospheric rivers and cool season extreme precipitation events in the Verde River basin of Arizona. *J. Hydrometeorol.* **2014**, *15*, 813–829. [CrossRef]
93. House, P.K.; Hirschboeck, K.K. Hydroclimatological and paleohydrological context of extreme winter flooding in Arizona, 1993. In *Storm-Induced Geological Hazards: Case Histories from the 1992–1993 Winter Storm in Southern California and Arizona. Reviews in Engineering Geology*; Larson, R.A., Slosso, J.E., Eds.; Geological Society of America: Boulder, CO, USA, 1997; Volume 11, pp. 1–24.
94. US Army Corps of Engineers. *Flood Damage Report, State of Arizona, Floods of 1993*; US Army Corps of Engineers: Los Angeles, CA, USA, 1994; p. 107.
95. Kopytkovskiy, M.; Geza, M.; McCray, J. Climate-change impacts on water resources and hydropower potential in the Upper Colorado River Basin. *J. Hydrol. Reg. Stud.* **2015**, *3*, 473–493. [CrossRef]
96. Molotch, N.P.; Fassnacht, S.R.; Bales, R.C.; Helfrich, S.R. Estimating the distribution of snow water equivalent and snow extent beneath cloud cover in the Salt-Verde River basin, Arizona. *Hydrol. Process.* **2004**, *18*, 1595–1611. [CrossRef]
97. Rahmstorf, S.; Coumou, D. Increase of extreme events in a warming world. *Proc. Natl. Acad. Sci. USA* **2011**, *108*, 17905–17909. [CrossRef]

water

MDPI

Article

Contribution of Glacier Runoff during Heat Waves in the Nooksack River Basin USA

Mauri S. Pelto [1,*], Mariama Dryak [2], Jill Pelto [3], Tom Matthews [4] and L. Baker Perry [5]

1 Environmental Science, Nichols College, Dudley, MA 01571, USA
2 Environmental Science, Shenandoah University, Winchester, VA 22601, USA; mdryak@su.edu
3 North Cascade Glacier Climate Project, Dudley, MA 01571, USA; pelto.jill@gmail.com
4 Geography, Kings College London, London WC2R 2LS, UK; tom.matthews@kcl.ac.uk
5 Geography and Planning, Appalachian State University, Boone, NC 28608, USA; perrylb@appstate.edu
* Correspondence: mspelto@nichols.edu

Abstract: The thirty-eight-year record (1984–2021) of glacier mass balance measurement indicates a significant glacier response to climate change in the North Cascades, Washington that has led to declining glacier runoff in the Nooksack Basin. Glacier runoff in the Nooksack Basin is a major source of streamflow during the summer low-flow season and mitigates both low flow and warm water temperatures; this is particularly true during summer heat waves. Synchronous observations of glacier ablation and stream discharge immediately below Sholes Glacier from 2013–2017, independently identify daily discharge during the ablation season. The identified ablation rate is applied to glaciers across the North Fork Nooksack watershed, providing daily glacier runoff discharge to the North Fork Nooksack River. This is compared to observed daily discharge and temperature data of the North Fork Nooksack River and the unglaciated South Fork Nooksack River from the USGS. The ameliorating role of glacier runoff on discharge and water temperature is examined during 24 late summer heat wave events from 2010–2021. The primary response to these events is increased discharge in the heavily glaciated North Fork, and increased stream temperature in the unglaciated South Fork. During the 24 heat events, the discharge increased an average of +24% (±17%) in the North Fork and decreased an average of 20% (±8%) in the South Fork. For water temperature the mean increase was 0.7 °C (±0.4 °C) in the North Fork and 2.1 °C (±1.2 °C) in the South Fork. For the North Fork glacier runoff production was equivalent to 34% of the total discharge during the 24 events. Ongoing climate change will likely cause further decreases in summer baseflow and summer baseflow, along with an increase in water temperature potentially exceeding tolerance levels of several Pacific salmonid species that would further stress this population.

Keywords: glacier ablation; North Cascade Range; climate change; salmon; glacier mass balance; heat wave

Citation: Pelto, M.S.; Dryak, M.; Pelto, J.; Matthews, T.; Perry, L.B. Contribution of Glacier Runoff during Heat Waves in the Nooksack River Basin USA. *Water* **2022**, *14*, 1145. https://doi.org/10.3390/w14071145

Academic Editors: Rajesh R. Shrestha and Mohammad Reza Najafi

Received: 29 January 2022
Accepted: 28 March 2022
Published: 2 April 2022

1. Introduction

Climate observations in the Pacific Northwest (United States) show an accelerated warming for the 1970–2012 time periods of approximately 0.2 °C per decade [1]. Analysis of key components of the alpine North Cascade hydrologic system indicate significant changes in glacier mass balance, terminus behavior, alpine snowpack, and alpine streamflow from 1950–2015 [2,3]. Glacier runoff is of particular importance to aquatic life late in the summer when other water sources are at a minimum, raising minimum streamflow and reducing maximum temperatures [4]. Contributions from groundwater, precipitation, and non-glacier snowmelt reach a minimum after 1 July [4]. Whereas annual glacier runoff peaks during the late summer and is highest in warm, dry summers and lowest during wet, cool summers [5].

Watersheds in the Pacific Northwest are comprised of pluvial streams that experience peak flow in winter due to the winter storm events [6], nival streams that peak in the May and June due to high snowmelt, and glacially fed streams which peak in July and August during peak glacier melt [5–8]. A comparison of hydrographs from glaciated and unglaciated basins indicates a similar progression through June when runoff is dominated by non-glacier snowmelt, followed by increasing divergence as glacier runoff minimizes declines in glaciated watersheds until October when the hydrographs converge again [7,8]. The loss of a glacier from a watershed reduces streamflow primarily during late summer minimum flow periods [5,8,9]. The volume of glacier runoff is the product of surface area and ablation rate [2]. Glacier volume loss contributes to changes in streamflow, leading to an increase in overall streamflow if the rate of volume loss is sufficiently large [7], or a decline in streamflow if the area of glacier cover declines sufficiently to offset any increase in ablation rate [2]. It is evident that glaciers have a substantially larger role than the area they cover in August based on the identifiable glacier fingerprint on hydrographs for a watershed where glacier cover exceeds 1% of total watershed area [8]. The amount of summer runoff generated per unit area in the Nooksack River was 0.036 $m^3s^{-1}km^{-2}$, in the unglaciated South Fork Nooksack (SFN) discharge was 0.045 $m^3s^{-1}km^{-2}$, increasing to 0.312 $m^3s^{-1}km^{-2}$ in the heavily glaciated North Fork Nooksack (NFN). This represents nearly a seven-fold increase in runoff from glacier versus non-glacier areas in the Nooksack Basin [2].

Climate change is altering late summer streamflow in the North Cascades. There has been a coherent shift toward earlier runoff in snow fed basins across the western US, including a 10–30-day earlier date of the center of mass for annual flow for each water year [10]. A reduction in summer streamflow in six North Cascade basins from 1956–2006 has been observed [2]. In the North Cascades glacier volume loss has contributed up to 6% of the total August–September streamflow [11]. The loss in glacier area in the North Cascades and British Columbia is greater than the increased rate of ablation, as a result peak runoff in these same regions has been reached and the dominant ongoing change in glacier runoff is a decline in summer streamflow due to glacier area reductions [2,7,12].

Thermal regimes in streams reflect the balance of numerous physical processes that cause heating or cooling. Rates of warming in the Pacific Northwest's rivers have been highest during the summer, an increase of 0.17 °C per decade [4]. Air temperature was the dominant factor in both long term and inter-annual variability for Pacific Northwest rivers [4]. Discharge and air temperature appear additive and the seasonal variation in stream warming rates is determined by the degree of concert between these two variables [4]. For example, the largest warming trend during the summer resulted from the effects of the largest air temperature increases added to the largest river discharge decreases. This is further supported by Luce et al. [13] who identified a pattern where water temperature in cold streams had low sensitivity to air temperature, while warm streams had a tendency for higher sensitivity to air temperature.

An important impact of changing glacier runoff in the Nooksack River is the stress of warming stream temperatures on salmon populations [13,14]. Temperature thresholds for changes in fish communities in the Fraser River region of British Columbia were noted as 12 °C and 19 °C [14]. The reduction of the glacial melt component augmenting summer low flows is already resulting in more low-flow days in the North Cascade region as has occurred other alpine regions with small glaciers [12,15]. In the Skykomish River from 1950–2013, there were 230 days during the summer melt season with discharge below 10% of mean annual flow (14 m^3s^{-1}); of these, 99% (228 days) had occurred since 1985 [12]. Of great concern for aquatic life is the occurrence of extended periods of low flow [14] that have increased in frequency.

Climate change is a growing threat that has caused and will cause increases in winter flow, earlier spring snowmelt, decreased summer baseflow, and increased maximum summer water temperature in North Cascade watersheds [14]. Without mitigating steps, climate change will increase the frequency of low flow conditions and water temperatures that exceed the salmon tolerance levels. The impact is most acute during summer heat waves that result in minimum flow conditions coincident with maximum stream temperatures. This research identifies the specific response in glaciers ablation, glacier runoff and the resultant evolving water temperature threat during summer heat waves in the

Nooksack River Basin. In this study, heat events are identified as any period of five or more consecutive days where the average daily temperature observed at the Middle Fork Nooksack SNOTEL site (MFN) exceeds 14 °C, and precipitation is less than 3 mm for the entire period. A shorter heat event due to lagging responses would not yield a robust measure of streamflow response. The temperature threshold is simply chosen based on relation to locally identified heat waves. More than 3 mm of precipitation could influence discharge and complicate understanding the contrasting response of SFN and NFN. This is accomplished by monitoring ablation and runoff directly from Sholes Glacier and examining the simultaneous United States Geological Survey (USGS) discharge and stream temperature record in the two of the three principal forks of the Nooksack River, having varying amounts of glacier cover, for 24 summer heat waves. This includes the most intense heat wave the region has experienced occurring at the end of June 2021. The exceptional nature of the June 2021 heat wave is identified using summer air temperature reanalysis using ERA5.

2. Study Area

The Nooksack River consists of the North, South, and Middle Fork which combine near Deming to create the main stem Nooksack River. The Nooksack River empties into Birch Bay near Bellingham, Washington. The Nooksack River Basin is a hybrid basin with the various sub-basins dominated by pluvial, nival, and glacial runoff contributions, resulting in differing seasonal timing of maximum discharge, reducing the magnitude and duration of the summer minimum flow period. The USGS has gaging stations on each of the three main forks and the main stem of the Nooksack Basin. There are no significant reservoirs or flow diversions upstream of the gaging locations. From October-March is a storage period characterized by precipitation exceeding discharge, whereas April-August is a period of excess runoff release [6,16]. In the Nooksack River basin, glacier runoff supplied 10–20% of summer streamflow in the late 20th century [16]. The primary focus is on glacier runoff for the North Fork Nooksack River with 6.1% glacier cover above the gaging station (Figure 1). The Nooksack River system is home to five species of Pacific salmon including: Chinook, Coho, Pink, Chum, and Sockeye, with Chinook listed as Threatened under the Endangered Species Act (ESA) [14]. In the last two centuries the numbers of salmon that return to spawn in the Nooksack watershed have greatly diminished due to substantial loss of habitat primarily from human-caused alteration [14].

Thirty-seven years of mass balance work in the basin identify glacier ablation that yields an average of 11 m^3s^{-1}–12 m^3s^{-1} from July–September [3]. This is 10–20% of the total summer discharge of the Nooksack River at Ferndale, Washington, depending on the specific year. The glaciated area coverage in 2015 was 6.1% in North Fork Nooksack River (NFN) basin, 0% in South Fork Nooksack River (SFN) basin, and 1.1% in the Nooksack River basin at Ferndale (Figure 1). This difference in glacier covered area allows assessment of the impact of glaciers on stream discharge and temperature.

The NFN is a 65-km long tributary, with salmon habitat extending to the base of Nooksack Falls (Figure 1). From 1985 to 2017 mean July-September discharge is 25.9 m^3s^{-1}. In 2015, the NFN watershed had a glacier area of 16.9 km^2, with 12.3 km^2 of that glacier area located above the USGS gage site. On the NFN at Nooksack Falls there is a run of river hydropower plant constructed in 1906 that is rated at a production of 3.5 MW. There is no reservoir for this plant, just a low weir diverting water into a penstock above the falls. The 1.25 km long penstock returns diverted water to the river below the falls.

The SFN is a 57-km long tributary with salmon habitat extending 52 km upstream of the junction of with the Nooksack River. From 1985 to 2017 mean July–September discharge is 8.8 m^3s^{-1}. There are currently no glaciers in the watershed. There are no hydropower plants in the basin.

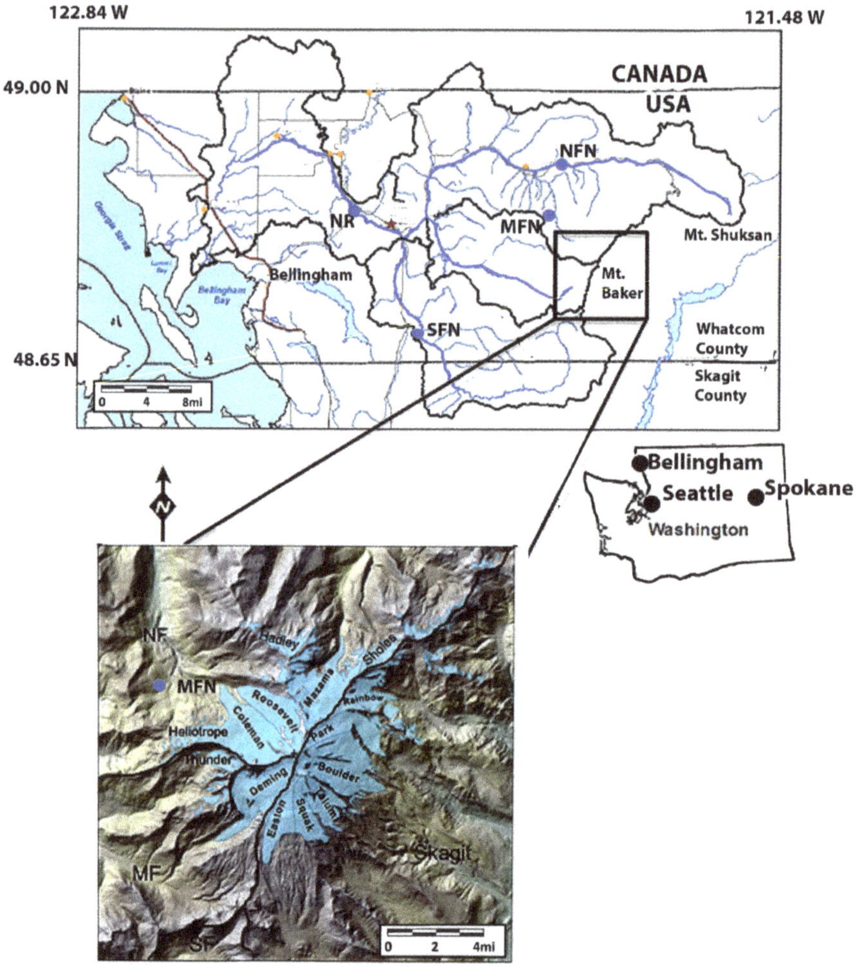

Figure 1. Map of Mount Baker glaciers and the Nooksack River Watershed. USGS gage locations for the Nooksack River (NR), North Fork Nooksack River (NFN), and South Fork Nooksack River (SFN) are indicated with blue dot. The Middle Fork Nooksack SNOTEL station (MFN) also indicated with blue dot, on inset also. Yellow dots mark cities, Red line is the I−5 highway and the red star is the Nooksack Indian Tribal Center.

From 1950–1980 the areal extent of glaciers in the NFN basin increased, with all Mount Baker glaciers advancing [17,18]. Since 1980, all glaciers in the basin have retreated significantly, with the retreat accelerating since 2000 [17,18]. On Mount Baker the average glacier retreat was 430 m during 1979–2015 [19]. Mass balance measurements indicate the cumulative loss as −17.3 m water equivalent (w.e.), equivalent to 20–30% of glacier volume lost during 1984–2015 [3].

3. Methods and Data Sources

3.1. Glacier Mass Balance

The North Cascade Glacier Climate Project (NCGCP) has monitored the annual mass balance during 1984–2021 on Lower Curtis and Rainbow Glacier in the Baker Lake watershed adjacent to the NFN and from 1990–2021 mass balance on Sholes Glacier in the NFN and Easton Glacier in the Baker Lake watershed [3,18,20]. Rainbow Glacier, which abuts Sholes Glacier, and Easton Glacier 8 km south in the Baker Lake Watershed are two of the 42 reference glaciers for the World Glacier Monitoring Service (WGMS). Accumulation and ablation measurements are completed yearly during the summer on each glacier at a density of over 100 pointskm^{-2}, and changes in glacier area are assessed every three to five years. The program relies on consistent methods applied to the same network of data points on the glacier's each year [20], with an uncertainty of 0.15 ma^{-1} falling in the typical range [21,22]. Direct measurement of ablation is accomplished using ablation stakes, changes in snow depth from repeat probing measurements and from snowline migration [19]. Ablation stakes are distributed across the entire elevation span of the glacier [3]. Stake measurement error over the shorter periods of observation determined here and in the Swiss Alps is 0.05 m [22,23], but less than 0.11 m to 0.14 m errors reported for annual stake observations [21,24]. These data, including the specific glacier area, are reported annually to the WGMS. Sholes Glacier had a mean elevation of 1840 m in 2015, while the mean elevation of glaciers in the NFN above the USGS gaging station was 1820 m in 2015 [19].

Overall Sholes Glacier has had a mass balance of −24.9 m w.e. during 1990–2021, this is a substantial loss for a glacier that averages 40–60 m in thickness [19]. The highest rate of loss occurred from 2013 to 2021 with a 13 m w.e. loss. The correlation in annual mass balance is from 0.96–0.98 between Sholes Glacier and Lower Curtis, Rainbow and Easton for 1990–2017, indicating the nearly identical response to annual climate conditions [3,20]. Glacier ablation measurements occurring simultaneously with discharge measurements below the glacier provide independent measures of glacier runoff. Here we report on observations during specific time periods that overlap with heat wave periods at ablation stakes.

The degree day function (DDF) is the most common means of modelling ablation on glaciers [25]. In this study both daily ablation and multi-day ablation observations are used to identify how much glacier runoff is produced. All daily ablation measurements from Easton, Sholes, and Rainbow Glacier completed during the summer have been used in combination with daily mean temperatures at MFN to derive a degree day factor (DDFs and DDFi) for daily snow and ice ablation respectively. In this study a specific heat event DDF is derived for snow and ice ablation for days where the temperatures average is 13 °C or above at MFN. This is 1 °C below the heat event threshold and expands the data set while maintaining the high temperature selection. The DDF Equation (1) is based on the average daily ablation (DA) at multiple sites between 1700 and 1900 m and the daily mean temperature (DT) at MFN. Neither ablation nor temperature is adjusted to the specific stake elevation. If there was a greater range in elevation of the ablation sites a lapse rate would be appropriate. Daily ablation measurements have been completed on 48 separate days, yielding 178 location specific observations when temperatures have exceeded this threshold. Of the 24 heat events we have collected ablation data throughout 10 of them. These data are used to generate a DDF model for ablation conditions during warm, dry periods.

$$DA = (DDFs \times DT) \text{ or } (DDFi \times DT) \quad (1)$$

3.2. USGS Stream Data

Both daily and monthly records of stream temperature and discharge are available from USGS stations at: North Fork Nooksack River at Glacier, South Fork Nooksack River at Saxon Bridge and Nooksack River at Ferndale and Cedarville [16,19]. Table 1 indicates the gage location characteristics, data type, and period of record utilized. This allows comparison of stream response to specific weather conditions and comparison between basins. The stream temperature records from the USGS only exist since 2008; hence no temporal trend analysis is performed.

Table 1. USGS stations characteristics and data records utilized.

Basin	USGS Station ID	Mean Elevation m a.s.l.	Basin Area km^2	Glacier Cover %	Discharge Records	Stream Temperature Records
Nooksack	12213100	800	2036	1.1	1970–2013	None
SF Nooksack	12210000	914	334	0	2008–2013	2008–2013
NF Nooksack	12205000	1311	272	6.1	1950–2013	2008–2013

3.3. SNOTEL Data

The United States Department of Agriculture-SNOTEL program has two stations in the Nooksack Basin that provide daily air temperature and precipitation. The Middle Fork Nooksack (MFN) station provides a consistent measure of hourly temperature and precipitation at an elevation 300 m below the glacier and 9 km west of Sholes Glacier, while Wells Creek is 600 m below the glacier and 6 km northwest.

3.4. ERA5 Data

There is no weather station above 1600 m in the area, below all but the lowest areas of a few glaciers. To understand the temperatures at higher elevations required use of an ERA5 dataset. Three-hourly air temperatures and geopotential height on pressure levels were obtained for the ablation season May–September from May 1979 to July 2021 from the $0.25 \times 0.25°$ ERA5 dataset, the latest and highest resolution reanalysis produced by the European Centre for Medium Range Weather Forecasting [26]. Both fields were interpolated to $-121.8°$ E, 48.8° N. The geopotential height, indicating the elevation of the pressure levels, was then used as the vertical coordinate to linearly interpolate temperature to the 3000 m contour. Daily maximum temperatures were computed from these interpolated data. The temperature lapse rates were defined as the slope coefficient in a regression of air temperature against the corresponding geopotential height [27]. We summarized the daily mean lapse rate as the arithmetic average of these (eight) slope coefficients computed on the three-hourly data. We caution that our use of ERA5 reanalysis in place of direct observations is a source of uncertainty, however given the lack of a long or consistent temperature record at elevation, this provides the most comparable data record for evaluating the significance of specific heat events. A good general agreement between station-based estimates of temperature and ERA5 in Western U.S. region, overlapping our region of study has been noted [28]. The highest seasonal correlation between stations and ERA5 for environmental lapse rate was 0.7 during the summer [28]. The highest correlations for specific temperature differences between stations and ERA was 0.9 for maximum temperatures [28]. This indicates that maximum temperatures in summer are one of the most reliable products of ERA5 in the Western US [28]. We also note that ERA5 is used here only to identify periods of highest temperatures, hence any mean biases will not affect our conclusions.

4. Results

4.1. Glacier Ablaiton

A few specific examples are reviewed below. In 2014 measurements of ablation daily during the 27 July–7 August period at a series of 12 stakes on the Sholes Glacier indicated a mean ablation rate for snow of 0.055 m w.e.d^{-1}. The ice melt was of the same thickness as noted for snow, but because of the greater density the water equivalent loss is higher, it indicated a mean ablation of 0.75 m w.e.d^{-1}. Between 29 July and 4 August 2015 measurements of ablation at a series of 6 stakes on the Sholes Glacier indicated a mean ablation rate for snow of 0.057 m w.e.d^{-1} and for ice of 0.078 m w.e.d^{-1}. In 2016, ablation measurement during the 12–21 August period at a series of 12 stakes on the Sholes Glacier yielded average snow ablation of 0.048 m w.e.d^{-1} and ice ablation of 0.070 m w.e.d^{-1}. In 2020, from 29 July to 5 August mean snow ablation at six stakes was 0.055 md^{-1} w.e.

4.2. Ablation Modelling

From 1990 to 2018, daily ablation measurements on Sholes Glacier and daily mean temperature reported at the Middle Fork Nooksack SNOTEL station, are utilized to generate a DDF for ablation. This model is generated from 148 days of observations of both ablation and air temperature (Figure 2). The focus on the observations for this study is specific daily to weekly observations of ablation during heat waves at the network of stakes spread across the glacier (Figure 3).

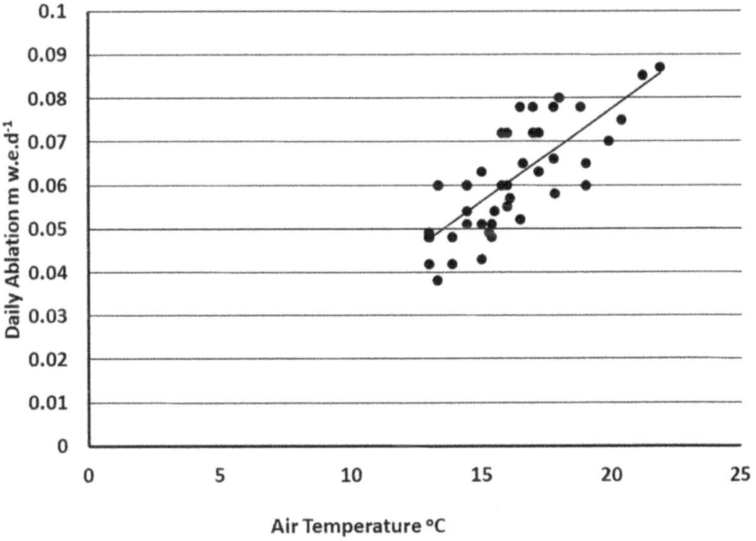

Figure 2. Daily mean air temperature at the Middle Fork Nooksack SNOTEL site and daily snow ablation measured on Sholes Glacier. The DDFs are derived from the linear regression slope coefficients.

The correlation coefficient between observed ablation and daily mean temperature is 0.82, for the entire ablation data set regardless of mean daily temperature. The overall DDFs for snow is 0.0035 m w.e. $°C^{-1}d^{-1}$. For ice, the DDFi is 0.0053 m w.e. $°C^{-1}d^{-1}$ [20].

This is similar to the reported relationship for nearby South Cascade Glacier during the 2003–2007 period; for snow was 0.0038 m w.e. $°C^{-1}d^{-1}$ and for ice was 0.0054 m w.e. $°C^{-1}d^{-1}$ [29]. Both ablation rates and DDF relationship in the limited elevation range of North Cascades glaciers have been found to be consistent from glacier to glacier [3,12,20,29]. The DDFs for Rainbow Glacier and Easton Glacier each with extensive mass balance records are between 0.0033 and 0.0039 m w.e. $°C^{-1}d^{-1}$. Both the annual balance and DDF relationships indicate it is reasonable to utilize Sholes Glacier ablation as a proxy for ablation on other glaciers in the watershed [3,19]. Sholes Glacier's mean elevation is also within ~20 m of the mean elevation of glaciers in the watershed in 2015.

For heat waves we have derived a separate DDF relationship based only on the 48 days when the average temperature at the Middle Fork Nooksack SNOTEL station exceeded 13 °C and we were measuring ablation. During heat waves, the DDF relationship changes yielding higher values with DDFs snow of 0.0043 m w.e. $°C^{-1}d^{-1}$. For ice, the DDFi is 0.0067 m w.e. $°C^{-1}d^{-1}$.This underscores the observation on other glaciers that incorporating weather types into a degree day model improves performance [30]

Figure 3. Limited retained snow cover on 8 August 2015 on Sholes Glacier above, with the discharge and weather station in foreground. Below is the stake network on Sholes Glacier annotated on this 2014 image.

The daily air temperature is scaled by the DDF to provide a daily value for ablation that can in turn be multiplied by the area of glacier ice and glacier snow to calculate the volume of runoff from Sholes Glacier and from all NFN glaciers. The model is further validated by comparison with periods of detailed ablation field observations (Table 2), yielding a mean daily ablation rate within 10% of observed ablation rates. Overall glacier runoff is the sum of the product of DDFs and snow-covered area, and DDFi and ice-covered area (Table 3). For each of the periods in Table 2, field work was completed during or within three days of the heat event allowing mapping of the snow and ice area. The area of snow and ice is a significant variable from year to year and through the melt season (Figure 3).

Table 2. Ablation rates determined from field measurements and degree day modeling on Sholes Glacier during portions of the 2012–2020 melt seasons.

Dates	Snow Ablation Rate-Measure (m w.e.d^{-1})	Snow Ablation Rate-Model (m w.e.d^{-1})	Ice Ablation Rate-Model (m w.e.d^{-1})	Ice Ablation Rate-Measure (m w.e.d^{-1})
8−6−2013 to 8−13−2013	0.049	0.045	0.064	0.073
7−27−2014 to 8−7−2014	0.055	0.053	0.075	0.077
7−29−2015 to 8−4−2015	0.057	0.053	0.075	0.078
7−25−2016 to 7−30−2016	0.053	0.054	None	None
8−12−2016 to 8−21−2016	0.048	0.050	0.070	0.067
7−31−2017 to 8−12−2017	0.056	0.060	0.084	0.078
8−5−2018 to 8−10−2018	0.061	0.051	0.072	None
8−4−2019 to 8−9−2019	0.051	0.051	0.072	0.073
7−29−2020 to 8−5−2020	0.055	0.053	0.075	None

Table 3. Impact of 24 heat events on North Fork Nooksack (NFN) and South Fork Nooksack discharge and temperature, glacier ablation, glacier runoff, and overall glacier contribution to flow of NFN.

Start Date	End Date	NFK Discharge (%)	SFK Discharge (%)	NFK Temp (°C)	SFK Temp (°C)	NFK Glacier Ablation (md^{-1})	Glacier Discharge (m^{3}s^{-1})	NFK Discharge (m^{3}s^{-1})	Glacier Runoff (%)
7/20/09	8/5/09	50%	−34%	1	4.7	0.058	8.46	28.05	30%
7/25/10	8/1/10	7%	−25%	0.3	1.0	0.048	7.00	41.7	17%
8/14/10	8/19/10	19%	−14%	0.7	1.8	0.055	8.02	29.2	27%
9/5/11	9/14/11	30%	−8%	0.4	1.2	0.055	8.02	24.2	33%
7/7/12	7/14/12	40%	−29%	0.2	1.5	0.057	8.31	88.6	9%
8/11/12	8/19/12	18%	−16%	0.3	1.5	0.063	9.18	35.3	26%
8/6/13	8/13/13	15%	−14%	0.6	0.7	0.045	6.56	25.5	26%
7/7/14	7/18/14	13%	−29%	2.1	4.7	0.054	7.87	45.9	17%
7/27/14	8/7/14	5%	−44%	0.8	2.7	0.053	7.72	25.3	31%
6/25/15	7/21/15	53%	−30%	0.7	3.9	0.051	7.26	18.8	39%
7/29/15	8/4/15	16%	−16%	0.6	1.4	0.053	7.54	13	58%
7/25/16	7/30/16	10%	−11%	0.8	1.6	0.054	7.69	24.2	32%
8/12/16	8/21/16	18%	−20%	0.7	1.7	0.05	7.12	17.5	41%
7/31/17	8/12/17	13%	−22%	0.5		0.06	8.26	21.8	38%
8/26/17	9/8/17	20%	−11%	1	0	0.055	7.57	16.4	46%
7/12/18	7/18/18	4%	−16%	0.6	1.4	0.053	7.30	29.3	25%
7/22/18	8/2/18	19%	−18%	1	4	0.057	7.85	25.6	31%
8/5/18	8/10/18	11%	−9%	0.6	2.1	0.059	8.12	22.4	36%
8/14/18	8/23/18	2%	−19%	0.5	1	0.052	7.16	19.3	37%
8/4/19	8/9/19	10%	−20%	0.6	1.2	0.051	7.02	17.3	41%
7/26/20	8/3/20	47%	−21%	0.6	2.6	0.053	7.30	26.6	27%
8/15/20	8/20/20	69%	−12%	1.5	4.2	0.059	8.13	17.2	47%
6/25/21	7/1/21	27%	−21%	0.8	2	0.072	9.92	73.1	14%
7/26/21	8/6/21	22%	−19%	0.3	1.2	0.055	7.58	26.3	34%
Avg.		24%	−20%	0.7	2.0	0.055	7.66	27.08	32%

4.3. Nooksack River Discharge and Stream Temperature

The 24 heat events are noted in Table 3. The change in discharge is reported as the percentage change in discharge from the start of the heat event to the maximum or minimum discharge during the event at the USGS gages in both NFN and SFN. For the NFN, discharge increased by more than 10% during 20 of the 24 periods, with a mean increase of 23%. In the SFN, discharge decreased by more than 10% during 22 of the 24 periods (Figure 4).

The stream temperature change is the difference between the daily stream temperature at the beginning of the period to the maximum daily temperature during the heat period at the USGS gage in both the NFN and SFN. Stream temperature rose by more than 1 °C during 5 of the 24 events in the NFN and during 21 of 23 events on SFN, temperature data were missing for one event for SFN (Figure 5). The mean stream temperatures change was 0.7 °C in NFN and 2.0 °C in SFN, quantifying the ameliorating impact of glaciers on stream temperature in NFN.

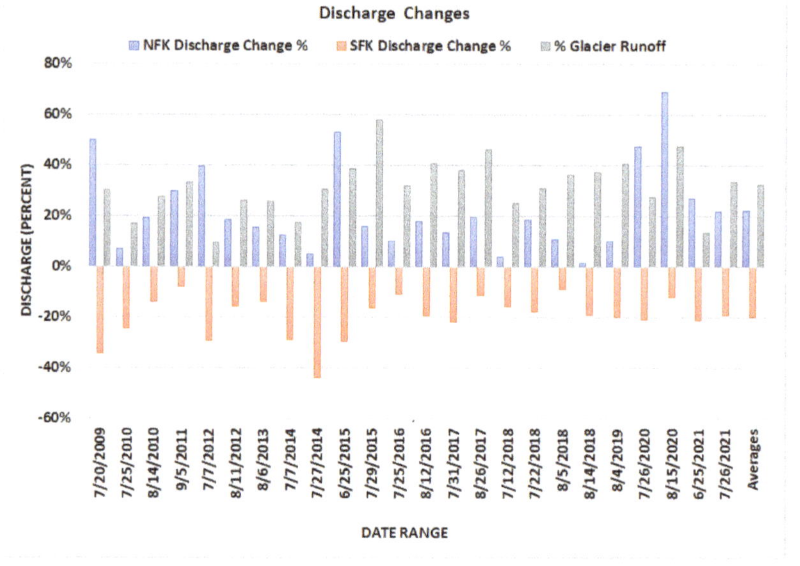

Figure 4. Change in discharge in the North Fork Nooksack (NFK) and South Fork Nooksack (SFK) during the 23 heat events. The percent of North Fork Nooksack discharge generated by glacier runoff is also indicated.

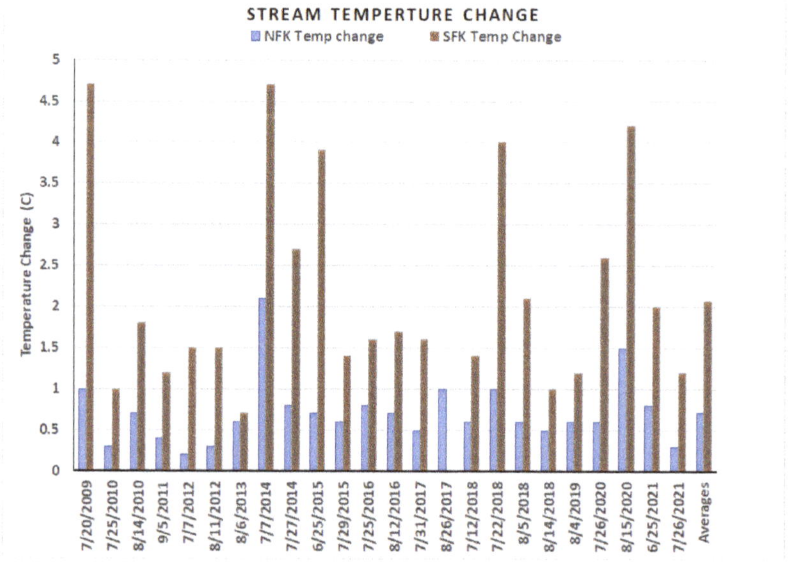

Figure 5. Change in daily stream temperature in the North Fork Nooksack (NFK) and South Fork Nooksack (SFK) during the 23 heat events from the beginning of the period to the maximum observed daily temperature.

The product of the observed or modelled daily glacier ablation and glacier area yields daily glacier discharge generated. Ablation measurements and modelling during these events indicate a mean daily ablation ranging from 0.045 md^{-1} w.e. to 0.063 md^{-1} w.e., with a mean rate of 0.055 md^{-1} w.e. The daily glacier discharge generated is then reported as a percent of the total observed NFN discharge. The daily melt is not all conveyed downstream to the gage the same day, hence the comparison to daily discharge is to glacier runoff generated. This analysis indicates that glacier runoff during these heat events generated discharge equivalent to more than 15% of total river flow during 22 of 23 events, with a mean of 34% (Figure 4; Table 3). Two of the events generated over 50% of the river discharge, both occurring during periods of particularly low flow.

4.4. ERA5 Temperature Data

The summer daily maximum temperature and daily lapse rate generated from ERA5 reanalysis for the 1979–2021 period yield a mean summer maximum temperature during 1979–2021 of 1.93 °C and a mean lapse rate of −5.73 °C km^{-1}. Maximum temperatures of greater than 10 °C approximates the 14 °C mean daily temperature threshold at the Middle Fork SNOTEL station for heat waves identification. From 1979–2021 there have been 378 days of the 6500 total days exceeding this threshold with an average lapse rate of −6.3 °C km^{-1}. During this period, 13 of the 20 highest maximum daily temperatures have been reported since 2015, with six of them (30%) occurring in 2021 (Table 4). This included a 141 h period from 25 June to 2 July 2021 where the temperature remained above 12 °C at 3000 m.

Table 4. The twenty highest maximum summer days with the highest maximum temperature at 3000 m from the ERA5 reconstruction for Mount Baker, Washington for May 1979 to July 2021. The lapse rate (°C km^{-1}) is also reported.

Date	Lapse Rate (°C km^{-1})	Maximum Temperature (°C)
6/30/2021	−5.99	18.88
6/29/2021	−7.89	18.66
6/28/2021	−7.47	17.55
7/1/2021	−4.12	15.86
9/5/1988	−6.71	15.53
7/13/2002	−6.54	15.33
6/27/2021	−6.53	15.33
9/4/1988	−7.87	14.95
7/22/2006	−7.10	14.86
9/3/1988	−7.70	14.49
9/7/2017	−6.14	14.39
6/28/2015	−7.67	14.36
9/5/2017	−7.53	14.33
6/26/2021	−5.87	14.29
9/6/2017	−7.13	14.12
8/10/2018	−6.66	14.04
9/23/2009	−6.54	13.81
6/27/2015	−6.50	13.70
5/29/1983	−8.29	13.64
7/31/2020	−7.24	13.36

5. Discussion

In a glaciated watershed, glaciers are important to maintaining sufficient discharge and stream temperature that are critical for salmon populations. This is illustrated in the NFN where the 24 heat events have led to an increased discharge and a stream temperature rise of less than 1 °C. The increased discharge in NFN during heat waves while SFN discharge decreased demonstrates the impact of glaciers on the NFN reversing the discharge trend during heat events. This is the result of increased glacier ablation during the heat waves. The continued loss of glacier area will lead to a decline in this mitigating effect of glaciers on NFN stream conditions. How will this impact fish species?

Some cold-water trout and salmon species are already constrained by warm water temperatures and additional warming will result in net habitat loss [4,14,26]. In the Fraser River and Thompson River, British Columbia fish community thresholds were observed for mean weekly average temperatures of about 12 °C and again above 19 °C [26]. Below 12 °C, the community were characterized by bull trout and some cold-water species, between 12 °C and 19 °C by salmonids and sculpins, and above 19 °C by minnows and some cold-water salmonids [31]. The temperature threshold above which mortality increases markedly for Pacific salmon in the region is 15 °C [32,33]. These thresholds indicated that small temperature changes can be expected to drive substantial changes in fish communities. During the 24 heat events noted in the North Fork only two events exceeded 12 °C, while in the South Fork 14 of the events exceeded 19 °C, which is well above the threshold where mortality increases [32,33]. This suggests that both rivers are near a threshold that could alter the fish community composition.

In Pacific Northwest rivers, air temperature drives 82–94% of the long-term stream temperature trends [4,34]. Summer discharge and air temperature both account for approximately half of the inter-annual variation in stream temperatures [34,35]. In spring, no temperature increase was observed and the rate of warming was highest during the summer at 0.17–0.22 °C increase in temperature per decade [4].

Nooksack River salmon begin and end their life cycle in the Nooksack River. The Washington Department of Fish and Wildlife [36] SalmonScape project maps the distribution of salmon in the Nooksack River basin. Each population is mapped separately for spawning, rearing, and presence. Chinook, Coho, and Chum salmon in the North Fork can migrate up to the base of Nooksack Falls 40 km upstream of the NFN-Nooksack River Junction. The SFN has the most extensive network of salmon streams with the presence of salmon extending 52 km upstream of the junction of SFN with the Nooksack River (Figure 6).

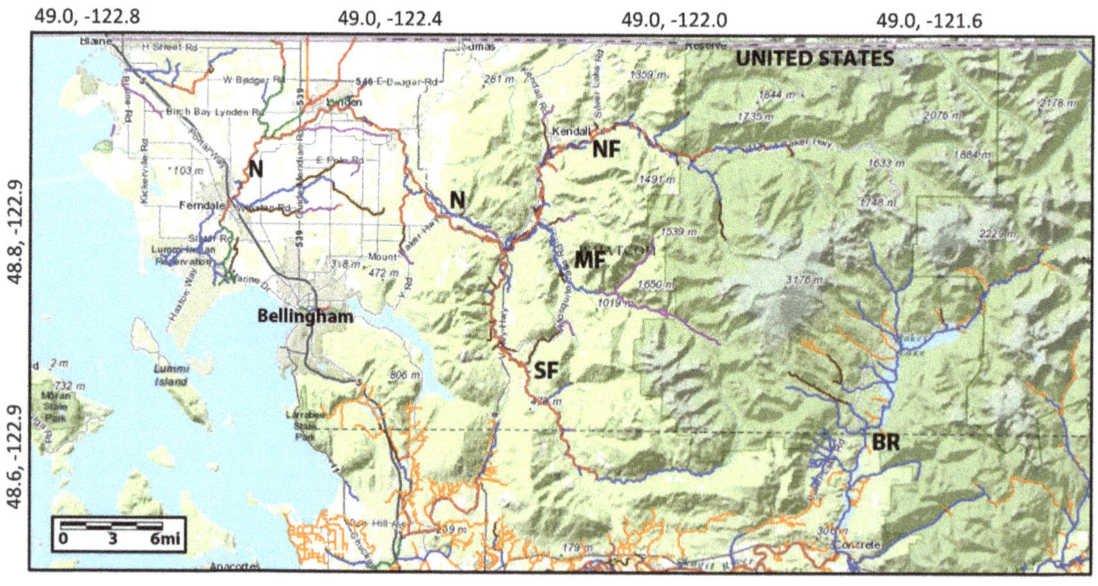

Figure 6. From the WDFW SalmonScape, this map indicates the extent of Chinook salmon in the Nooksack (N) and Baker River (BR) watersheds. Including the Nooksack sub-basins; MF = Middle Fork, NF = North Fork, SF = South Fork. Red = documented spawning, Blue = Documented presence, Green = Documented rearing, Yellow = Modelled presence, Purple = Blocked.

Chinook salmon surveys in the Nooksack River are conducted annually by the Washington Department of Fish and Wildlife. In the NFN Chinook spawn mainly in a 30 km stretch from Mosquito Lake Road to Wells Creek at the base of Nooksack Falls [36]. In the NFN, the number of returning Chinook is divided into natural and hatchery spawned salmon; the WDFW [37] report that 88% of recent spawning Chinook salmon originate from the Kendall Creek Hatchery at the junction of Kendall Creek and the NFN. From 2000 to 2011 the number of Chinook released in the Middle Fork and NFN watershed averaged 1,036,000 sub-yearling fish [37]. Though overall populations and escapements increased as a result, natural-origin spawning Chinooks have not increased from 1995 to 2016 and remained threatened in the NFN River [37].

For salmon, both their riverine and marine environments are experiencing physical changes due to climate change, compounding human alteration of the aquatic habitat. This is a consistent stress throughout their life cycle [14,32,33]. In late summer of 2021, the SFN experienced a ~2500 Chinook die off from warm water lowering resistance to columnaris disease [38]. There was no die off in the NFN potentially indicative of the ameliorating impact of glaciers on stressful stream conditions.

6. Conclusions

The increasing frequency and intensity of Pacific Northwest heat waves underscores the need to quantify the impact on all alpine watersheds; in this case the Nooksack Basin a glaciated alpine watershed. Alpine glaciers in the NFN drive an increase in discharge during heat events averaging 24%, while limiting water temperature rise to a mean of 0.7 C. This contrasts with the unglaciated SFN where during the same heat events discharge declined 20% and temperatures increased 2.1 C. During the heat events increased ablation drove an increase in glacier runoff and the importance of glacier runoff to overall river discharge. Heat events are of importance, because the low discharge and high temperatures that characterize heat events are stressful for salmon populations.

Mass balance losses in the basin are driving glacier area decline [3], that has already led to a declining glacier runoff [3,8,19]. The result of continued glacier area loss will be a reduction in the enhanced discharge, leading to reduced flow during warm-dry low flow events. Ongoing loss of glacier area will also lead to a greater increase in overall stream temperature in NFN. The summer of 2021 brought the highest observed air temperatures to the region further highlighting the importance of this issue [39]. The ERA5 maximum temperatures identified that the three hottest summer days of the 1979–2021 period were 28, 29 June, and 30 June 2021.

This study is the first detailed quantification of glacier ablation, glacier runoff, and consequent alpine river discharge during heat waves in this region. The study highlights the importance of completing additional ablation measurements of bare ice surfaces and consistent repeat mapping of the distribution of snow-covered area on these glaciers using remote sensing products to effectively apply melt models.

Author Contributions: Data from field observations of glacier mass balance and ablation observation completed by M.S.P., J.P. and M.D. All weather station data, USGS discharge and steam temperature data were analyzed by M.S.P. ERA5 temperature reconstruction and lapse rate calculation was performed by T.M. and L.B.P.; writing—original draft preparation, M.S.P.; writing—review and editing by T.M. and M.D. All authors have read and agreed to the published version of the manuscript.

Funding: This research received no external funding.

Data Availability Statement: All USGS data used is available for North Fork Nooksack River is from: https://waterdata.usgs.gov/nwis/uv?site_no=12205000. South Fork Nooksack River data is from https://waterdata.usgs.gov/nwis/uv?site_no=12210000. Temperature data for the USDA Snotel site Middle Fork Nooksack is at: https://www.nwrfc.noaa.gov/snow/snowplot.cgi?MNOW1. Mass balance data is reported to the World Glacier Monitoring Service at: https://wgms.ch/data_databaseversions/ (all accessed on 28 January 2022).

Acknowledgments: This work has benefitted from the support of the Nooksack Indian Tribe. This project has been sustained by over 60 field assistants including several who spent at least five field seasons with the project Tom Hammond, Ben Pelto, and Bill Prater. The leadership in glacier monitoring and data reporting of the World Glacier Monitoring Service has been integral to the ongoing work of the North Cascade Glacier Climate Project field program. The long-term discharge measurements by the USGS and weather data from the USDA SNOTEL system have been indispensable.

Conflicts of Interest: The authors declare no conflict of interest.

References

1. Abatzoglou, J.; Rupp, D.; Mote, P. Seasonal climate variability and change in the Pacific Northwest of the United States. *J. Clim.* **2014**, *27*, 2125–2142. [CrossRef]
2. Pelto, M.S. Impact of climate change on North Cascade alpine glaciers, and alpine runoff. *Northwest Sci.* **2008**, *82*, 65–75. [CrossRef]
3. Pelto, M.S. How Unusual Was 2015 in the 1984–2015 Period of the North Cascade Glacier Annual Mass Balance? *Water* **2018**, *10*, 543. [CrossRef]
4. Isaak, D.J.; Wollrab, S.; Horan, D.; Chandler, G. Climate change effects on stream and river temperatures across the northwest U.S. from 1980–2009 and implications for salmonid fishes. *Clim. Change* **2012**, *113*, 499–524. [CrossRef]
5. Fountain, A.; Tangborn, W.V. The effect of glaciers on streamflow variations. *Water Resour. Res.* **1985**, *21*, 579–586. [CrossRef]
6. Dery, S.; Stahl, K.; Moore RWhitfield, W.; Menounos, B.; Burford, J.E. Detection of runoff timing changes in pluvial, nival, and glacial rivers of western Canada. *Water Resour. Res.* **2009**, *45*, W04426. [CrossRef]
7. Stahl, K.; Moore, D. Influence of watershed glacier coverage on summer streamflow in British Columbia, Canada. *Water Resour. Res.* **2006**, *42*, W06201. [CrossRef]
8. Moore, R.; Pelto, B.; Menounos, B.; Hutchinson, D. Detecting the Effects of Sustained Glacier Wastage on Streamflow in variably Glacierized Catchments. *Front. Earth Sci.* **2020**, *8*, 136. [CrossRef]
9. Nolin, A.W.; Phillippe, J.; Jefferson, A.; Lewis, S.L. Present-day and future contributions of glacier runoff to summertime flows in a Pacific Northwest watershed: Implications for water resources. *Water Resour. Res.* **2010**, *46*, W12509. [CrossRef]
10. Hidalgo, H.G.; Das, T.; Dettinger, M.D.; Cayan, D.R.; Pierce, D.W.; Barnett, T.P.; Bala, G.; Mirin, A.; Wood, A.W.; Bonfils, C.; et al. Detection and attribution of streamflow timing changes to climate change in the western United States. *J. Clim.* **2009**, *22*, 3838–3855. [CrossRef]
11. Granshaw, F.; Fountain, A. Glacier change (1958–1998) in the North Cascades National Park Complex, Washington, USA. *J. Glaciol.* **2006**, *52*, 251–256. [CrossRef]
12. Pelto, M.S. Skykomish River, Washington: Impact of ongoing glacier retreat on streamflow. *Hydrol. Process.* **2011**, *25*, 3356–3363. [CrossRef]
13. Luce, C.; Staab, B.; Kramer, M.; Wenger, S.; Isaak, D.; McConnell, C. Sensitivity of summer stream temperatures to climate variability in the Pacific Northwest. *Water Resour. Res.* **2014**, *50*, 3428–3443. [CrossRef]
14. Grah, O.; Beaulieu, J. The effect of climate change on glacier ablation and baseflow support in the Nooksack River basin and implications on Pacific salmonid species protection and recovery. In *Climate Change and Indigenous Peoples in the United States*; Springer: Cham, Switzerland, 2013. [CrossRef]
15. Huss, M.; Bookhagen, B.; Huggel, C.; Jacobsen, D.; Bradley, R.S.; Clague, J.J.; Vuille, M.; Buytaert, W.; Cayan, D.R.; Greenwood, G.; et al. Toward mountains without permanent snow and ice. *Earth's Future* **2017**, *5*, 418–435. [CrossRef]
16. Bach, A.J. Snowshed contributions to the Nooksack River watershed, North Cascades range, Washington. *Geogr. Rev.* **2002**, *92*, 192–212. [CrossRef]
17. Harper, J.T. Glacier Terminus Fluctuations on Mount Baker, Washington, USA, 1940–1990, and Climatic Variations. *Arct. Alp. Res.* **1993**, *25*, 332–340. [CrossRef]
18. Pelto, M.S.; Hedlund, C. The terminus behavior and response time of North Cascade glaciers, Washington, USA. *J. Glaciol.* **2001**, *47*, 497–506. [CrossRef]
19. Pelto, M.S. Climate Driven Retreat of Mount Baker Glaciers and Changing Water Resources. In *Briefs in Climate Studies*; Springer: Cham, Switzerland, 2015; ISBN 978-3-319-22605-7. [CrossRef]
20. Pelto, M.S.; Brown, C. Mass balance loss of Mount Baker, Washington glaciers 1990–2010. *Hydrol. Process.* **2012**, *26*, 2601–2607. [CrossRef]
21. Cogley, J. Effective sample size for glacier mass balance. *Geogr. Ann. Ser. A Phys. Geogr.* **1999**, *81*, 497–507. [CrossRef]
22. Pelto, M. The impact of sampling density on glacier mass balance determination. *Hydrol. Process.* **2000**, *14*, 3215–3225. [CrossRef]
23. Huss, M.; Dhulst, L.; Bauder, A. New long-term mass-balance series for the Swiss Alps. *J. Glaciol.* **2015**, *61*, 551–562. [CrossRef]
24. Thibert, E.; Blanc, R.; Vincent, C.; Eckert, N. Glaciological and volumetric mass-balance measurements: Error analysis over 51 years for Glacier de Sarennes glacier, French Alps. *J. Glaciol.* **2008**, *54*, 522–532. [CrossRef]
25. Hock, R. Glacier melt: A review of processes and their modelling. *Prog. Phys. Geogr.* **2005**, *29*, 362–391. [CrossRef]
26. Hersbach, H.; Bell, B.; Berrisford, P.; Hirahara, S.; Horányi, A.; Muñoz-Sabater, J.; Nicolas, J.; Peubey, C.; Radu, R.; Schepers, D.; et al. The ERA5 global reanalysis. *Q. J. R. Meteorol. Soc.* **2020**, *146*, 1999–2049. [CrossRef]
27. Matthews, T.; Perry, L.B.; Koch, I.; Aryal, D.; Khadka, A.; Shrestha, D.; Abernathy, K.; Elmore, A.C.; Seimon, A.; Tait, A.; et al. Going to Extremes: Installing the World's Highest Weather Stations on Mount Everest. *Bull. Am. Meteorol. Soc.* **2020**, *101*, E1870–E1890.
28. Dutra, E.; Muñoz Sabater, J.; Boussetta, S.; Komori, T.; Hirahara, S.; Balsamo, G. Environmental lapse rate for high-resolution land surface downscaling: An application to ERA5. *Earth Space Sci.* **2020**, *7*, e2019EA000984. [CrossRef]

29. Bidlake, W.R.; Josberger, E.G.; Savoca, M.E. *Modelled and Measured Glacier Change and Related Glaciological, Hydrological, and Meteorological Conditions at South Cascade Glacier, Washington, Balance and Water Years 2006 and 2007*; Science Investigations Report; U.S. Geological Survey: Reston, VA, USA, 2010.
30. Matthews, T.; Hodgkins, R.; Wilby, R.L.; Guðmundsson, S.; Pálsson, F.; Björnsson, H.; Carr, S. Conditioning temperature-index model parameters on synoptic weather types for glacier melt simulations. *Hydrol. Process.* **2015**, *29*, 1027–1045. [CrossRef]
31. Parkinson, E.A.; Lea, E.V.; Nelitz, M.A.; Knudson, J.M.; Moore, R.D. Identifying Temperature Thresholds Associated with Fish Community Changes in British Columbia, Canada, to Support Identification of Temperature Sensitive Streams. *River Res. Appl.* **2016**, *32*, 330–347. [CrossRef]
32. Atlas, W.; Seitz, K.M.; Jorgenson, J.; Millard-Martin, B.; Housty, W.; Ramos-Espinoza, D.; Burnett, N.; Reid, M.; Moore, J. Thermal sensitivity and flow-mediated migratory delays drive climate risk for coastal sockeye salmon. *Facets* **2021**, *6*, 71–89. [CrossRef]
33. Crozier, L.; McClure, M.; Beechie, T.; Bograd, S.; Boughton, D.; Carr, M.; Cooney, T.D.; Dunham, J.B.; Greene, C.M.; Haltuch, M.A.; et al. Climate vulnerability assessment for Pacific salmon and steelhead in the California Current Large Marine Ecosystem. *PLoS ONE* **2019**, *14*, e0217711. [CrossRef]
34. Tennant, D.L. Instream Flow Regimens for Fish, Wildlife, Recreation, and Related Environmental Resources. In *Instream Flow Needs*; Osborne, J., Allman, C., Eds.; American Fisheries Society, Western Division: Bethesda, MD, USA, 1976; Volume 2, pp. 359–373.
35. Rieman, B.; Isaak, D.; Adams, S.; Horan, D.; Nagel, D.; Luce, C.; Myers, D. Anticipated climate warming effects on bull trout habitats and populations across the interior Columbia River basin. *Trans. Am. Fish. Soc.* **2007**, *136*, 1552–1565. [CrossRef]
36. Washington Department of Fish and Wildlife. Salmon Scape. Available online: http://apps.wdfw.wa.gov/salmonscape/map.html (accessed on 10 February 2018).
37. Washington Department of Fish and Wildlife. Salmon Score. Available online: https://fortress.wa.gov/dfw/score/ (accessed on 10 February 2018).
38. Ryan, J. Heat-loving bacteria kills thousands of Washington salmon. KUOW, 10/15/2021. Available online: https://www.kuow.org/stories/heat-loving-bacteria-kills-thousands-of-washington-salmon (accessed on 20 November 2021).
39. Philip, S.Y.; Kew, S.F.; van Oldenborgh, G.J.; Anslow, F.S.; Seneviratne, S.I.; Vautard, R.; Coumou, D.; Ebi, K.L.; Arrighi, J.; Singh, R.; et al. Rapid attribution analysis of the extraordinary heatwave on the Pacific Coast of the US and Canada June 2021. *Earth Syst. Dynam. Discuss.* **2021**, preprint. [CrossRef]

water

MDPI

Article

Uncertainties in Riverine and Coastal Flood Impacts under Climate Change

Shuyi Wang [1], Mohammad Reza Najafi [1,*], Alex J. Cannon [2] and Amir Ali Khan [3]

[1] Department of Civil and Environmental Engineering, Western University, London, ON N6A 5B9, Canada; swang737@uwo.ca
[2] Climate Research Division, Environment and Climate Change Canada, Victoria, BC V8N 1V8, Canada; alex.cannon@canada.ca
[3] Water Rights, Investigations and Modelling Section, Department of Environment, Climate Change and Municipalities, St. John's, NL A1B 4J6, Canada; akhan@gov.nl.ca
* Correspondence: mnajafi7@uwo.ca

Abstract: Climate change can affect different drivers of flooding in low-lying coastal areas of the world, challenging the design and planning of communities and infrastructure. The concurrent occurrence of multiple flood drivers such as high river flows and extreme sea levels can aggravate such impacts and result in catastrophic damages. In this study, the individual and compound effects of riverine and coastal flooding are investigated at Stephenville Crossing located in the coastal-estuarine region of Newfoundland and Labrador (NL), Canada. The impacts of climate change on flood extents and depths and the uncertainties associated with temporal patterns of storms, intensity–duration–frequency (IDF) projections, spatial resolution, and emission scenarios are assessed. A hydrologic model and a 2D hydraulic model are set up and calibrated to simulate the flood inundation for the historical (1976–2005) as well as the near future (2041–2070) and far future (2071–2100) periods under Representative Concentration Pathways (RCPs) 4.5 and 8.5. Future storm events are generated based on projected IDF curves from convection-permitting Weather Research and Forecasting (WRF) climate model simulations, using SCS, Huff, and alternating block design storm methods. The results are compared with simulations based on projected IDF curves derived from statistically downscaled Global Climate Models (GCMs). Both drivers of flooding are projected to intensify in the future, resulting in higher risks of flooding in the study area. Compound riverine and coastal flooding results in more severe inundation, affecting the communities on the coastline and the estuary area. Results show that the uncertainties associated with storm hyetographs are considerable, which indicate the importance of accurate representation of storm patterns. Further, simulations based on projected WRF-IDF curves show higher risks of flooding compared to the ones associated with GCM-IDFs.

Keywords: climate change; uncertainty; riverine flooding; coastal flooding; compound flooding; projected IDF curves; design storm; Stephenville Crossing

Citation: Wang, S.; Najafi, M.R.; Cannon, A.J.; Khan, A.A. Uncertainties in Riverine and Coastal Flood Impacts under Climate Change. *Water* **2021**, *13*, 1774. https://doi.org/10.3390/w13131774

Academic Editor: Laurens M. Bouwer

Received: 15 May 2021
Accepted: 24 June 2021
Published: 27 June 2021

1. Introduction

Flooding accounts for 43% of natural disasters, affecting around 2.3 billion people worldwide between 1995 and 2015 [1]. With $673 million estimated annual costs, floods comprise the highest proportion (75%) of extreme weather-related expenses in Canada [2]. In Newfoundland and Labrador (NL), over 600 flood events were recorded during 1950–2011. These events were associated with intense rainfall (72%), coastal flooding (17%), ice jams, and snowmelt (7%), as well as other factors [3]. NL's estuaries and coastal lands are commonly considered as flood-prone areas affected by both inland (riverine and pluvial flooding) and coastal extreme water levels associated with high tides, storm surges, and waves.

It is widely recognized that climate change can affect the drivers of flood events, including heavy rainfall [4–7], river discharge [8–10], and coastal water levels [11–14]. These flood risks are expected to increase across Canada because of more intense rainfall events, warmer temperatures that can cause sudden snowmelt, and sea-level rise, among others [15]. Nonetheless, there are considerable uncertainties associated with GCMs, future emission scenarios, downscaling, and hydrologic/hydraulic modeling [16–20] that can challenge the design and planning of communities and infrastructure systems in a changing climate.

Traditionally, research on flood impacts has been mostly focused on individual flood types, including pluvial [21–23], riverine [24,25], and coastal flooding [14,26,27]. However, compound flooding, caused by multiple flood drivers such as the concurrent occurrence of river overflows and extreme coastal water levels, can lead to more severe impacts, especially in densely populated low-elevation coastal zones [28,29]. In recent years, several studies have been conducted to assess compound flood effects at regional [30–34], continental [35], and global scales [36]. Statistical and process-based models are developed and applied to assess the characteristics and impacts of such events, including the simultaneous occurrence of river overflows and extreme water levels globally [37], and in different regions around the world including Canada [38], Australia [39], the U.S. [40–43], and Asia [44], among others. Kumbier et al. (2018) [39] investigated compound flood effects on an Australian estuarine environment by considering the storm surge and extreme river discharge using the Delft3D hydrodynamic model. Pasquier et al. (2019) [45] integrated the 1D-2D HEC-RAS model to assess the sensitivity of different drivers of flooding in the UK coastal regions, and found that storm surge is likely the main driver.

A few studies have investigated the impacts of climate change on compound coastal and fluvial flooding [46–49]. However, there are major gaps in understanding the combined effects of sea-level rise and future changes in the pattern and intensity of precipitation associated with climate change. Further, comprehensive assessments of the contribution of GCMs, design storm methods, hydrodynamic models, and projected intensity–duration–frequency (IDF) curves to the overall uncertainties are lacking. Besides, in an engineering context, projected IDF curves provide essential information on short-duration rainfall events. Statistically downscaled GCMs, however, may not accurately represent such events [50]. To address these research gaps, we assess the individual and compound effects of riverine and coastal flooding by setting up and calibrating a hydrologic and a two-dimensional hydrodynamic model for Stephenville Crossing located on the west coast of Newfoundland. The uncertainties associated with GCMs, design storm methods, projected IDF curves, and hydrodynamic modeling (i.e., terrain data, model structure, and roughness coefficient) are analyzed. Further, we investigate the flood characteristics based on projected IDF curves generated using downscaled GCMs as well as high-resolution, convection-permitting, WRF simulations.

In the remainder of this paper, the study area and data are presented in Section 2. Section 3 discusses the hydrologic and hydraulic model setup, calibration, and validation. The sensitivity analysis, design storm methods, and projected IDF curves are further discussed in this section. The results are provided in Section 4, followed by conclusions in Section 5.

2. Study Area and Data

Stephenville Crossing is located on the western coast of Newfoundland Island at 48°31′ N latitude and 58°27′ W longitude (Figure 1). It has a total area of 80.8 km^2, with most of the population (~1700 people) living close to the coastline and along Harry's river [51]. There are many properties and commercial premises along the coastline and the mouth of the river. The average monthly temperature varies between around −7 °C and 16 °C, and the annual average relative humidity is 81%. The lowest and highest temperatures occur in February (−10 °C) and August (20 °C), respectively. The annual precipitation rate is 1340 mm based on the 1981 to 2020 normal climate at Stephenville

Airport's station. Precipitation in March, April, and May is lower compared to the other months. The region has experienced a slight increasing trend in both temperature and precipitation from the period of 1961–1990 to 1981–2010. During the winter, winds are stronger than in other seasons, and the maximum wind gust can reach approximately 140 km/h. Stephenville has been frequently affected by riverine and coastal flooding based on historical flood records [52]. The frequency of 50-year water level events is projected to double in Newfoundland due to around 10 cm of sea-level rise [53].

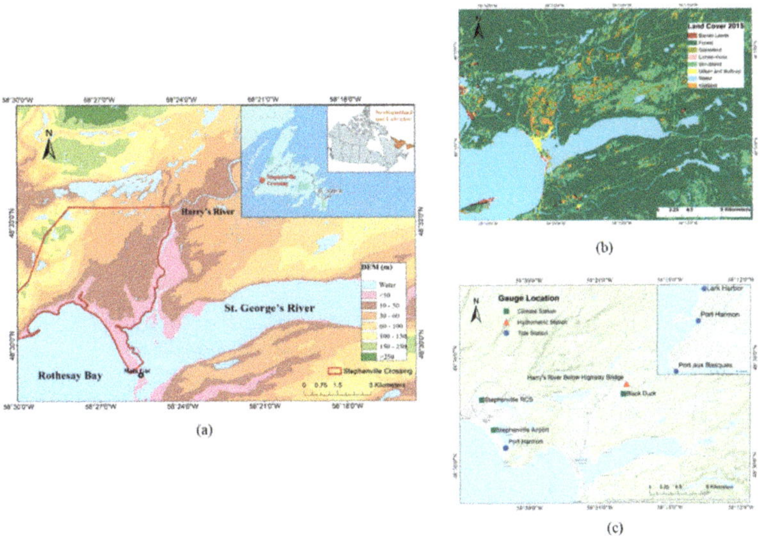

Figure 1. (**a**) Study area including the Town of Stephenville Crossing, (**b**) land cover map, and (**c**) location of available climate (green), hydrometric (red), and tide (blue) gauges.

Harry's river discharges into St. George's River from the north, which then flows westward into Rothesay Bay through a narrow channel called Main Gut (Figure 1). The drainage area corresponding to the upstream gauge (Harry's River below Highway Bridge) is 640 km^2 [52]. There are two bridges across the Main Gut to link the town of Stephenville Crossing with other communities. The study area is mainly covered by forest, followed by shrubland and wetland across the river system. The land cover map is classified into eight types (Figure 1) and the corresponding roughness values are provided in Table S1 [54]. Only a small part of the domain between the bay and the estuary of Harry's River is developed.

Flooding has repeatedly impacted this area in the past. In late December 1951, coastal flooding affected the area, resulting in the displacement of ~600 people. The severe storm caused high-speed winds of 177 km per hour that swept through the railway station and destroyed 15 surrounding electrical poles. In addition to seawater overtopping the coastal area of Stephenville Crossing, heavy rainfall resulted in Harry's River overflow, which inundated the streets. Many fishermen lost their boats and tools, and some stores and house interiors were damaged. In December 1977, another coastal flood event forced families to evacuate and caused house damages. High winds and tides caused flooding and washed out the roads and streets [3]. Other examples include flood events in March 2003 and November 2014 generated by high river flows and gusty winds of up to 110 km per hour respectively, which resulted in bridge damage, inundated pavements, highway closure, and basement flooding [3].

2.1. Data

We used three Digital Elevation Models (DEMs), including Shuttle Radar Topography Mission (SRTM), Canadian Digital Elevation Model (CDEM), and TanDEM-X. CDEM is a pan-Canadian product provided by Natural Resources Canada. In areas south of 68° N latitude, the spatial resolution is 0.75 arc-seconds (~20 m). The measured altimetric accuracy of CDEM in the study area is within a range of 5–10 m. SRTM, produced by the National Aeronautics and Space Administration (NASA), provides global elevation data at three arc-seconds (~90 m) and one arc-second (30 m) resolution. We use the 30 m SRTM data covering Stephenville Crossing, which has an absolute vertical accuracy of below 16 m and absolute horizontal accuracy of less than 20 m. The German Aerospace Center's TanDEM-X is a synthetic aperture radar mission that provides global elevation data at three arc-seconds spatial resolution. The absolute horizontal and vertical accuracies are below 10 m within a 90% confidence interval. We interpolated 46 detailed cross-sections, surveyed along Harry's River in 2010, to generate the river bathymetry. The bathymetry was then fused into the DEMs for hydraulic simulations.

Sub-daily ground-based precipitation records at Stephenville Airport, available for 1953–present, and streamflow data from the hydrometric station located at Harry's River Below Highway Bridge, available for 1968–present, were used for hydrologic and hydraulic model simulations. The climate station data at Stephenville Airport were also used by ECCC to generate the historical IDF curves. There are no gauges within the simulation area for calibration, except the one used as the upstream boundary. Therefore, we used water level measurements along the river channel available for 25 September 2010 and 3 November 2010 [52] to calibrate and validate the hydraulic model.

Tides are the cyclic rise and fall of seawater caused by the gravitational attraction between the moon, the sun, and the Earth. Oceans are expected to experience two high tides and low tides every tidal period, moving westwards. However, continents block the water movement, causing different tidal patterns at each location. Two major tide patterns are observed in the Canadian shoreline: semidiurnal tides along the eastern coastline and mixed-semidiurnal tides along the western coastline. A semi-diurnal tidal cycle represents two high tides and two low tides each day, while a mixed-semi-diurnal tidal cycle shows different tide sizes. Daily tide prediction data is available at the tide station, Port Harmon, which is the nearest station located between the towns of Stephenville and Stephenville Crossing.

2.1.1. Climate Projections

We considered nine GCMs that participated in the Coupled Model Intercomparison Project Phase 5 (CMIP5), following Perez et al. (2014) [55], who evaluated the performance of GCMs over the northwestern Atlantic region, including Stephenville Crossing. They applied the scatter index and relative entropy to assess the skill of GCM datasets to reproduce synoptic situations, historical seasonal variability, and the consistency of future projections. The selected GCMs include: ACCESS1.0, HadGEM2-CC, HadGEM2-ES, GFDL-CM3, MPI-ESM-LR, HadGEM-AO, CSIRO-Mk3.6.0, GFDL-ESM2G, and CanESM-2 (Table S2). Statistically downscaled daily minimum and maximum temperatures from 1950 to 2100 under Representative Concentration Pathways (RCPs) 4.5 and 8.5 are provided by the Pacific Climate Impacts Consortium. The downscaled climate data is created based on the Bias Correction/Constructed Analogues with Quantile mapping reordering version 2 (BCCAQ-V2) and is available at 300 arc-seconds (roughly 10 km).

2.1.2. Intensity–Duration–Frequency (IDF) Curves

IDF curves are essential for the design and maintenance of sewers, stormwater ponds, and catchment basins, among other various types of engineering infrastructures. We used the 2007 IDF curve corresponding to the weather station at Stephenville Airport, which is generated based on in situ data from 1967 to 2007. The 24 h extreme rainfall events with return periods of 25 and 100 years for the historical period (1976–2005) and two future periods of 2041–2070 (2050s) and 2071–2100 (2080s) are considered in this study.

IDF curves generated by Environment and Climate Change Canada are based on Bernard's equation:

$$\text{Bernard's formula: } i(t) = \frac{a}{t^b} \tag{1}$$

where i (mm/h) is the rainfall intensity at time t (hour) and a, b are parameters for each return period.

2.1.3. Projected IDF Curves

Traditionally, IDF curves are generated based on historical rainfall observations, assuming that the historical variations can represent the future climate system. However, this stationarity assumption might not be valid because the future rainfall patterns are projected to change [56,57]. It is important to consider the impacts of climate change on IDF curves for future infrastructure design and planning, and management of water resources.

Statistically downscaled General Circulation Models (GCMs) have been used to assess the projected impacts of climate change on hydrological processes. However, GCM resolution is too coarse to represent small-scale physical processes, and the short-duration rainfall extremes may not be adequately represented in the downscaled data. In this study, we compare the flood characteristics based on IDFs generated from statistically downscaled GCMs [58] and a high-resolution climate model to assess the corresponding uncertainties. In the first approach, the observed sub-daily maximum rainfall data (from 5 min to 24 h) and GCM simulated daily maximum rainfall from historical and future GCMs are extracted. Generalized Extreme Value distribution (GEV) is fitted to the sub-daily/daily maxima using the L-moments method. Next, an equidistant quantile-matching approach is applied to downscale precipitation data by establishing a direct statistical relationship between daily maximum precipitation simulated by GCM (at the reference period) and sub-daily historical observations. Similarly, the relationship between maximum rainfall for historical and future GCM simulations is established. The relative change in simulated precipitation between GCM baseline and future scenarios is calculated and applied on the established relationship between observations and historical GCM simulations to generate the projected IDF curves [58].

The second approach to generate projected IDF curves is based on the Weather Research and Forecasting (WRF) system, which is a numerical weather prediction model designed to simulate meteorological processes, and provide weather forecasting and climate change analyses based on actual atmospheric or idealized conditions, across scales from tens of meters to thousands of kilometers. WRF model simulations used here were conducted by Rasmussen (2017) [59] to assess the impacts of climate change on convective population and thermodynamic environments over the North American domain at a relatively high resolution of 4 km. The adopted WRF model explicitly characterizes convective precipitation events. WRF CTRL (control) represents the historical control run, forced with ERA-Interim boundary conditions, and PGW represents future climate simulations based on a Pseudo-Global Warming approach, which uses boundary conditions that superimpose coarse-resolution differences between present and GCM-projected warming conditions overtop ERA-Interim data [60]. Several studies have assessed the WRF model's convective and non-convective rainfall simulations, and the results show that it can adequately represent the features of rainfall events [61–63]. In the context of IDF curves specifically, Cannon et al. (2019) [64] used sub-daily precipitation outputs from the WRF CTRL and PGW simulations to investigate future changes in IDF curves over North America. A novel parsimonious Generalized Extreme Value Simple Scaling (GEVSS) model was fitted to annual maxima from 1 to 24 h duration, and future changes in the resulting IDF curve parameters were estimated [64]. The study showed an increase in the scaling exponent of the GEVSS parameter, indicating that the return levels corresponding to the short-duration rainfall events can increase to a larger extent compared to the ones associated with longer duration events (e.g., 24 h). Further, this approach is not bound by the stationarity assumption made by Simonovic et al. (2016) [58] above—the estimated scaling factors can change for events with different durations.

Cannon et al. (2019) [64] expressed projected relative changes in sub-daily precipitation extremes from the WRF simulations for different return levels based on temperature changes, and assessed the adherence to the theoretical Clausius–Clapeyron (CC) relation. Under theoretical CC relation, the atmosphere can "hold" approximately 7% more moisture for every 1 K warming of air temperature [65]. The temperature scaling rate, defined as the percent change of precipitation rate per degrees Celsius, is determined for different return periods and rainfall durations. In this study, we used the scaling rates determined by Cannon et al. (2019) [64] to perturb the observed IDF curve at Stephenville Crossing based on projected temperature changes from the GCMs. We first found the average temperature of the region over the historical and future periods based on downscaled GCMs. The scaling factor per degree Celsius was then applied to the temperature changes between future and historical periods to estimate the projected increases in rainfall events at different durations. Then, the final change rate of precipitation in the future period was used to update historical IDF curves. Depending on the design storm method, the scaling rate is either directly applied to the total rainfall amount calculated from IDF curves or rainfall intensity obtained from IDF equations, with a constant rate of change at each time step of the storm event.

2.1.4. Coastal Components

We assessed the individual and compounding effects of coastal and riverine flooding considering tidal effects as well as changes in storm surge, waves, and sea-level rise. Glacier melt and thermal expansion of seawater due to climate change are expected to increase the global sea level. Besides, the population and economic growth in the low-lying coastal areas make the cities and communities more vulnerable to coastal flooding. Batterson et al. (2010) [66] studied the historical and future sea-level changes in Newfoundland and Labrador considering the effects of land subsidence and global sea-level rise. They showed that sea level is projected to increase by 30 and 80 cm by 2050 and 2099, respectively [66]. The projected local ground subsidence rate is 2 mm/year for the main area of Newfoundland Island [67]. Probability density functions of water levels due to astronomic tides and atmospheric forcing are combined to generate a new frequency distribution of water levels corresponding to tide, surge, and wave [52]. High tide levels obtained from tide predictions of Port Harbor station are used to generate the tidal probability density function. Although Port Harmon is the nearest tide station, it does not have sufficient observation data for surge analysis, therefore the observed water levels obtained from the Lark Harbor gauge were used to conduct a surge frequency analysis. Surge is calculated based on the difference between water level observation and tide prediction at each time. Wave analysis involves the frequency analysis of wind data and wind hindcast (Table 1). We considered the worst-case scenario by applying a triangular shape hydrograph on tide prediction graphs, assuming that the peaks of surge and tide occur at the same time, consistent with Karim and Mimura (2008) [68]. Figure 2 shows the downstream boundary condition estimated by imposing the triangular shape of super-elevation and constant future SLR on tide predictions.

Table 1. The components considered in coastal flood assessments under climate change.

Coastal Components	Scenarios	
Storm surge and wave (m)	25-year event	100-year event
	5.25	6.34
Sea-level rise (m)	2050s period	2080s period
	0.3	0.8

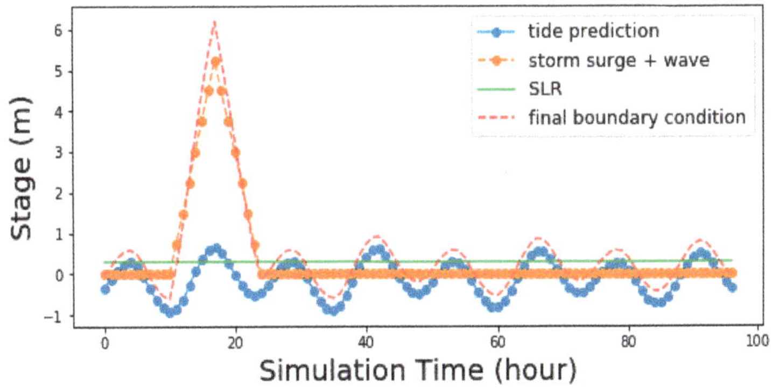

Figure 2. Coastal boundary condition with the tide, storm surge, wave, and future sea-level rise (SLR) corresponding to a 25-year event by the 2050s.

2.1.5. Satellite Imagery

The Sentinel-1 mission by the European Space Agency (ESA) provides enhanced revisit frequency and coverage with interferometry capability. The satellite covers the entire world's land at different frequencies, i.e., bi-weekly for sea and ice zones, and daily frequency for European coastal regions. The first and second Sentinel satellites were launched in 2014 and 2016 respectively, and the corresponding imagery is used to evaluate the flood event on 14 January 2018.

Long et al. (2014) [69] proposed a change detection and thresholding approach to extract the flood extents using Sentinel-1 images. The method identifies the brightness changes between the flood event image and the imagery corresponding to the normal conditions (i.e., the reference image; Table S3). River volume generally varies between seasons, and therefore the reference images are selected for the same season of the flood event (from 8 January 2017 to 20 January 2019). The HH polarization of the transmitter-receiver is preferred to other polarizations [70]. A reference image is generated by taking the median of all available selected images. A speckle filter is applied for both reference and flood images to remove speckle and improve the smoothness of the image with reduced resolution and blurred features. Speckle noise is a granular interference in synthetic aperture radar (SAR) images due to random interference [71]. Senthilnath et al. (2013) [72] evaluated different speckle filters (Lee filter, Frost filter, and Gamma MAP filter) in flood extent extraction from the Sentinel-1 C band image. Accordingly, the Gamma MAP filter, based on Bayesian analysis and Gamma distribution, is considered in this study as it filters more speckles and provides relatively better performance. Google Earth Code Editor is used for image collection, reference image calculation, and image generation. In addition, speckle removal is completed through multiple types of filters in the Sentinel Application Platform toolbox (SNAP). The difference image is filtered based on a threshold in ArcGIS to identify the actual flooded area.

3. Methodology

This study characterizes the riverine and coastal flooding in Stephenville Crossing using the Hydrologic Engineering Center-Hydrologic Modeling System (HEC-HMS) hydrologic model and Hydrologic Engineering Center-River Analysis System (HEC-RAS) hydraulic model. Both models, developed by the U.S. Army Corps of Engineers (USACE), have been widely applied for flood hazard modeling across the world [73–75], including the study of Hurricane Ike 2008 [41] and Typhoon Maemi 2008 [44]. The hydraulic model is driven by the observed and simulated streamflow at the upstream and (coastal) water levels downstream. The calibrated hydrological model [52] was applied to simulate the hydrological response of the river system to short-duration extreme rainfall events. Further, the two-dimensional hydraulic model was calibrated and validated against water level observations and compared with simulation results of a calibrated one-dimensional model. A sensitivity analysis of the hydraulic model was conducted considering different terrain data, simulation cell size, and roughness coefficients.

3.1. Design Storms

We considered three methods to generate design storms, i.e., Soil Conservation Service (SCS), Huff, and Alternative Block Method (ABM), to assess the corresponding uncertainties in flood inundation modeling [76]. The required input parameters and procedures to generate hyetographs, the corresponding features and limitations, and their effects on model simulations are discussed.

3.1.1. Method of SCS

The method of Soil Conservation Service (SCS) is widely used in engineering designs of dams and urban facilities, among others, which uses standardized rainfall intensities arranged to maximize the peak runoff at a given storm depth. The SCS rainfall distribution was developed in 1986 and applied for a single storm event with 6 or 24 h duration across the U.S. Four different distribution types were generated based on the data in multiple areas. Considering that Stephenville Crossing is on the Atlantic coast, the SCS curve Type III is applied to generate the design storm. The curves are applied for storm events for up to 24 h. The required information includes storm duration (24 h), design return periods (25 and 100 years), distribution type (Type III is used for the Atlantic coast), and total rainfall amount (using the IDF curves). The hyetograph was generated by first estimating the total precipitation amount for a given duration and return period. The SCS curve was then applied to generate the cumulative precipitation, followed by determining the increments between each time step and plotting the precipitation amount vs. time.

3.1.2. Method of Huff

The Huff method is similar to the SCS method, as they both use a standardized distribution type to describe rainfall patterns. However, the method of Huff provides more flexibility because there is no restriction in the duration of design storms. The Huff method was developed based on approximately 300 storms with durations ranging from 3 to 48 h. Four types of distribution curves describe the relationship between the cumulative fraction of precipitation and time, with the timing of peak intensity varying between each type. The distribution is selected based on the duration of the design storm with Type III used for 12 to 24 h storm duration. The drawback of the Huff method is that the generated hyetograph may lose the rainfall features such as extreme peak intensity because it flattens the peak of precipitation during an event.

3.1.3. Alternating Block Method (ABM)

The precipitation pattern produced by the Alternating Block Method maximizes the rainfall depth for different storm durations using the IDF curve functions. The duration of the storm event and the time step of hyetographs are first selected. Methods of Huff and SCS have variations in the time of peak rainfall by choosing different distribution curves, however, the ABM method always generates the peak rainfall in the middle of the storm event. The required information includes storm duration (24 h), design return periods (25 and 100 years), time interval (1 h increment for the 24 h event), and equation expression of IDF curves. To generate the design storm patterns based on ABM, the precipitation amount (mm) for a specific duration is determined based on the corresponding rainfall intensity (mm/h). The increments of precipitation amount between each time interval are calculated and the highest precipitation increment (maximum block) is placed in the middle of the hyetograph. The second-highest increment is placed to the right of the maximum block, and the third-highest increment to the left of the maximum block, and so on until the last block is located. In this study, design storms based on projected WRF-IDF curves were updated in two ways, resulting in two types of hyetographs generated through

ABM. The first approach applies a constant temperature scaling rate to the entire event (ABM1), and the other approach applies different temperature scaling rates at different time steps (ABM2).

3.2. Hydrologic and Hydraulic Model Setup and Calibration

3.2.1. HEC-HMS

The HEC-HMS model represents the drainage basin of Harry's River up to Black Duck Siding, which consists of 33 sub-basins, 10 river reaches, and 17 junctions, including the hydrometric gauge of 02YJ001, Harry's River, below the highway bridge. For each reach, the required inputs of channel characteristics, which include the length, shape, and slope of the channel, and Manning's n coefficient are determined. All reaches are represented by trapezoidal cross-sections, and the longitudinal slopes vary between 0.001 and 0.025, with a Manning's n value of 0.04. The loss, transform, base-flow, and routing methods are represented by the U.S. Soil Conservation Service (SCS) Curve Number, SCS Unit Hydrograph, Constant Monthly rates, and Muskingum-Cunge routing, respectively. A weighted Curve Number is estimated for each sub-basin based on soil group and land use types. An area reduction factor of 0.9 is applied over the precipitation inputs to drive the model. The model is calibrated using measured hydrographs corresponding to the 1990 event (December 8–9) and validated based on the events on 8 June 1995 and 26–27 September 2005. During calibration and validation, base flow is estimated from flow records at the hydrometric gauge (02YJ001) before the date of the simulation event. Further details are provided in the Hydrotechnical study of Stephenville Crossing (2012) by the Atlantic Canadian Adaptation Solutions Association (Government of Newfoundland and Labrador, 2012).

3.2.2. HEC-RAS

The HEC-RAS 1D model simulates river flow from the downstream of Harry's River to the Main Gut (Government of Newfoundland and Labrador, 2012). Eleven surveyed bathymetric cross-sections across the reach are used to describe the channel geometry and floodplains (Figure S1). Roughness coefficients of the channel and floodplain are estimated based on the type of channel and overbank. The model is driven by the flow hydrograph as the upstream and stage hydrograph as the downstream boundary conditions. The 1D HEC-RAS model was calibrated based on several water level measurements at cross-sections of 10–12, 14, and 16–17 during 25 to 28 September 2010 and validated for 3–7 November (Figure S1; [52]).

Further, we set up and calibrated the two-dimensional HEC-RAS model, which represents floodplain flow as a 2D cell, by assuming that the third dimension of water depth is relatively shallow. The conservation of mass and momentum equations are expressed as follows:

$$\textit{Mass Conservation}: \frac{\partial H}{\partial t} + \frac{\partial (hu)}{\partial x} + \frac{\partial (hv)}{\partial y} + q = 0 \tag{2}$$

where t is time, x and y represent spatial dimensions, the 2D vector (u,v) represents the velocity components in two dimensions, q is flux, H is water surface elevation, and h is water depth [77].

Momentum Conservation:

$$\frac{\partial u}{\partial t} + u\frac{\partial u}{\partial x} + v\frac{\partial u}{\partial y} = -g\frac{\partial H}{\partial x} + v_t\left(\frac{\partial^2 u}{\partial x^2} + \frac{\partial^2 u}{\partial y^2}\right) - c_f u + fv \tag{3}$$

$$\frac{\partial v}{\partial t} + u\frac{\partial v}{\partial x} + v\frac{\partial v}{\partial y} = -g\frac{\partial H}{\partial y} + v_t\left(\frac{\partial^2 v}{\partial x^2} + \frac{\partial^2 v}{\partial y^2}\right) - c_f v + fu \tag{4}$$

where g is the gravitational acceleration, c_f represents the bottom friction, f is the Coriolis parameter, and v_t is the horizontal eddy viscosity coefficient [77].

DEM, channel bathymetry, and land cover map with spatially varied roughness coefficients were used to set up the 2D model. The 20 m resolution Canadian Digital Elevation Model (CDEM) was used to represent the terrain's topography. Considering that the DEM does not include the bathymetric details under the water surface, surveyed cross-sections are interpolated into a surface profile and then fused into the topography data (Figure S1). The 30 m resolution Canada's Land Cover map was used to generate spatially varied Manning's n values for each cell. Table S1 lists all types of land cover in the study region with the corresponding roughness coefficients. A value of 0.035 was considered for the reach along Harry's River. We set up the 2D model considering a 20×20 m cell size consistent with the 20 m resolution DEM. Break-lines are added along the river centerline and right and left of the overbank. The cell size around the break-line includes smaller irregular meshes for a more accurate simulation of the channel and overbank area.

The 2D model is driven using the upstream flow hydrographs at Harry's River below the Highway Bridge and coastal water levels as the downstream boundary condition. The flow hydrographs of the upstream boundary were obtained from HEC-HMS at the hydrometric station of Harry's River below the Highway Bridge. The coastal boundary condition was constructed based on hourly tidal records, which were collected from the tide gauge at Port Harmon, an active station close to St. George's Bay.

4. Results and Discussion

4.1. Model Performance

The roughness coefficients in the channel and floodplain were calibrated based on water surface elevation (WSE) measurements at specific points along the channel on 27 September 2010 (Figure 3). 2D model simulations are consistent with the results of the 1D model, including the peak values, and both models represent the observed levels quite well. At the lowest levels, the maximum difference between 2D- and 1D-model simulations is about 0.1 m. The results are further evaluated based on observations on 3–7 November 2010 (Figure S2).

Further, we assessed the sensitivity of the simulations to changes in the cell size, DEM product, and the Manning's n roughness factors in the HEC-RAS 2D model. The sensitivity analysis was conducted for the November 2010 event with higher peak flow rates (80 m^3/s) than the September 2010 event (30 m^3/s). Sensitivity analysis results show that DEM can considerably affect the model simulations. The effects of the DEMs are most pronounced at the upstream reach, and the distinction between inundated areas based on different DEMs gradually decreases from upstream to downstream. Similarly, the comparison of simulations based on different cell sizes shows the considerable effects of spacing (Figure 4). Run 4 (20 m in 2D area and 15 m around break-line) has the largest simulated inundation area. Further, we investigate the sensitivity to Manning's n values for river channel and floodplain. It was found that the lower part of the reach in the HEC-RAS 2D model is less sensitive to Manning's n values.

4.2. GCM-IDF vs. WRF-IDF Simulations

4.2.1. Uncertainties in Storm Patterns

A total of 432 hyetographs (288 for WRF-IDF curves and 144 for GCM-IDF curves) were generated for Stephenville Crossing, based on projected IDF curves, three design storm methods, Representative Concentration Pathways (RCP) 4.5 and 8.5, and two future periods of 2041–2070 (2050s) and 2071–2100 (2080s) (Table S4). As discussed before, nine GCMs are selected in climate change analysis using WRF-IDF curves. Six of those models were available for the GCM-IDF curve assessment (using IDF-Tools). Common GCMs were used to compare WRF- and GCM-simulated results. The hyetographs based on the historical and future IDF curves were then used to drive the HEC-HMS model based on the three design storm methods, including Soil Conservation Service (SCS), Huff, and

Alternative Block Method (ABM). Figure S5 shows an example of hyetographs generated from three methods, which result in different magnitudes and timing of peak rainfall.

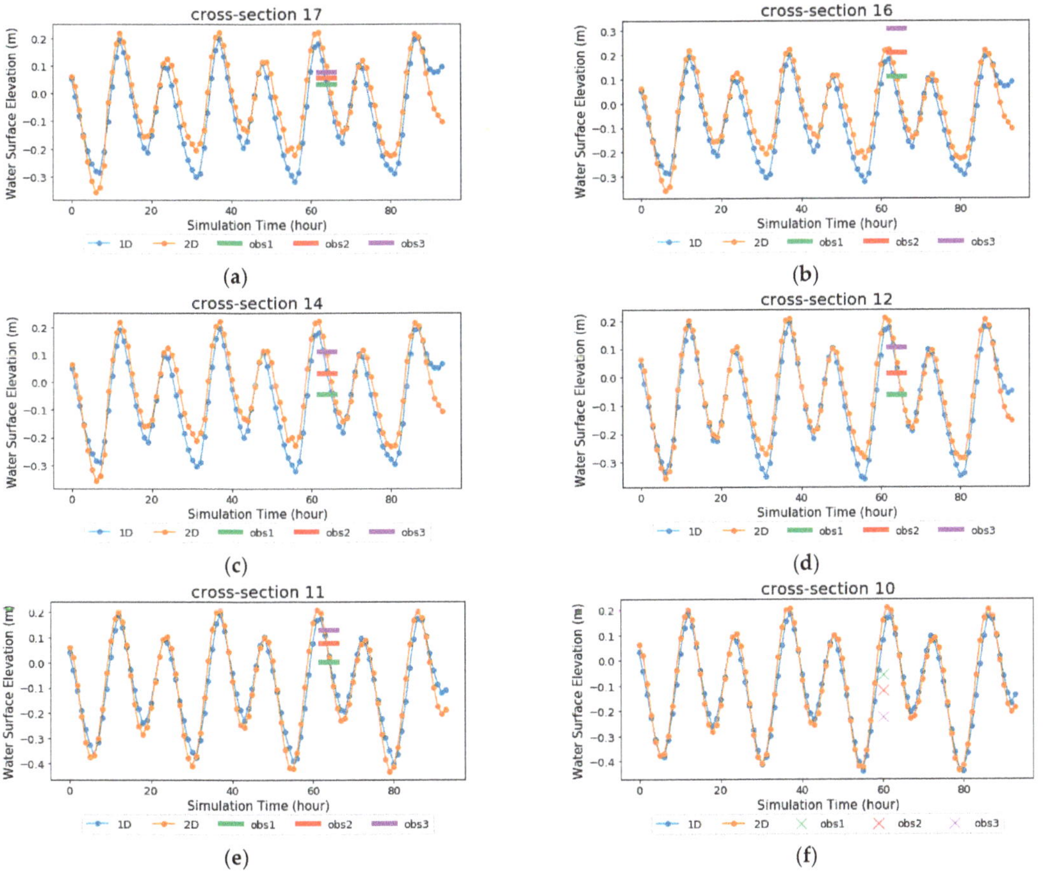

Figure 3. Observed Water Surface Elevations on 27 September 2010 and the corresponding HEC-RAS 1D and 2D model simulations (from 25 to 28 September). Orange represents HEC-RAS 2D results, blue represents HEC-RAS 1D results; obs1, 2, and 3 represent observations along the cross-section during different times. The horizontal line represents the duration of taking the measurements from 3 to 7 pm. Results are shown for different cross-sections (**a–f** representing crosssections 17, 16, 14, 12–10, respectively) along the river (Figure 1 and Figure S1). The exact time of the measurements is available for Cross-section 10 (at 1 pm).

The variations of total rainfall amount between GCM- and WRF-IDFs are shown in Table 2. For a 25-year event, WRF-IDF generates higher rainfall amounts. The upper bound of WRF-IDF curves is similar to GCM-IDF curves, however, the lower bound is much higher than GCM-IDF curves (26% higher for the RCP8.5 scenario in the 2080s). For a 100-year event, WRF-IDF generates a lower rainfall amount (except for RCP8.5 in the 2080s) with a narrower uncertainty range than GCM-IDF for future scenarios.

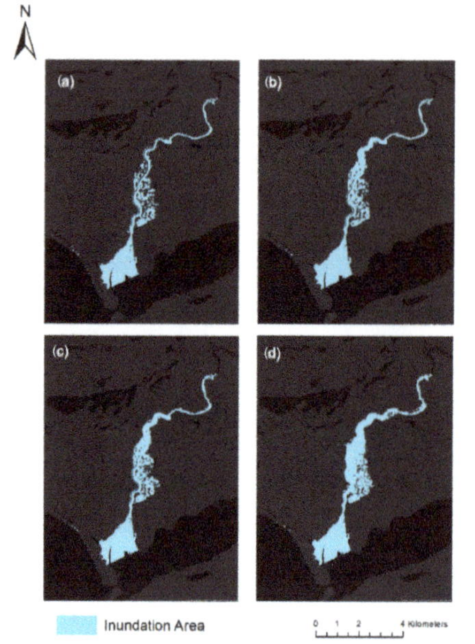

Figure 4. Comparison between 2D simulated flood inundation extents using different mesh sizes (around break line): (**a**) 100 m (70 m), (**b**) 50 m (30 m), (**c**) 30 m (30 m), (**d**) 20 m (15 m).

Table 2. Comparison of 24 h rainfall (mm) amounts generated based on WRF- and GCM-IDF curves based on six GCMs.

Return Period (Years)	Historical (mm)	Future (mm)			
		Period	RCP	The Multi-Model Ensemble Average of GCMs (Minimum and Maximum)	
				GCM-IDF	WRF-IDF
25	107.94	2050s	4.5	131.44 (111.38, 152.31)	137.17 (121.80, 147.35)
			8.5	135.75 (118.8, 169.00)	143.03 (126.65, 154.14)
		2080s	4.5	129.93 (105.00, 153.94)	141.62 (127.04, 151.94)
			8.5	142.50 (113.86, 176.36)	163.05 (145.5, 176.54)
100	142.79	2050s	4.5	184.82 (144.33, 233.31)	169.73 (150.15, 182.70)
			8.5	185.32 (149.58, 241.31)	177.20 (156.33, 191.35)
		2080s	4.5	181.76 (124.37, 237.56)	175.40 (156.83, 188.55)
			8.5	200.00 (133.85, 333.34)	202.70 (180.34, 219.89)

The resulting hyetographs generated based on three design storm methods for a 25-year event over the historical and future (2050s corresponding to the RCP4.5 emission scenario) periods are compared in Figure 5. The figure shows the average, minimum, and maximum values of hyetographs based on multiple GCMs. The peak rainfall occurs at around the 11th hour for both ABM and SCS design storms, however, the peak rainfall corresponding to the Huff design storms occurs at around the 14th hour. Design hyetographs based on ABM have the highest peak rainfall and peak intensity, followed by the hyetographs based on SCS. In general, the peak precipitation values generated from Huff are relatively low, with smaller variations in magnitude. The overall rainfall pattern in the Huff method is more even and flat compared to the other two methods. This can cause an underestimation of the peak flood volume in the hydraulic model simulation. The overall pattern of rainfall hyetographs is similar between ABM-1 and ABM-2, however,

ABM2 generated by different scaling rates shows slightly higher peak values. Overall, the peak values corresponding to the GCM- and WRF-IDFs for the 25-year event in the 2050s are close. However, differences are more distinguishable for larger events. Based on Figure S6, the lowest hyetograph peak generated by WRF-IDF curves is higher than that generated by GCM-IDF curves (for a 100-year event in the 2080s), contrary to the highest values. The uncertainty range of hyetographs based on GCM-IDF curves is larger than that of the WRF-IDFs because of larger variations in precipitation projections between GCMs compared to the corresponding temperature simulations, used for temperature scaling in WRF-IDFs.

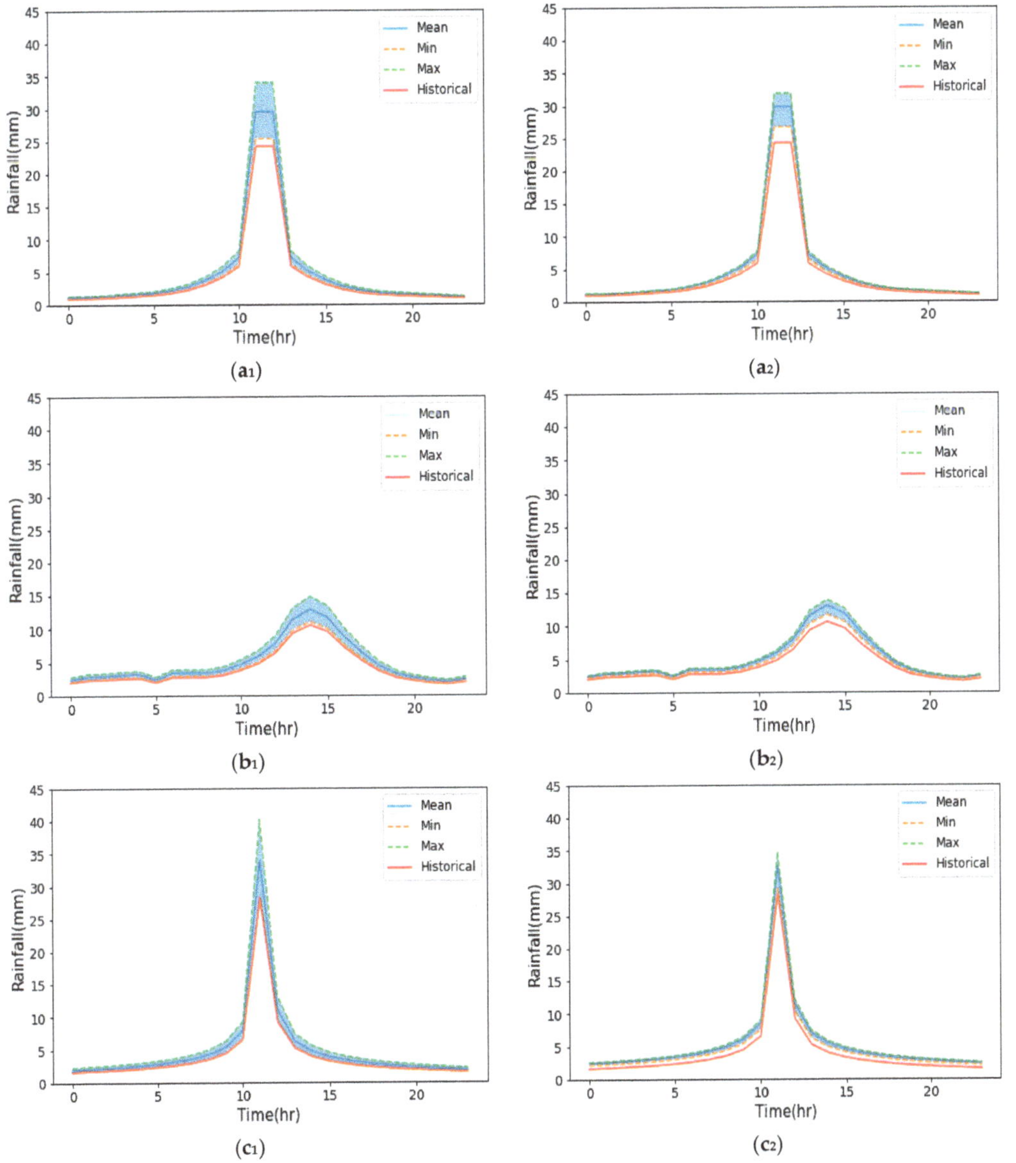

Figure 5. *Cont.*

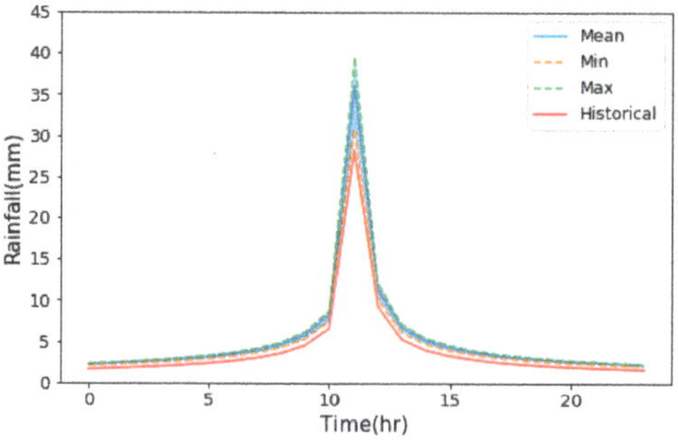

(c3)

Figure 5. Rainfall hyetographs corresponding to a 25-year event for the historical and future (2050s, RCP4.5) periods. Hyetographs are generated based on (**a$_1$**) HUFF, (**b$_1$**) SCS, and (**c$_1$**) ABM design storm methods for GCM-IDFs, and (**a$_2$**) SCS, (**b$_2$**) HUFF, (**c$_2$**) ABM-1, and (**c$_3$**) ABM-2 for WRF-IDFs.

Differences between GCM- and WRF-IDF effects are more pronounced when investigating the individual GCMs. The resulting design storms for CanESM2 based on the RCP4.5 emission scenario are shown in Figure S7. All hyetographs (based on ABM, SCS, and Huff methods) are defined with a one-hour time interval and a total storm duration of 24 h. Results show a considerable difference in rainfall patterns based on different approaches. In ABM, high rainfall intensity is maximized within a short duration, which occurs in the midst of the event, for example, the peak rainfall intensity always occurs at the 12th hour during the 24 h event. Differences between ABM hyetographs in the 2080s are generally larger than those in the 2050s. The peak rainfall value of the ABM2 hyetograph is always higher than the one in the ABM1 hyetograph because a shorter duration provides a higher scaling rate, and the difference in the two approaches of WRF-IDF curve in the ABM varies with RCP scenarios and future periods. The overall pattern of hyetographs generated by the SCS method is very similar to ABM hyetographs, however, SCS hyetographs generate a longer time of maximum rainfall. The timing of peak rainfall value in the hyetographs generated by the Huff method is about 3 h later than the peak time of ABM and SCS hyetographs. In addition, the magnitude of maximum precipitation of Huff hyetographs is considerably smaller than the hyetographs generated by the other two methods. The differences in the peak rainfall can be as high as three times among design storm methods. The ABM hyetographs have the maximum precipitation peak, followed by SCS and Huff hyetographs. The maximum rainfall amount in a 25-year event during the future period of the 2050s ranges from 13 mm, based on the Huff approach, to 39 mm based on ABM2. Within a 24 h duration storm, the peak rainfall intensities are the largest in ABM and SCS hyetographs, while Huff hyetographs provide relatively low rainfall intensities that are distributed over an extended period. Consequently, the variations of rainfall patterns are highly dependent on the choice of design storm methods. The relative differences between the projected IDF curves (GCM vs. WRF precipitation simulations) based on CanESM2 under two future periods and return levels are also shown in Figure S7. Considering the RCP4.5 scenario, there are slight differences in the 25-year rainfall event between the hyetographs generated by GCM-IDF and WRF-IDF curves. For simulations based on CanESM2, the peak rainfall in design storms based on the GCM-IDF curve is higher than that based on WRF-IDF curves, particularly for a 100-year event. However, it is not always the case for all GCMs, for example in HadGEM-AO (AO), WRF-IDF curves can generate higher peak rainfall in design hyetographs than that based on GCM-IDF curves (Figure S8). Compared with Huff hyetographs, the differences between the two updated IDF curves are more in ABM and SCS hyetographs. Although the differences in peak values between the methods of design storms and projected IDF curves are not large, they can cause major effects in hydrological simulations.

4.2.2. Uncertainties in Flow Hydrographs

The hyetographs generated based on SCS, Huff, and ABM methods, corresponding to projected WRF- and GCM-IDF curves, are used as inputs to the HEC-HMS hydrological model to simulate the upstream basin's hydrological response (i.e., flow discharge). The variations of simulated discharge rates among the two types of updated IDF curves and different design storm methods are shown in Figure 6. Overall, the uncertainties corresponding to different design storm methods are considerable compared to other major sources of uncertainties, such as GCMs. Based on WRF-IDF curves, the uncertainties between design storm methods increase from the 2050s to the 2080s, and from RCP4.5 to RCP8.5. Based on GCM-IDF curves, the peak discharge rates in the future periods are consistent for RCP4.5, and the rates further increase in the 2080s considering the RCP8.5 emission scenario. Overall, the uncertainties corresponding to the design storm methods are larger in GCM-IDF curves compared to the WRF simulations. The hyetographs generated from SCS provide the highest peak discharge simulation for future scenarios and two projected IDF curves, while the method of Huff provides the lowest rates for all cases. The highest river discharge rates occur in the 2080s under the RCP8.5 emission scenario, while the lowest values are in the 2050s corresponding to RCP4.5, which indicates more intense flood events in the future periods under climate change. The simulated rates from WRF-IDF curves are larger than the ones corresponding to the GCM-IDF curves for 100-year events in the 2080s under the high emission scenario of RCP8.5. In other scenarios, the overall peak discharge rates are relatively close between the multi-model means corresponding to the two projected IDF curves, however, differences in upper quantiles are relatively large. Further, GCM-IDFs show larger uncertainty ranges for 100-year events and project higher rates in the upper bounds (Figure 6).

According to WRF-IDF curves, CSIRO-Mk3.6.0 (CSIRO), GFDL-ESM2G (ESM2G), and MPI-ESM-LR (MPI) provide relatively lower results, whereas the discharge rates are close for other GCMs. However, for GCM-IDF, except for HadGEM2-ES (ES) that shows the highest peak discharge rates, the projections of other GCMs vary among different future periods. GFDL-ESM2G (ESM2G) simulates a low peak flow rate in both projected IDF curves. The performances of GFDL-CM3 (CM3) and HadGEM-AO (AO) are distinct between projected IDF curves (Figure S9).

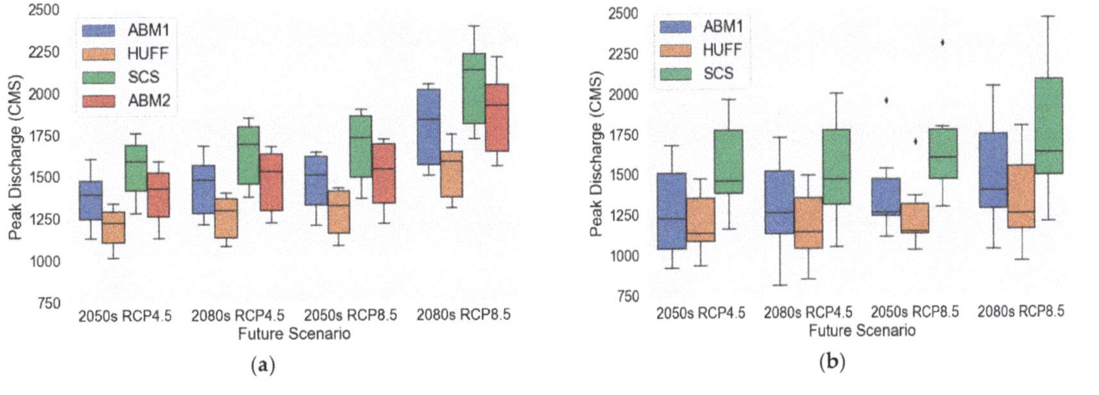

Figure 6. *Cont.*

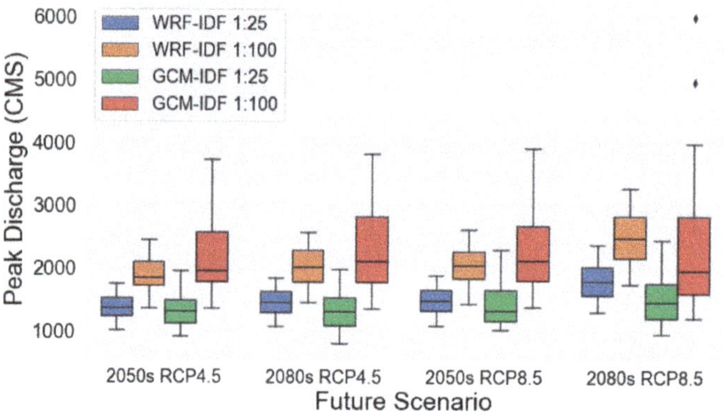

(c)

Figure 6. Simulated peak discharge rates corresponding to a 25-year event based on different design storm methods corresponding to (**a**) WRF-IDF and (**b**) GCM-IDF curves. ABM1 (Alternative Block Method) represents constant temperature scaling rate applied to the entire event, and ABM2 shows varied temperature scaling rates applied at each time step. (**c**) Simulated peak discharge rates between WRF- and GCM-IDFs corresponding to 25- and 100-year events.

Further, the uncertainties between design storms and GCMs are compared for a 25-year event during the 2050s under the RCP8.5 emission scenario (Figure 6 and Figure S8). The means of peak flow rates among the three design storm methods range from 1300 to 1700 m^3/s for WRF-IDF curves and from 1125 to 1475 m^3/s for GCM-IDF curves, while the mean peak discharges among GCMs vary from 1150 to 1650 m^3/s for WRF-IDF curves and from 1100 to 1900 m^3/s for GCM-IDF curves. The uncertainties from the choice of design storm methods are slightly larger than the uncertainties brought by GCMs when using WRF-IDF curves, however, different GCMs have larger variations than design storm methods in using GCM-IDF curves.

The flow hydrographs for a 100-year event in the 2050s corresponding to the RCP8.5 emission scenario were compared between three design storm methods (Figure S10). The figure shows the average values of hydrographs generated based on nine GCMs and the corresponding minimum and maximum values. The overall pattern of simulated hydrographs generated based on the three design storm methods is similar, however, the magnitude and timing of peak discharge rates are different. The peak discharge occurs at around the 16th hour for both ABM and SCS design storms, however, peak discharge of Huff design storms occurs around the 19th hour. The 3 h time lag is the same as the time lag of peak rainfall between Huff hyetographs and the other two hyetographs. Simulated peak runoff by SCS hyetographs exceeds the peak discharge by ABM hyetographs. The peak discharge rates simulated by Huff hyetographs are much smaller, with less variation in the magnitude. Relatively low rainfall intensities evenly distributed over the event allow the watershed more time to respond, and thus, the simulated results of Huff hyetographs result in lower peak runoff. Consequently, the estimated flow discharge is much smaller, and it may cause an underestimation in peak flood values in the hydraulic model simulations. The overall pattern and magnitude of peak runoff are similar in ABM-1 and ABM-2. However, the ABM2 hyetographs generated by varied scaling rates have more variations in peak flow, as there is a slightly wider higher uncertainty range.

The hydrological response of the two projected IDF curves (WRF- and GCM-IDFs) are shown in Figure 7 and Figure S10. The hydrographs corresponding to the two projected IDF curves for the 25-year flood event have a consistent pattern and average peak values, but GCM-IDF simulations show larger variations between different GCMs, resulting in differences between lower and higher quantiles. The peak flow corresponding to the GCM-IDF curve ranges from 900 to 1600 m^3/s, while WRF-IDF simulations range between 1100 and 1500 m^3/s. The 100-year flood event simulations show similar behavior. Compared with the 25-year event, the results of a 100-year event based on GCM-IDF hyetographs have larger variations, ranging from approximately 1600 to 5500 m^3/s. Therefore, the hyetographs based on the GCM-IDF curve are very sensitive to the choice of GCM. The multi-model means of the peak discharge values, based on two future IDF curves, is around 1250 m^3/s, corresponding to the 25-year event in the 2050s under RCP4.5. This value increases to approximately 2500 m^3/s for a 100-year event in the 2080s under RCP8.5. The uncertainties corresponding to ABM-1 and ABM-2 IDF methods are relatively low compared to the uncertainties between other design storm methods and projected IDF curves, especially in 100-year flood event simulation.

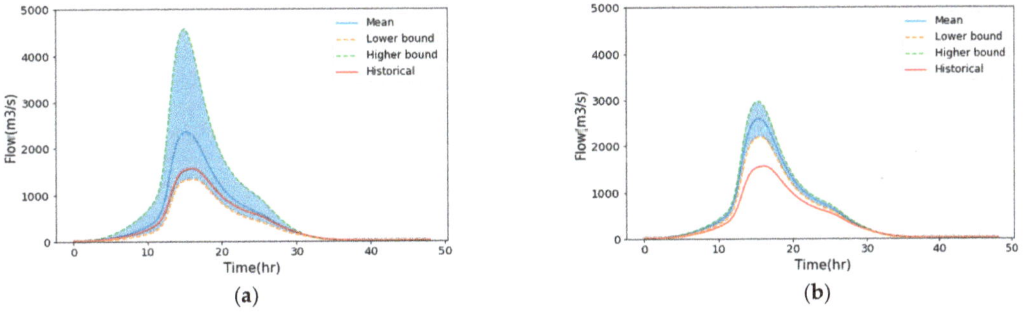

Figure 7. Flow hydrographs at the gauge of Harry's River below the Highway Bridge for a 100-year event corresponding to the historical and future (2080s, RCP8.5) periods. Hyetographs are generated based on (**a**) GCM-IDFs and ABM method, and (**b**) WRF-IDFs and ABM2 method. The lower and higher bounds represent the minimum and maximum values of six GCMs.

4.3. Projected Changes in Flood Characteristics and the Corresponding Uncertainties

The maximum flood extent areas corresponding to each design storm are summarized in Table 3. The Huff method results in the lowest flood inundation area, indicating that it can be considered as the lower bound of flood risk estimates in floodplain management and planning. The ABM and SCS approaches result in the largest inundation areas for the 25- and 100-year events, respectively. These results highlight the importance of characterizing the uncertainties in design storms for flood risk analysis.

Table 3. Inundation area (square kilometers) for different design storms in the historical period.

Design Storm	25-Year Event	Difference from Huff Method	100-Year Event	Difference from Huff Method
ABM	6.221	0.057	6.427	0.178
SCS method	6.21	0.046	6.431	0.182
Huff method	6.164	0	6.249	0

Relative changes of the simulated maximum flood depths between the three design storm methods and the corresponding mean values are shown in Figure 8. For the future period of the 2050s under RCP8.5, the SCS method provides the most conservative 100-year flood estimate, while Huff shows the lowest impacts. ABM2, which applies different scaling rates at different time steps, provides slightly higher values than the ones from ABM1.

We assessed the projected changes of the maximum flood depths corresponding to 25- and 100-year events based on WRF- and GCM-IDF curves (Figures 9 and 10). The projected changes of flood depths for a 25-year event based on GCM-IDFs are relatively small for both future periods of the 2050s and 2080s under RCP4.5 (Figure 9 and Figure S11), with changes slightly higher at the middle region of Harry's River under RCP8.5. Results from WRF-IDF curves are relatively similar for future scenarios, except in the 2080s under RCP8.5, which shows larger inundation at the upstream and middle of the river. Overall, GCM-IDF under RCP4.5 provides the lowest projected changes of flood depth, while the WRF-IDF curve results in the highest values in the 2080s under RCP8.5. The inundation areas of the upstream are projected to increase for a 100-year event, however, WRF-IDF under RCP4.5 shows milder changes in the 2050s. Further, the coastal regions are inundated based on the two projected IDF curves in the 2050s and 2080s under RCP8.5, however GCM-IDFs project lower inundation extents in the 2080s under RCP8.5. The multi-model mean peak discharge corresponding to GCM-IDF curves is considerably lower than simulations from WRF-IDF curves. Therefore, the relative changes of projected flood depths are considerably different between the two IDF approaches under a high emission scenario of RCP8.5 in the 2080s (Figure 10).

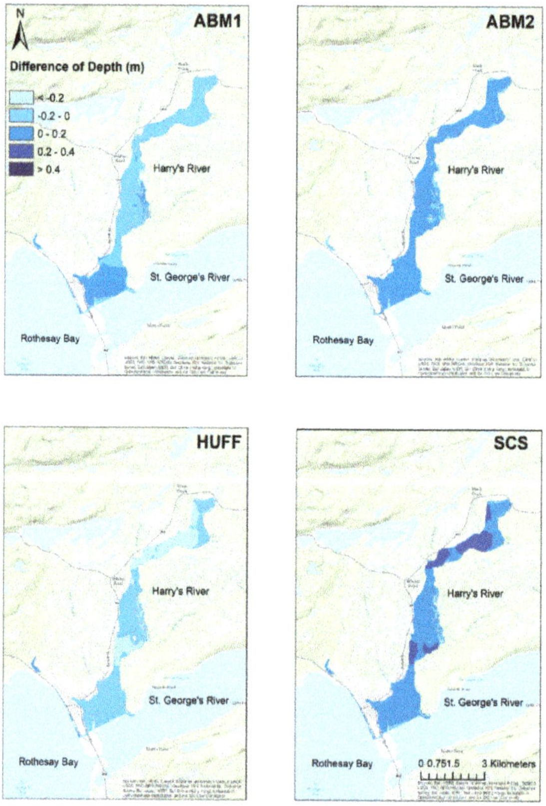

Figure 8. Relative changes in simulated maximum flood depths (m) between different design storm methods and the average of maximum depths from all methods. Results correspond to a 100-year event in the 2050s based on WRF-IDF under the RCP8.5 emission scenario.

Although differences between the mean rainfall amount corresponding to a 100-year event in the 2080s under RCP8.5 are not very large (Table 2), GCM-simulations show relatively large variations in rainfall values in both the upper and lower bounds. Therefore, differences in the simulated hydrographs and corresponding peak flows become relatively large, which translate into considerable changes between estimated flood depths (Figure 10). These results suggest that the uncertainties associated with GCMs contribute considerably to the total uncertainties in future flood risk analyses, particularly for GCM-IDFs.

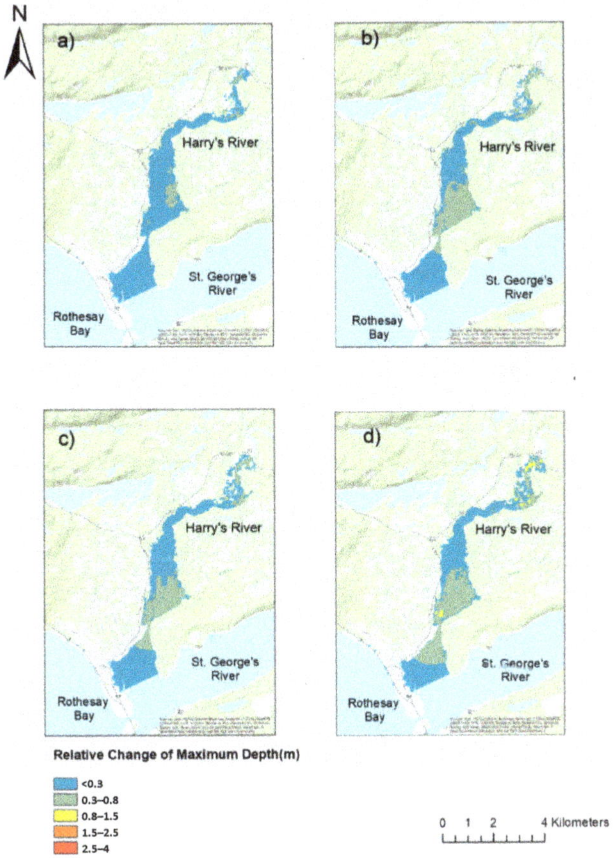

Figure 9. Projected changes in maximum flood depths corresponding to a 25-year event between the future (2080s) and historical periods: (**a**) GCM-IDF and RCP4.5, (**b**) WRF-IDF and RCP4.5, (**c**) GCM-IDF and RCP8.5, (**d**) WRF-IDF and RCP8.5.

4.4. Compound Flood Assessment

The simulation of fluvial flooding is conducted by considering historical tide estimates as the downstream boundary condition and projected flow hydrographs generated based on future design storms as the upstream boundary condition. As discussed previously, increases in both rainfall intensity and coastal water levels, associated with climate change, can lead to higher risks of flooding in the low-lying areas.

The compounding effects of riverine and coastal flooding can result in severe damages to communities and infrastructure. We quantify the return periods of such events by developing the joint distribution of both flood drivers and characterizing the dependencies using copula functions [33,57,78–81]. A set of 41 copulas are considered in this study and the best function is selected using the AIC criterion. Results show that the probability of such events, as suggested by the estimated joint return periods, is higher if the dependencies between the variables are considered compared to the conventional approach, which assumes that different flood drivers are independent. For example, the joint return period of a 10-year riverine and a 10-year coastal flood event is 88 years (considering the dependence structure), as opposed to 100 years (based on the independence assumption). Such an event has a 70-year return period considering extreme rainfall and coastal events. The joint return periods are 520 years (vs. 625 years considering independence) and 416 years for 25-year riverine and coastal, and 25-year rainfall and coastal extremes, respectively.

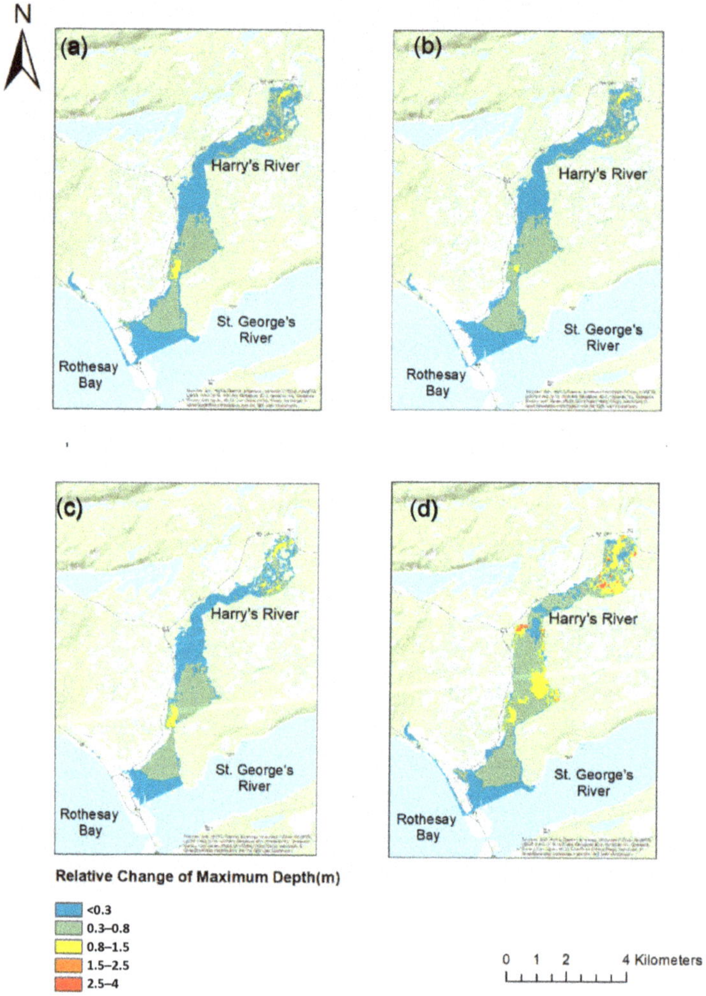

Figure 10. Projected changes in maximum flood depths corresponding to a 100-year event between future (2080s) and historical periods: (**a**) GCM-IDF and RCP4.5, (**b**) WRF-IDF and RCP4.5, (**c**) GCM-IDF and RCP8.5, (**d**) WRF-IDF and RCP8.5.

Further, we quantified the return period of the occurrence of either extreme discharge/rainfall or coastal events considering the dependence structure of the flood drivers. Results show that in the OR scenario, the return period is almost half of the univariate return periods, for example, the return period of a 100-year fluvial OR 100-year coastal flooding is ~50 years. This indicates that assessments of different flood types in isolation can result in a major underestimation of their impacts.

We added the effects of projected coastal flood drivers (storm surge, wave, and sea-level rise) and assessed compound flooding under climate change. We assume that the peak of the stage hydrograph coincides with the peak of flow hydrographs, which is a conservative assumption. Table 4 lists the simulated flood inundation areas corresponding to rainfall-only and compound flooding simulations under future climate scenarios. In all scenarios, the compound flooding simulation estimates a larger flooded area compared to the rainfall-only analysis, which increases from RCP4.5 to RCP8.5 and from the 2050s to the 2080s.

Table 4. Projected flood inundation extents (square meter) based on WRF-IDF curves (multi-model means of six GCMs).

Return Level	RCP	Future Period	Fluvial Flood Scenario	Compound Flood Scenario
			Mean	Mean
25-year event	4.5	2050s	6.16	6.66
		2080s	6.29	6.87
	8.5	2050s	6.28	6.75
		2080s	6.78	7.3
100-year event	4.5	2050s	6.97	7.63
		2080s	7.88	8.7
	8.5	2050s	7.81	8.49
		2080s	8.98	9.78

A comparison between the impacts of fluvial flooding (25-year event, 2050s RCP4.5) and the compound scenario is shown through a flood inundation map of the estuarine area (Figure 11). The blue area represents the simulation under the changes of future extreme rainfall events. With the addition of the coastal components (i.e., storm surge, wave, and local sea-level rise), the inundation extent increases considerably in coastal areas, which extends further upstream of Harry's River, affecting the urban zone between the coastline and the estuary area. The results show that the upstream area of Harry's River suffers more from riverine flooding, while the estuary and the mouth of the river are mainly affected by both coastal and riverine flooding.

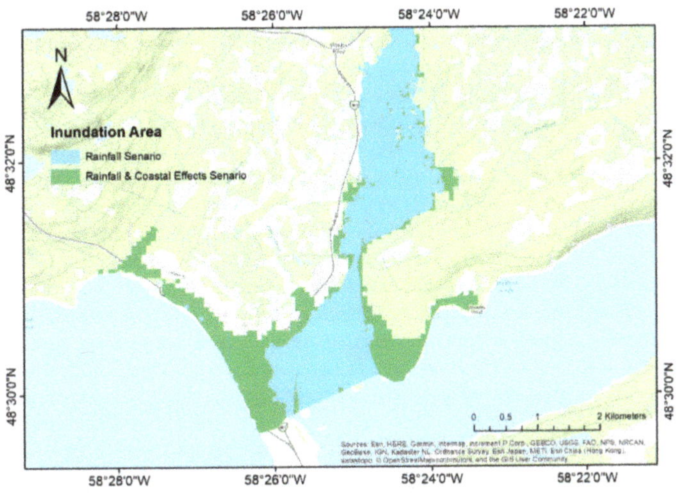

Figure 11. Flood inundation areas corresponding to fluvial flooding (blue) and compound fluvial and coastal flooding (blue and green).

5. Conclusions

In this study, the individual and compounding effects of riverine and coastal flooding were analyzed over Stephenville Crossing on the west coast of Newfoundland. The area is located between St. George's River estuary and Rothesay Bay. In the past, this community suffered from floods due to storm surge, high river flows caused by heavy rainfall, and their combination. With increases in extreme rainfall events and sea-level rise associated with climate change, such impacts are expected to be exacerbated. A two-dimensional hydraulic model (HEC-RAS 2D) was set up and coupled with a hydrologic model (HEC-HMS) to simulate the historical and projected changes in flood events and analyze the corresponding uncertainties. The 2D model is driven by the flow hydrographs as the upstream boundary condition and coastal water levels at the downstream boundary. The model was validated using water surface elevation (WSE) measurements at surveyed locations along the river. Further, Sentinel-1 satellite imagery was used to assess simulated inundation extents.

Identifying different sources of uncertainties and understanding their influences are crucial for floodplain management in a changing climate. In this study, the uncertainties associated with GCMs (ACCESS1.0, HadGEM2-CC, HadGEM2-ES, GFDL-CM3, MPI-ESM-LR, HadGEM-AO, CSIRO-Mk3.6.0, GFDL-ESM2G, and CanESM2), future scenarios (RCPs 4.5 and 8.5), design storms (SCS, Huff, ABM), and projected IDF curves (WRF- and GCM-IDF) were investigated. Results suggest that all components have a major contribution to the uncertainties in flood risk assessments. The uncertainties in design storms can be as large as the ones associated with GCMs in climate change impact assessments. The results show that the Huff method can underestimate the peak flood volume, which is consistent with a study of design storms on urban flood simulation conducted by Pan (2017). The differences between the two ways of applying WRF-IDF temperature scales in the Alternating Block Method (ABM1 and ABM2) were relatively small in our analyses, and the corresponding means and uncertainty ranges of hydrographs were almost the same during the two future periods.

Further, analyses show larger uncertainties corresponding to GCM-IDFs compared to those corresponding to WRF-IDFs, including higher variations in estimated hydrographs and flood depths. GCMs have limitations in simulating convectional rainfall, and the uncertainties of simulated short-duration rainfall extremes can translate from projected GCM-IDF curves into flood modeling analysis. Consequently, analyses show inconsistent trends between projected WRF- and GCM-IDFs from RCP4.5 to RCP8.5. In some cases, this results in an underestimation of projected flood impacts, which can undermine future adaptation plans.

The differences in flood extents for historical and future climate conditions are considerable, with more inundation in the estuarine area. Projected coastal water levels were estimated by overlaying the storm surge values to future sea level rise. Future analyses should quantify the non-stationarity of the storm surge as well as changes in the mean sea level [82–84]. The analyses show positive dependencies between fluvial and coastal flooding over the region, suggesting that the corresponding compound effects should be considered in developing mitigation and adaptation measures. While the riverine flooding mainly affects the inundation area upstream of the study reach, coastal flooding combined with river overflows can significantly impact the areas close to Harry's River mouth and the upstream regions. Further, areas close to the estuary are vulnerable to compound flooding caused by river overflows, storm surge, wave, and sea-level rise. Future urbanization growth and population increases in urban low-lying areas can increase the flood risks.

Supplementary Materials: The following are available online at https://www.mdpi.com/article/10.3390/w13131774/s1, Table S1: Roughness value (manning's n) for different land cover types, Table S2: Characteristics of the selected GCMs, Table S3: List of satellite images including the reference images and the flood image, Table S4: List of scenarios and the simulations, Table S5: Peak Rainfall (mm) values corresponding to WRF- and GCM-IDF curves based on CanESM2 simulations in the 2050s, Table S6: Peak Rainfall (mm) values corresponding to WRF- and GCM-IDF curves based on CanESM2 simulations in the 2080s, Figure S1: (a) Survey cross sections in the HEC-RAS model,

(b) Additional surveyed cross-sections (red line) with bathymetry-fused DEM, Figure S2: HEC-RAS 1D & 2D model evaluation for 3 November at 8pm 7 November 2010 at 4 pm. Orange represents 1D HEC-RAS results, blue represents 2D HEC-RAS results; obs represents the measurements at 4 pm, 6 November 2010, Figure S3: Comparison between 2D simulated flood inundation extents based on different roughness values for channel and floodplain: (a) 0.033 and 0.05; (b) 0.045 and 0.05; (c) 0.033 and 0.08, Figure S4: The detected flood inundated area based on Sentinel-1 imagery (14 January 2018) compared with HEC-RAS 2D model simulations, Figure S5: Hyetographs corresponding to a 25-year rainfall event generated by three design methods for the historical and future (2050s; RCP 4.5) periods, Figure S6: Similar to Figure 5 but for 100-year event and future period of 2080s corresponding to RCP 8.5 emission scenario (a) GCM-IDF (ABM) and (b) WRF-IDF (ABM-2), Figure S7: Rainfall hyetographs corresponding to CanESM2 simulations for 2050s under RCP4.5, Figure S8: Rainfall hyetographs corresponding to HadGEM-AO (AO) simulations for 2080s under RCP8.5, Figure S9: Simulated peak discharge rates corresponding to WRF- and GCM-IDFs for (a) 25-yr event and (b) 100-yr event. Both simulations correspond to 2050s (2041-2070) under RCP 8.5. emission scenario. The participating GCMs include: HadGEM-AO (AO), GFDL-CM3 (CM3), CSIRO-Mk3.6.0 (CSIRO), HadGEM2-ES (ES), GFDL-ESM2G (ESM2G), and CanESM2 (CAN), Figure S10: Projected HEC-HMS hydrographs corresponding to the 100-year rainfall event for the historical and future (2050s; RCP8.5) conditions. The results correspond to the WRF-IDF curves based on a. ABM1, b. ABM2, c. Huff, d. SCS design storm methods, Figure S11: Flow hydrographs at the gauge of Harry's River below Highway Bridge (see location in Figure 1) for a 25-year event corresponding to the historical and future (2050s, RCP4.5) periods. Hyetographs are generated based on the HUFF method and a) GCM-IDFs b) WRF-IDFs, Figure S12: Projected changes in maximum flood depths corresponding to a 25-year event between future (2050s) and historical periods; (a) GCM-IDF & RCP 4.5, (b) WRF-IDF & RCP 4.5, (c) GCM-IDF & RCP 8.5, (d) WRF-IDF & RCP 8.5, Figure S13: Projected changes in maximum flood depths corresponding to a 100-year event between future (2050s) and historical periods; (a) GCM-IDF & RCP 4.5, (b) WRF-IDF & RCP 4.5, (c) GCM-IDF & RCP 8.5, (d) WRF-IDF & RCP 8.5, Figure S14: Projected changes in 25-year flood inundation corresponding to RCP 4.5 in 2050s compared to the historical condition (based on the SCS design storm method).

Author Contributions: Conceptualization, M.R.N. and A.J.C.; methodology, M.R.N., S.W., A.J.C. and A.A.K.; software, S.W., M.R.N. and A.A.K.; validation, S.W.; formal analysis, S.W. and M.R.N; investigation, S.W. and M.R.N.; resources, M.R.N. and A.A.K.; data curation, S.W., A.J.C. and M.R.N.; writing—original draft preparation, S.W.; writing—review and editing, M.R.N., A.J.C. and A.A.K.; visualization, S.W.; supervision, M.R.N.; project administration, M.R.N.; funding acquisition, M.R.N. All authors have read and agreed to the published version of the manuscript.

Funding: This project was supported by NSERC CRD, grant number CRDPJ 523924-18.

Institutional Review Board Statement: Not applicable.

Informed Consent Statement: Not applicable.

Acknowledgments: TanDEM data was obtained from Geoservices under the German Aerospace Center (https://geoservice.dlr.de/web/dataguide/tdm90/ (accessed on 26 June 2021)). SRTM DEM (https://www2.jpl.nasa.gov/srtm/statistics.html (accessed on 26 June 2021)) was downloaded from U.S. Geological Survey (USGS) EarthExplorer. Further, tide predictions and coastal water levels were available from Fisheries and Oceans Canada. We thank the Water Rights, Investigations and Modelling Section in the Department of Environment, Climate Change and Municipalities in Newfoundland and Labrador for providing access to the data and the calibrated HEC-HMS hydrological model.

Conflicts of Interest: The authors declare no conflict of interest and the funders had no role in the design of the study, analyses, or interpretation of data and results.

References

1. United Nations Office for Disaster Risk Reduction; Centre for Research on the Epidemiology of Disaster. *The Human Cost of Natural Disasters: A Global Perspective*; UN Office for Disaster Risk Reduction: Geneva, Switzerland; Centre for Research on the Epidemiology of Disasters: Brussels, Belgium, 2015.
2. Office of the Parliamentary Budget Officer. *Estimate of the Average Annual Cost for Disaster Financial Assistance Arrangements Due to Weather Events*; Office of the Parliamentary Budget Officer: Ottawa, ON, Canada, 2016.

3. Atlantic Climate Adaption Solutions Association. Flood Risk and Vulnerability Analysis Project. 2012. Available online: https://atlanticadaptation.ca/en/islandora/object/acasa%253A446 (accessed on 26 June 2021).
4. Zhou, Q.; Mikkelsen, P.S.; Halsnæs, K.; Arnbjerg-Nielsen, K. Framework for economic pluvial flood risk assessment considering climate change effects and adaptation benefits. *J. Hydrol.* **2012**, *414*, 539–549. [CrossRef]
5. Kaspersen, P.S.; Ravn, N.H.; Arnbjerg-Nielsen, K.; Madsen, H.; Drews, M. Comparison of the impacts of urban devel-opment and climate change on exposing European cities to pluvial flooding. *Hydrol. Earth Syst. Sci.* **2017**, *21*, 4131–4147. [CrossRef]
6. Pregnolato, M.; Ford, A.; Glenis, V.; Wilkinson, S.; Dawson, R. Impact of Climate Change on Disruption to Urban Transport Networks from Pluvial Flooding. *J. Infrastruct. Syst.* **2017**, *23*, 04017015. [CrossRef]
7. Evans, B.; Chen, A.S.; Djordjević, S.; Webber, J.; Gómez, A.G.; Stevens, J. Investigating the Effects of Pluvial Flooding and Climate Change on Traffic Flows in Barcelona and Bristol. *Sustainability* **2020**, *12*, 2330. [CrossRef]
8. Wilby, R.L.; Beven, K.J.; Reynard, N. Climate change and fluvial flood risk in the UK: More of the same? *Hydrol. Process.* **2008**, *22*, 2511–2523. [CrossRef]
9. Van PD, T.; Popescu, I.; Van Griensven, A.; Solomatine, D.P.; Trung, N.H.; Green, A. A study of the climate change im-pacts on fluvial flood propagation in the Vietnamese Mekong Delta. *Hydrol. Earth Syst. Sci.* **2012**, *16*, 4637–4649. [CrossRef]
10. Eccles, R.; Zhang, H.; Hamilton, D. A review of the effects of climate change on riverine flooding in subtropical and trop-ical regions. *J. Water Clim. Chang.* **2019**, *10*, 687–707. [CrossRef]
11. Purvis, M.J.; Bates, P.D.; Hayes, C.M. A probabilistic methodology to estimate future coastal flood risk due to sea level rise. *Coast. Eng.* **2008**, *55*, 1062–1073. [CrossRef]
12. Thompson, K.R.; Bernier, N.B.; Chan, P. Extreme sea levels, coastal flooding and climate change with a focus on Atlantic Canada. *Nat. Hazards* **2009**, *51*, 139–150. [CrossRef]
13. Garner, A.J.; Mann, M.E.; Emanuel, K.A.; Kopp, R.E.; Lin, N.; Alley, R.B.; Horton, B.P.; DeConto, R.M.; Donnelly, J.P.; Pollard, D. Impact of climate change on New York City's coastal flood hazard: Increasing flood heights from the preindustrial to 2300 CE. *Proc. Natl. Acad. Sci. USA* **2017**, *114*, 11861–11866. [CrossRef]
14. Didier, D.; Baudry, J.; Bernatchez, P.; Dumont, D.; Sadegh, M.; Bismuth, E.; Bandet, M.; Dugas, S.; Sévigny, C. Multihazard simulation for coastal flood mapping: Bathtub versus numerical modelling in an open estuary, Eastern Canada. *J. Flood Risk Manag.* **2019**, *12*, 12505. [CrossRef]
15. Zhang, X.; Flato, G.; Kirchmeier-Young, M.; Vincent, L.; Wan, H.; Wang, X.; Rong, R.; Fyfe, J.; Li, G.; Kharin, V.V. Changes in Temperature and Precipitation across Canada. In *Canada's Changing Climate Report*; Bush, E., Lemmen, D.S., Eds.; Government of Canada: Ottawa, ON, USA, 2019; Chapter 4; pp. 112–193.
16. Abbasian, M.S.; Abrishamchi, A.; Najafi, M.R.; Moghim, S. Multi-site statistical downscaling of precipitation using generalized hier-archical linear models: A case study of the imperilled Lake Urmia basin. *Hydrol. Sci. J.* **2020**, *65*, 2466–2481. [CrossRef]
17. Abbasian, M.S.; Najafi, M.R.; Abrishamchi, A. Increasing risk of meteorological drought in the Lake Urmia basin under climate change: Introducing the precipitation–temperature deciles index. *J. Hydrol.* **2021**, *592*, 125586. [CrossRef]
18. Kay, A.L.; Davies, H.N.; Bell, V.A.; Jones, R. Comparison of uncertainty sources for climate change impacts: Flood frequency in England. *Clim. Chang.* **2009**, *92*, 41–63. [CrossRef]
19. Najafi, M.R.; Moradkhani, H.; Jung, I.W. Assessing the uncertainties of hydrologic model selection in climate change impact studies. *Hydrol. Process.* **2011**, *25*, 2814–2826. [CrossRef]
20. Najafi, M.R.; Zwiers, F.W.; Gillett, N.P. Attribution of Observed Streamflow Changes in Key British Columbia Drainage Basins. *Geophys. Res. Lett.* **2017**, *44*, 11–12. [CrossRef]
21. Falconer, R.; Cobby, D.; Smyth, P.; Astle, G.; Dent, J.; Golding, B. Pluvial flooding: New approaches in flood warning, mapping and risk management. *J. Flood Risk Manag.* **2009**, *2*, 198–208. [CrossRef]
22. Yin, J.; Yu, D.; Yin, Z.; Liu, M.; He, Q. Evaluating the impact and risk of pluvial flash flood on intra-urban road network: A case study in the city center of Shanghai, China. *J. Hydrol.* **2016**, *537*, 138–145. [CrossRef]
23. Maksimović, Č.; Prodanović, D.; Boonya-Aroonnet, S.; Leitão, J.P.; Djordjević, S.; Allitt, R. Overland flow and pathway analysis for modelling of urban pluvial flooding. *J. Hydraul. Res.* **2009**, *47*, 512–523. [CrossRef]
24. Yu, D.; Lane, S. Urban fluvial flood modelling using a two-dimensional diffusion-wave treatment, part 1: Mesh resolution effects. *Hydrol. Process.* **2004**, *20*, 1085–1092. [CrossRef]
25. Xue, X.; Wang, N.; Ellingwood, B.R.; Zhang, K. The impact of climate change on resilience of communities vulnerable to riverine flooding. In *Climate Change and Its Impacts*; Springer: Cham, Switzerland, 2018; pp. 129–144.
26. Bates, P.D.; Dawson, R.J.; Hall, J.W.; Horritt, M.S.; Nicholls, R.J.; Wicks, J.; Hassan, M.A.A.M. Simplified two-dimensional numerical modelling of coastal flooding and example applications. *Coast. Eng.* **2005**, *52*, 793–810. [CrossRef]
27. Didier, D.; Bernatchez, P.; Boucher-Brossard, G.; Lambert, A.; Fraser, C.; Barnett, R.L.; Van-Wierts, S. Coastal flood as-sessment based on field debris measurements and wave runup empirical model. *J. Mar. Sci. Eng.* **2015**, *3*, 560–590. [CrossRef]
28. Ganguli, P.; Merz, B. Extreme Coastal Water Levels Exacerbate Fluvial Flood Hazards in Northwestern Europe. *Sci. Rep.* **2019**, *9*, 1–14. [CrossRef] [PubMed]
29. Najafi, M.R.; Zhang, Y.; Martyn, N. A flood risk assessment framework for interdependent infrastructure systems in coastal envi-ronments. *Sustain.Cities Soc.* **2021**, *64*, 102516. [CrossRef]

30. Serafin, K.A.; Ruggiero, P.; Parker, K.; Hill, D.F. What's streamflow got to do with it? A probabilistic simulation of the competing oceanographic and fluvial processes driving extreme along-river water levels. *Nat. Hazards Earth Syst. Sci.* **2019**, *19*, 1415–1431. [CrossRef]
31. Lian, J.; Xu, K.; Ma, C. Joint impact of rainfall and tidal level on flood risk in a coastal city with a complex river net-work: A case study of Fuzhou city, China. *Hydrol. Earth Syst. Sci.* **2013**, *17*, 679. [CrossRef]
32. Zheng, F.; Westra, S.; Sisson, S.A. Quantifying the dependence between extreme rainfall and storm surge in the coastal zone. *J. Hydrol.* **2013**, *505*, 172–187. [CrossRef]
33. Wahl, T.; Jain, S.; Bender, J.; Meyers, S.; Luther, M.E. Increasing risk of compound flooding from storm surge and rainfall for major US cities. *Nat. Clim. Chang.* **2015**, *5*, 1093–1097. [CrossRef]
34. Zhang, Y.; Najafi, M.R. Probabilistic Numerical Modeling of Compound Flooding Caused by Tropical Storm Matthew Over a Da-ta-Scarce Coastal Environment. *Water Resour. Res.* **2020**, *56*, e2020WR028565. [CrossRef]
35. Bevacqua, E.; Maraun, D.; Vousdoukas, M.I.; Voukouvalas, E.; Vrac, M.; Mentaschi, L.; Widmann, M. Higher probability of compound flooding from precipitation and storm surge in Europe under anthropogenic climate change. *Sci. Adv.* **2019**, *5*, eaaw5531. [CrossRef] [PubMed]
36. Couasnon, A.; Eilander, D.; Muis, S.; Veldkamp, T.I.E.; Haigh, I.D.; Wahl, T.; Winsemius, H.C.; Ward, P.J. Measuring compound flood potential from river discharge and storm surge extremes at the global scale. *Nat. Hazards Earth Syst. Sci.* **2020**, *20*, 489–504. [CrossRef]
37. Ikeuchi, H.; Hirabayashi, Y.; Yamazaki, D.; Muis, S.; Ward, P.J.; Winsemius, H.C.; Verlaan, M.; Kanae, S. Compound simulation of fluvial floods and storm surges in a global coupled river-coast flood model: Model development and its application to 2007 C yclone S idr in B angladesh. *J. Adv. Modeling Earth Syst.* **2017**, *9*, 1847–1862. [CrossRef]
38. Jalili Pirani, F.; Najafi, M.R. Recent trends in individual and multivariate compound flood drivers in Canada's coasts. *Water Re-sour. Res.* **2020**, *56*, e2020WR027785. [CrossRef]
39. Kumbier, K.; Carvalho, R.C.; Vafeidis, A.T.; Woodroffe, C.D. Investigating Compound Flooding in an Estuary Using Hydrodynamic Modelling: A Case Study from the Shoalhaven River, Australia. *Nat. Hazards Earth Syst. Sci.* **2018**, *18*, 463–477. [CrossRef]
40. Bunya, S.; Dietrich, J.C.; Westerink, J.J.; Ebersole, B.A.; Smith, J.M.; Atkinson, J.H.; Jensen, R.; Resio, D.T.; Luettich, R.A.; Dawson, C.; et al. A High-Resolution Coupled Riverine Flow, Tide, Wind, Wind Wave, and Storm Surge Model for Southern Louisiana and Mississippi. Part I: Model Development and Validation. *Mon. Weather. Rev.* **2010**, *138*, 345–377. [CrossRef]
41. Ray, T.; Stepinski, E.; Sebastian, A.; Bedient, P.B. Dynamic Modeling of Storm Surge and Inland Flooding in a Texas Coastal Floodplain. *J. Hydraul. Eng.* **2011**, *137*, 1103–1110. [CrossRef]
42. Bacopoulos, P.; Tang, Y.; Wang, D.; Hagen, S.C. Integrated Hydrologic-Hydrodynamic Modeling of Estuarine-Riverine Flooding: 2008 Tropical Storm Fay. *J. Hydrol. Eng.* **2017**, *22*, 04017022. [CrossRef]
43. Bakhtyar, R.; Maitaria, K.; Velissariou, P.; Trimble, B.; Mashriqui, H.; Moghimi, S.; Abdolali, A.; van der Westhuysen, A.J.; Ma, Z.; Clark, E.P.; et al. A new 1D/2D coupled modeling approach for a riverine-estuarine system under storm events: Application to Delaware River Basin. *J. Geophys. Res. Oceans* **2020**, *125*, e2019JC015822. [CrossRef]
44. Lee, C.; Hwang, S.; Do, K.; Son, S. Increasing flood risk due to river runoff in the estuarine area during a storm landfall. *Estuar. Coast. Shelf Sci.* **2019**, *221*, 104–118. [CrossRef]
45. Pasquier, U.; He, Y.; Hooton, S.; Goulden, M.; Hiscock, K.M. An integrated 1D–2D hydraulic modelling approach to assess the sensitivity of a coastal region to compound flooding hazard under climate change. *Nat. Hazards* **2018**, *98*, 915–937. [CrossRef]
46. Ganguli, P.; Paprotny, D.; Hasan, M.; Güntner, A.; Merz, B. Projected Changes in Compound Flood Hazard from Riverine and Coastal Floods in Northwestern Europe. *Earth's Future* **2020**, *8*, e2020EF001752. [CrossRef]
47. Garcia, E.S.; Loáiciga, H.A. Sea-level rise and flooding in coastal riverine flood plains. *Hydrol. Sci. J.* **2013**, *59*, 204–220. [CrossRef]
48. Ghanbari, M.; Arabi, M.; Kao, S.; Obeysekera, J.; Sweet, W. Climate Change and Changes in Compound Coastal-Riverine Flooding Hazard Along the U.S. Coasts. *Earth's Future* **2021**, *9*, e2021EF002055. [CrossRef]
49. Moftakhari, H.R.; Salvadori, G.; AghaKouchak, A.; Sanders, B.; Matthew, R.A. Compounding effects of sea level rise and fluvial flooding. *Proc. Natl. Acad. Sci. USA* **2017**, *114*, 9785–9790. [CrossRef] [PubMed]
50. Maraun, D.; Wetterhall, F.; Ireson, A.; Chandler, R.; Kendon, E.J.; Widmann, M.; Brienen, S.; Rust, H.W.; Sauter, T.; Themeßl, M.; et al. Precipitation downscaling under climate change: Recent developments to bridge the gap between dynamical models and the end user. *Rev. Geophys.* **2010**, *48*, RG3003. [CrossRef]
51. Canada, S. *2016 Census of Canada: Technical Report*; Statistics Canada: Ottawa, ON, Canada, 2016.
52. Government of Newfoundland and Labrador. *Hydrotechnical Study of Stephenville Crossing/Black Duck Siding*; Water Resources Management Division, Department of Environment and Conservation: St. John's, NL, Canada, 2012.
53. Vitousek, S.; Barnard, P.L.; Fletcher, C.H.; Frazer, N.; Erikson, L.; Storlazzi, C.D. Doubling of coastal flooding frequency within decades due to sea-level rise. *Sci. Rep.* **2017**, *7*, 1399. [CrossRef] [PubMed]
54. Chow, V.T. *Open Channel Hydraulics*; McGraw-Hill Interamericana: New York, NY, USA, 1959.
55. Perez, J.; Menendez, M.; Mendez, F.J.; Losada, I.J. Evaluating the performance of cmip3 and cmip5 global climate models over the north-east Atlantic region. *Climat. Dyn.* **2014**, *43*, 2663–2680. [CrossRef]
56. Mahmoudi, M.H.; Najafi, M.R.; Singh, H.; Schnorbus, M. Spatial and temporal changes in climate extremes over northwestern North America: The influence of internal climate variability and external forcing. *Clim. Chang.* **2021**, *165*, 1–19. [CrossRef]

57. Singh, H.; Najafi, M.R.; Cannon, A.J. Characterizing non-stationary compound extreme events in a changing climate based on large-ensemble climate simulations. *Clim. Dyn.* **2021**, *56*, 1389–1405. [CrossRef]
58. Simonovic, S.P.; Schardong, A.; Sandink, D.; Srivastav, R. A web-based tool for the development of Intensity Duration Frequency curves under changing climate. *Environ. Model. Softw.* **2016**, *81*, 136–153. [CrossRef]
59. Rasmussen, K.L.; Prein, A.F.; Rasmussen, R.M.; Ikeda, K.; Liu, C. Changes in the convective population and thermodynamic environ-ments in convection-permitting regional climate simulations over the United States. *Climat. Dyn.* **2020**, *55*, 383–408. [CrossRef]
60. Kimura, F.; Kitoh, A. *Downscaling by Pseudo Global Warming Method*; The Final Report of ICCAP; Research Institute for Humanity and Nature: Kyoto, Japan, 2007; pp. 43–46.
61. Mugume, I.; Basalirwa, C.P.; Waiswa, D.; Ngailo, T. Spatial variation of WRF model rainfall prediction over Uganda. *J. Environ. Chem. Ecol. Geol. Geophys. Eng.* **2017**, *11*, 553–557.
62. Innocenti, S.; Mailhot, A.; Frigon, A.; Cannon, A.J.; LeDuc, M. Observed and Simulated Precipitation over Northeastern North America: How Do Daily and Subdaily Extremes Scale in Space and Time? *J. Clim.* **2019**, *32*, 8563–8582. [CrossRef]
63. Prein, A.F.; Rasmussen, R.M.; Ikeda, K.; Liu, C.; Clark, M.P.; Holland, A.F.P.R.M.R.K.I.C.L.M.P.C.G.J. The future intensification of hourly precipitation extremes. *Nat. Clim. Chang.* **2017**, *7*, 48–52. [CrossRef]
64. Cannon, A.J.; Innocenti, S. Projected intensification of sub-daily and daily rainfall extremes in convection-permitting climate model simulations over North America: Implications for future intensity–duration–frequency curves. *Nat. Hazards Earth Syst. Sci.* **2019**, *19*, 421–440. [CrossRef]
65. Pall, P.; Allen, M.R.; Stone, D.A. Testing the Clausius–Clapeyron constraint on changes in extreme precipitation under CO_2 warming. *Climat. Dyn.* **2007**, *28*, 351–363. [CrossRef]
66. Batterson, M.; Liverman, D. *Past and Future Sea-Level Change in Newfoundland and Labrador: Guidelines for Policy and Planning: Current Research*; Report 10–1; Newfoundland and Labrador Department of Natural Resources Geological Survey: St. John's, NL, Canada, 2010.
67. Tarasov, L.; Peltier, W. A geophysically constrained large ensemble analysis of the deglacial history of the North American ice-sheet complex. *Quat. Sci. Rev.* **2004**, *23*, 359–388. [CrossRef]
68. Karim, M.; Mimura, N. Impacts of climate change and sea-level rise on cyclonic storm surge floods in Bangladesh. *Glob. Environ. Chang.* **2008**, *18*, 490–500. [CrossRef]
69. Long, S.; E Fatoyinbo, T.; Policelli, F. Flood extent mapping for Namibia using change detection and thresholding with SAR. *Environ. Res. Lett.* **2014**, *9*, 035002. [CrossRef]
70. Twele, A.; Cao, W.; Plank, S.; Martinis, S. Sentinel-1-based flood mapping: A fully automated processing chain. *Int. J. Remote Sens.* **2016**, *37*, 2990–3004. [CrossRef]
71. Pasmurov, A.; Zinoviev, J. *Radar Imaging and Holography*; Bibliovault OAI Repository, the University of Chicago Press: Chicago, IL, USA, 2005. [CrossRef]
72. Senthilnath, J.; Shenoy, H.V.; Rajendra, R.; Omkar, S.N.; Mani, V.; Diwakar, P.G. Integration of speckle de-noising and image segmentation using Synthetic Aperture Radar image for flood extent extraction. *J. Earth Syst. Sci.* **2013**, *122*, 559–572. [CrossRef]
73. Razi, M.; Ariffin, J.; Tahir, W.; Arish, N. Flood Estimation Studies using Hydrologic Modeling System (HEC-HMS) for Johor River, Malaysia. *J. Appl. Sci.* **2010**, *10*, 930–939. [CrossRef]
74. Halwatura, D.; Najim, M. Application of the HEC-HMS model for runoff simulation in a tropical catchment. *Environ. Model. Softw.* **2013**, *46*, 155–162. [CrossRef]
75. Tassew, B.G.; Belete, M.A.; Miegel, K. Application of HEC-HMS Model for Flow Simulation in the Lake Tana Basin: The Case of Gilgel Abay Catchment, Upper Blue Nile Basin, Ethiopia. *Hydrology* **2019**, *6*, 21. [CrossRef]
76. Chow, V.T. *Applied Hydrology: International Edition*; MacGraw-Hill, Inc.: New York, NY, USA, 1988; p. 149.
77. US Army Corps of Engineers. *Hydrologic Modeling System (HEC-HMS) Application Guide*; Version 3.1.0; Institute for Water Resources: Davis, CA, USA, 2008.
78. Singh, H.; Najafi, M.R. Evaluation of gridded climate datasets over Canada using univariate and bivariate approaches: Implications for hydrological modelling. *J. Hydrol.* **2020**, *584*, 124673. [CrossRef]
79. Singh, H.; Pirani, F.J.; Najafi, M.R. Characterizing the temperature and precipitation covariability over Canada. *Theor. Appl. Clim.* **2019**, *139*, 1543–1558. [CrossRef]
80. Wu, W.; Westra, S.; Leonard, M. Estimating the probability of compound floods in estuarine regions. *Hydrol. Earth Syst. Sci.* **2021**, *25*, 2821–2841. [CrossRef]
81. Zscheischler, J.; Westra, S.; Hurk, B.J.J.M.V.D.; Seneviratne, S.I.; Ward, P.J.; Pitman, A.; AghaKouchak, A.; Bresch, D.N.; Leonard, M.; Wahl, T.; et al. Future climate risk from compound events. *Nat. Clim. Chang.* **2018**, *8*, 469–477. [CrossRef]
82. Lowe, J.A.; Gregory, J.M.; Flather, R.A. Changes in the occurrence of storm surges around the United Kingdom under a future climate scenario using a dynamic storm surge model driven by the Hadley Centre climate models. *Clim. Dyn.* **2001**, *18*, 179–188. [CrossRef]
83. Rahmstorf, S. Rising hazard of storm-surge flooding. *Proc. Natl. Acad. Sci. USA* **2017**, *114*, 11806–11808. [CrossRef] [PubMed]
84. Yang, J.-A.; Kim, S.; Son, S.; Mori, N.; Mase, H. Correction to: Assessment of uncertainties in projecting future changes to extreme storm surge height depending on future SST and greenhouse gas concentration scenarios. *Clim. Chang.* **2020**, *162*, 425–442. [CrossRef]

water

MDPI

Article

Nonstationary Design Flood Estimation in Response to Climate Change, Population Growth and Cascade Reservoir Regulation

Yuzuo Xie [1], Shenglian Guo [1,*], Lihua Xiong [1], Jing Tian [1] and Feng Xiong [2]

1 State Key Laboratory of Water Resources and Hydropower Engineering Science, Wuhan University, Wuhan 430072, China; yuzuoxie@whu.edu.cn (Y.X.); xionglh@whu.edu.cn (L.X.); jingtian@whu.edu.cn (J.T.)
2 Bureau of Hydrology, Changjiang Water Resources Commission, Wuhan 430010, China; fxiong07@whu.edu.cn
* Correspondence: slguo@whu.edu.cn

Abstract: The hydrologic data series are nonstationary due to climate change and local anthropogenic activities. The existing nonstationary design flood estimation methods usually focus on the statistical nonstationarity of the flow data series in the catchment, which neglect the hydraulic approach, such as reservoir flood regulation. In this paper, a novel approach to comprehensively consider the driving factors of non-stationarities in design flood estimation is proposed, which involves three main steps: (1) implementation of the candidate predictors with trend tests and change point detection for preliminary analysis; (2) application of the nonstationary flood frequency analysis with the principle of Equivalent Reliability (ER) for design flood volumes; (3) development of a nonstationary most likely regional composition (NS-MLRC) method, and the estimation of a design flood hydrograph at downstream cascade reservoirs. The proposed framework is applied to the cascade reservoirs in the Han River, China. The results imply that: (1) the NS-MLRC method provides a much better explanation for the nonstationary spatial correlation of the flood events in Han River basin, and the multiple nonstationary driving forces can be precisely quantified by the proposed design flood estimation framework; (2) the impacts of climate change and population growth are long-lasting processes with significant risk of flood events compared with stationary distribution conditions; and (3) the swift effects of cascade reservoirs are reflected in design flood hydrographs with lower peaks and lesser volumes. This study can provide a more integrated template for downstream flood risk management under the impact of climate change and human activities.

Keywords: cascade reservoirs; design flood; nonstationary conditions; equivalent reliability; most likely regional composition; dependence structure

Citation: Xie, Y.; Guo, S.; Xiong, L.; Tian, J.; Xiong, F. Nonstationary Design Flood Estimation in Response to Climate Change, Population Growth and Cascade Reservoir Regulation. *Water* **2021**, *13*, 2687. https://doi.org/10.3390/w13192687

Academic Editors: Rajesh R. Shrestha and Mohammad Reza Najafi

Received: 16 August 2021
Accepted: 23 September 2021
Published: 28 September 2021

1. Introduction

Traditional design flood for hydraulic structures such as reservoirs is based on a stationary hypothesis, meaning that the driving factors (e.g., climate change, urbanization and reservoir flood regulation) act to generate the hydrological variables in the same way as in the past, present and likely the future [1–5]. However, the statistical characteristics of flood series might alter due to a changing environment [6,7]. If hydrological engineers do not fully consider the nonstationarity of the hydrological series, the results of the conventional stationary flood frequency analysis would be inaccurate in practice [8]. López and Francés [9] used climate and reservoir indices as external covariates in nonstationary flood frequency analysis. Yan et al. [10] considered climate change and population growth to explain the nonstationary properties of hydrological time series. Global warming, the primary factor that drives climate change, has altered the meteorological regimes in some regions [11,12].

Population growth will not only lead urbanization but also lead to increased water consumption. Rapid urbanization over recent decades has significantly changed the regional hydrological characteristics of catchments [13–15]. The river network systems have been obviously influenced by the process of urbanization, aggravating the hazards of floods and water degradation [16,17]. Water consumption is broadly reflected in daily life and productions, which can also affect the hydrological regime in the catchment. Reservoirs, one of the effective facilities for flood control, hydropower generation and other social functions, have gradually formed cascade reservoir systems [18,19]. The great impact brought by reservoir flood regulation has tremendously altered the hydrological characteristics in rivers [20]. To sum up, it is difficult to find a river basin that is not impacted by global warming and anthropogenic activities, particularly in rapidly developing China.

Milly et al. have elucidated the challenges about how to deal with design floods and water resources management under nonstationary conditions [1]. The commonly used technique to gain the changing knowledge of flood regimes is nonstationary flood frequency analysis [9,21]. Strupczewski et al. [22–24] presented a nonstationary approach to at-site flood frequency analysis, in which the distribution parameters are represented as the functions of explanatory variables to explain the nonstationary hydrological series [25,26]. Stasinopoulos and Rigby proposed "Generalized Additive Models for Location, Scale and Shape" (GAMLSS), which is a powerful implement for nonstationary frequency analysis of time series [27]. The meteorological factors (such as precipitation), population growth and reservoir index (RI) are widely used as explanatory variables incorporated in GAMLSS as covariates [9,28,29]. Taking RI for an example, López and Francés [9] considered the effects of RI from two aspects: the data used for flood frequency analysis are observed daily flow series' that have been affected by upstream reservoirs; some of the reservoir characteristic parameters, such as the catchment area and the reservoir total storage capacity, are integrated to form RI. However, the RI series are piecewise constants in spite of reservoir operation rules [30]. Although RI has been improved (by a hand of studies) for greater performance in nonstationary model fitting, the strategies of reservoir operation are hard to be quantified with the GAMLSS-based nonstationary flood frequency analysis framework [8,31]. Therefore, the covariates grounded in (or modified by) RI are unable to accurately consider the impact of cascade reservoir regulation.

The flood regional composition (FRC) combined with reservoir operation rules [32–34] can overcome this drawback. The aim of the FRC is to study the flood generation mechanism at the investigated downstream site. The inflow of the target reservoir consists of the first upstream reservoir inflow and all interval inflows between adjacent reservoirs. Then, the outflow of the downstream reservoir can be obtained through the reservoir operation rules. Among all possible compositions based on the water balance equation, an appropriate FRC needs to be selected. Guo et al. [33] proposed the most likely regional composition (MLRC) method and derived theoretical formula for triple cascade reservoirs. The MLRC method presumes that FRCs are diverse with their occurrence probabilities, which can be quantified by the multivariate probability density function (PDF) of flood events occurring at all sub-basins, and the FRC with the largest occurrence probability should be chosen for representing the actual spatial correlation of flood events. With a rigorous statistical basis, the MLRC method has been successfully applied in the cascade reservoirs in the upper Yangtze River [35]. In contrast to the nonstationary flood frequency analysis applied with RI, the natural flow data restored by the observed data is employed in the FRC framework. Although the MLRC-based approach takes into full consideration both FRC and reservoir operation rules, it is based upon a stationary assumption condition so that the other variables such as climate change and population growth are ignored.

Under nonstationary conditions, the copulas with time-varying dependence structures have been primarily applied for modeling coincidence probabilities such as the joint return periods [8,36–38]. In fact, the changing environments might also alter the statistical correlation of FRC, and the dependence structure of the MLRC-based method is unable to catch the nonstationary spatial correlation. Therefore, we propose a nonstationary MLRC (NS-MLRC) method and then compare it with the MLRC method under stationary distribution conditions. Equivalent Reliability (ER) [39] is employed in this study to concatenate the stationary and nonstationary design criteria grounded in a given return period [40].

In a word, this study focuses on three objectives: (1) to propose a nonstationary design flood estimation framework; (2) to develop and verify the NS-MLRC method; and (3) to estimate design flood hydrographs at downstream sites. The rest of the paper is organized as follows: Section 2 describes the methodology used in this study. Section 3 briefly introduces the study area and data acquisition. Section 4 analyzes the nonstationary design flood estimation results. Section 5 discusses the nonstationary characteristics and the worst regional flood composition. Finally, Section 6 ends with conclusions.

2. Methodology

The flowchart of nonstationary design flood estimation for cascade reservoirs is shown in Figure 1, of which there are three main modules. The first module preliminarily analyzes the nonstationarity of the flood volume series. The second module calculates the design flood volumes based on nonstationary flood frequency analysis and ER criteria. The last module solves the flood regional compositions using the NS-MLRC method and derives the design flood hydrograph regulated by cascade reservoirs.

2.1. Preliminary Analysis Module

This module aims to manipulate the dataset, determine the candidate univariate and multivariate distributions and covariates, and analyze the nonstationarity preliminarily.

2.1.1. Candidate Distribution Functions, Copulas and Covariates

For nonstationary univariate frequency analysis, the type of the probability distribution $f_Y(\cdot)$ determines which form of frequency curves will be generated. Table 1 lists five probability distributions, namely Gamma, Weibull, Log normal, Gumbel, and Pearson type III, which are the alternative distributions for flood frequency analysis with the different link functions [41–43]. On the other hand, the Archimedean copulas are widely used in hydrological researches [8,33]. Three mono-parameter Archimedean copulas, i.e., Gumbel-Hougaard, Frank and Clayton copulas (Table 2) are selected as the candidate copulas for the time-varying dependence structure modelling.

The multiple covariates (i.e., explanatory variables, predictors or addictive terms) indicating both population growth and climate change are incorporated to explain the variation in nonstationary time series. The impact of urbanization and water consumption on the hydrologic flood series is considered by the population (Pop) [37,43], while the annual total precipitation (Prcp), one of the important hydro-climate factors, is applied to quantify the effects of climate change [10,44].

2.1.2. Trend Test and Change Point Detection

The Mann–Kendall trend test [45–47] and Pettitt's test [48] are widely used for trend test and change point detection, which can be applied to analyze the nonstationarity of time series [43,44]. A nonparametric multiple change point analysis approach designed by Matteson and James [49] is used for change point analysis of multivariate time series in this study.

2.2. Hydrologic Design Module

This module is responsible for nonstationary flood frequency analysis and design flood volume estimation.

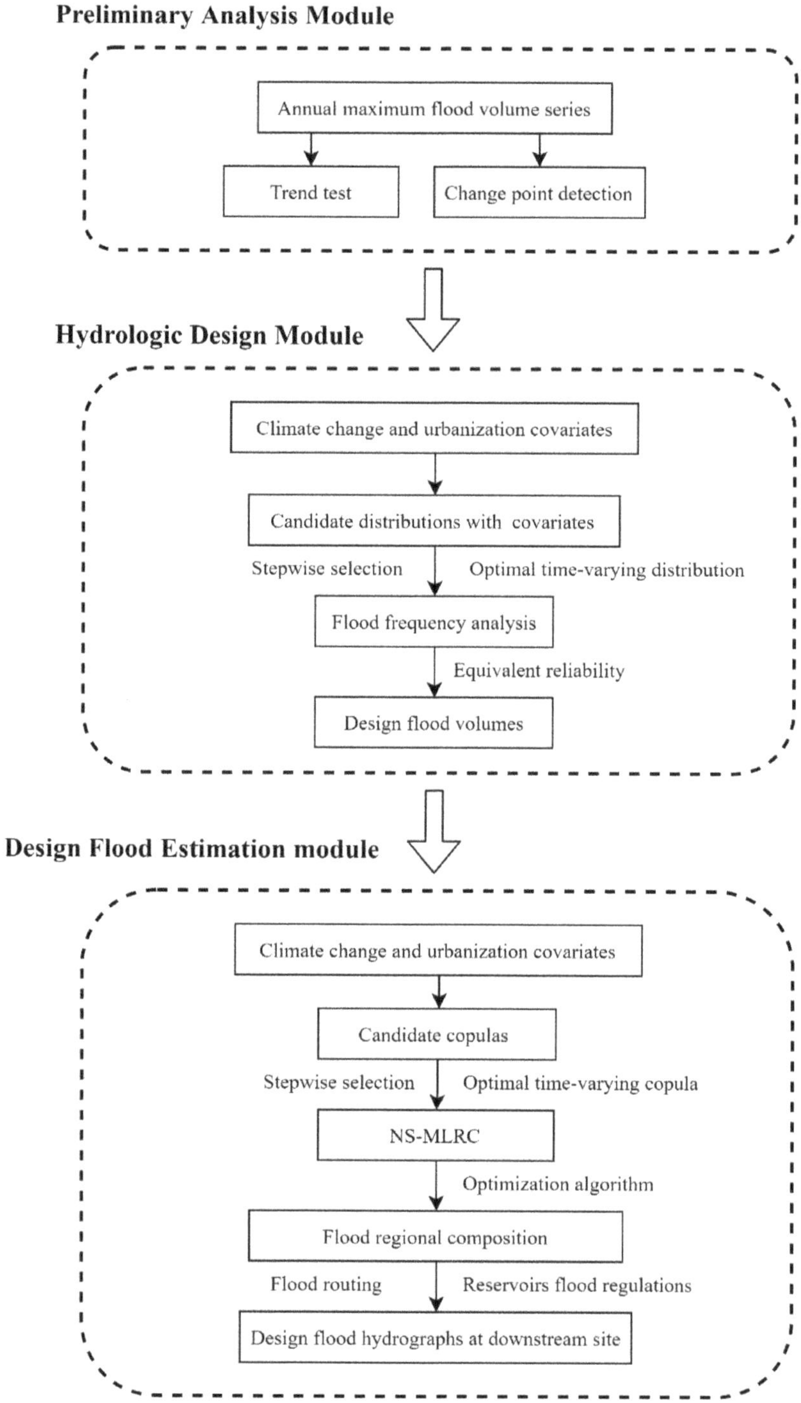

Figure 1. Flowchart of nonstationary design flood estimation in response to climate change, population growth and cascade reservoir regulation.

Table 1. The univariate distributions used to model the flood series in this study.

Distribution	PDF	Distribution Moments	Link Functions
Gamma	$f_Z(z\|\mu,\sigma) = \frac{1}{(\mu\sigma^2)^{1/\sigma^2}} \frac{z^{(1/\sigma^2-1)}e^{-z/\mu\sigma^2)}}{\Gamma(1/\sigma^2)}$ $z > 0, \mu > 0, \sigma > 0$	$E(Z) = \mu$ $\mathrm{Var}(Z) = \mu^2\sigma^2$	$g_1(\mu) = \ln(\mu)$ $g_2(\sigma) = \ln(\sigma)$
Weibull	$f_Z(z\|\mu,\sigma) = \frac{\sigma z^{\sigma-1}}{\mu^\sigma} \exp\left[-\left(\frac{z}{\mu}\right)^\sigma\right]$ $z > 0, \mu > 0, \sigma > 0$	$E(Z) = \mu\Gamma(1/\sigma+1)$ $\mathrm{Var}(Z) = \mu^2\left\{\Gamma(2/\sigma+1) - [\Gamma(1/\sigma+1)]^2\right\}$	$g_1(\mu) = \ln(\mu)$ $g_2(\sigma) = \ln(\sigma)$
Log-normal	$f_Z(z\|\mu,\sigma) = \frac{1}{\sqrt{2\pi}\sigma}\frac{1}{z}\exp\left[-\frac{(\log(z)-\mu)^2}{2\sigma^2}\right]$ $z > 0, \mu > 0, \sigma > 0$	$E(Z) = \omega^{1/2}e^{\mu}$ $\mathrm{Var}(Z) = \omega(\omega-1)e^{2\mu}$ $\omega = \exp(\sigma^2)$	$g_1(\mu) = \mu$ $g_2(\sigma) = \ln(\sigma)$
Gumbel	$f_Z(z\|\mu,\sigma) = \frac{1}{\sigma}\exp\left\{-\left(\frac{z-\mu}{\sigma}\right) - \exp\left[-\left(\frac{z-\mu}{\sigma}\right)\right]\right\}$ $-\infty < z < \infty, -\infty < \mu < \infty, \sigma > 0$	$E(Z) = \mu + \gamma\sigma$ $\mathrm{Var}(Z) = \frac{\pi}{\sqrt{6}}\sigma$	$g_1(\mu) = \mu$ $g_2(\sigma) = \ln(\sigma)$
Pearson type III	$f_Z(z\|\mu,\sigma,\nu) = \frac{1}{\|\sigma\|\Gamma(\nu)}\left(\frac{z-\mu}{\sigma}\right)^{\nu-1}\exp\left[-\left(\frac{z-\mu}{\sigma}\right)\right]$ $z > 0, \sigma \neq 0, \nu > 0$	$E(Z) = \mu + \nu\sigma,\ Var(Z) = \nu\sigma^2$ $Cv = \frac{2\sigma}{\|\sigma\|\nu^{1/2}}, Cs = \frac{\sigma\nu^{1/2}}{\|\sigma\|(\nu+\mu\sigma)}$	$g_1(\mu) = \ln(\mu)$ $g_2(\sigma) = \ln(\sigma)$ $g_3(\kappa) = \nu$

Table 2. Three Archimedean copulas used for the time-varying dependence structure modelling in this study.

Copula	Cumulative Distribution Function	Parameter	Link Function
Gumbel-Hougaard	$C(u,v\|\theta_c) = \exp\left(-\left((-\ln u)^{\theta_c} + (-\ln v)^{\theta_c}\right)^{1/\theta_c}\right)$	$\theta_c > 1$	$g_c(\theta_c) = \ln(\theta_c - 1)$
Frank	$C(u,v\|\theta_c) = -\frac{1}{\theta_c}\ln\left(1 + \frac{(e^{-u\theta_c}-1)\times(e^{-v\theta_c}-1)}{e^{-\theta_c}-1}\right)$	$\theta_c \neq 0$	$g_c(\theta_c) = \theta_c$
Clayton	$C(u,v\|\theta_c) = \left((u)^{-\theta_c} + (v)^{-\theta_c} - 1\right)^{-1/\theta_c}$	$\theta_c > 0$	$g_c(\theta_c) = \ln(\theta_c)$

2.2.1. Time-Varying Moments

Generally, the nonstationary flood frequency analysis can be established by the time-varying moments [50,51] based on the GAMLSS model [27]. Furthermore, the time-varying moments can be improved by replacing time with other physical covariates which have physical meanings [8,10]. The generalized linear model (GLM) [52] is implemented to establish the function between distribution parameters and their predictors. To facilitate understanding, a three-parameter time-varying distribution is taken as an example in the remainder of this literature.

For the time-varying probability distribution function $f_Y(y^t|\mu^t,\ \sigma^t,\ \nu^t)$, let y^t be the response variable at time t $(t = 1,2,\ldots,n)$ and the vector $[\mu^t,\ \sigma^t,\ \nu^t]$ be the three time-varying parameters. Then each parameter can be expressed as a function of the covariates x_i^t $(i = 1,2,\ldots,m)$ via a link function as follows:

$$\begin{cases} g_1(\mu^t) = \alpha_{10} + \sum\limits_{i=1}^{m} \alpha_{1i} x_i^t \\ g_2(\sigma^t) = \alpha_{20} + \sum\limits_{i=1}^{m} \alpha_{2i} x_i^t \\ g_3(\nu^t) = \alpha_{30} + \sum\limits_{i=1}^{m} \alpha_{3i} x_i^t \end{cases} \tag{1}$$

where $g_i(\cdot)$ are the link functions of distribution parameters for restricting the sample space [8]; $\alpha_{ki}(k = 1,2,3;\ i = 0,1,2,\ldots,m)$ are the GLM parameters. The time-varying moments are fitted by maximum penalized likelihood estimation [27,53]. The centile curve [54] and worm plot [55] are used for the goodness-of-fit test of univariate nonstationary distributions.

2.2.2. Selection of Time-Varying Distributions

The selection of the covariates and the type of probability distribution are the two steps for seeking the best time-varying moment model. For one specific time-varying distribution, the selection of covariates contains two aspects:

(1) The selection of the best GLM for a single parameter, which is conducted by the forward procedure [27]. In the forward procedure, all variables not currently in the model are considered for addition at each step, while all variables currently in the model are individually considered for deletion.

(2) For a given distribution, there exist diverse strategies to select the GLMs used to model all the parameters, i.e., μ, σ, and ν. The strategy proposed by Stasinopoulos and Rigby [27] is executed for GLMs selection of all the distribution parameters.

Since the shape parameter ν is too sensitive to the data series, it is often assumed as constant as other studies have done [10,56]. In practice, only the first two moments (i.e., μ and σ) are taken to consider nonstationary data series. The optimal probability distribution is selected from these five distributions mentioned in Table 1 based upon the Bayesian information criterion (BIC) [57]:

$$\text{BIC} = -2\text{ML} + \ln(n) \times \text{df} \tag{2}$$

where ML denotes the log-likelihood within a likelihood-based inferential procedure; df represents the total effective degrees of freedom and; n is the length of the data series. Comparing with the widely used Akaike information criterion (AIC) [58], the BIC has a larger penalty on the over-fitting phenomenon.

After the above selection procedures is performed for each probability distribution, then the distribution corresponding to the smallest BIC value is selected as the optimal univariate distribution.

2.2.3. Equivalent Reliability (ER)

The hydrologic design criteria according to the definition of return period under stationary conditions might be modified to adapt nonstationarity [43,59]. To ensure that the nonstationary flood frequency analysis is closely relevant to the practical condition, the flood events can be linked with the design lifespan of hydraulic projects [39]. The project reliability in the no flood event exceeds the specific design value z_p during the life period $T_1 - T_2$ of the project, which is defined as follows [59]:

$$\begin{cases} \mathrm{RE}^{S}_{T_1\sim T_2} = \prod\limits_{t=T_1}^{T_2} (1-p_t) = \prod\limits_{t=T_1}^{T_2} F_Z(z_p|\mu_0,\ \sigma_0,\ \nu_0) \\ \mathrm{RE}^{NS}_{T_1\sim T_2} = \prod\limits_{t=T_1}^{T_2} F_Z(z_p|\mu^t,\ \sigma^t,\ \nu^t) \end{cases} \tag{3}$$

where p_t is the exceedance probability at year t for design value z_p; the superscript S and NS represent stationary and nonstationary conditions, respectively.

If a project is designed to withstand a flood event that occurs once in m years under stationary conditions, then the reliability within the design life period $T_1 - T_2$ of the project is given by

$$\mathrm{RE}^{S}_{T_1\sim T_2} = \prod_{t=T_1}^{T_2} (1-p_t) = (1-\frac{1}{m})^{T_2-T_1+1} \tag{4}$$

The ER supposes that the hydrological design values derived using stationary and nonstationary flood control design standard should have the same reliability during the life time of hydraulic structures [39]. Assuming that $\mathrm{RE}^{S}_{T_1\sim T_2} = \mathrm{RE}^{NS}_{T_1\sim T_2}$, the nonstationary design value for a specific return period m is denoted by $z_{T_1\sim T_2}(m)$, which is measured by the following equation:

$$\prod_{t=T_1}^{T_2} F_Z(z_{T_1\sim T_2}(m)|\mu^t,\ \sigma^t,\ \nu^t) = (1-\frac{1}{m})^{T_2-T_1+1} \tag{5}$$

For given design life period $T_1 - T_2$ and return period m, the stationary ER criteria is calculated by Equation (4) firstly, and then the nonstationary design value $z_{T_1\sim T_2}(m)$ corresponding to m can be obtained eventually by solving Equation (5).

2.3. *Design Flood Estimation Module*

This module is used to establish the time-varying copula and derive the nonstationary design flood hydrographs based on the NS-MLRC method.

2.3.1. Time-Varying Dependence Parameter of Copulas

Traditional joint hydrological frequency analysis assumes that the parameters of both marginal distributions and copula functions are constant. Nevertheless, either the individual series or the dependence structure between the multiple series might be also nonstationary under changing environments [8,38]. To consider such possibility, a general form of the joint distribution of multiple random variables $(Y_1^t,\ Y_2^t,\ \ldots,\ Y_n^t)$ at any time t, is built. According to Sklar's theorem [60], the joint PDF can be expressed in terms of its marginal distributions and the associated dependence function, i.e., the time-varying copula can be expressed as follows:

$$\begin{aligned} H_{NS}(y_1^t, y_2^t,\ \ldots,\ y_n^t) &= C[F_1(y_1^t|\mu_1{}^t,\ \sigma_1{}^t,\ \nu_1{}^t),\ \ldots,\ F_n(y_n^t|\mu_n{}^t,\ \sigma_n{}^t,\ \nu_n{}^t)] \\ &= C(u_1^t, u_2^t,\ \ldots,\ u_n^t|\theta_c^t) \end{aligned} \tag{6}$$

where $H_{NS}(\cdot)$ denotes the cumulative distribution function (CDF) that defines the dependence structure of multiple variables; $C(\cdot)$ represents the CDF of copula function; $F(\cdot)$ represents the CDF of the hydrological random variables; $[\mu_i{}^t,\ \sigma_i{}^t,\ \nu_i{}^t]$ $(i = 1, 2, \ldots, n)$ are the time-varying marginal distribution parameters; θ_c^t denotes the time-varying copula parameter; and u_1^t , u_2^t, …, u_n^t are the marginal CDFs of the time-varying copula, which should range on [0,1].

Like the univariate distribution parameters expressed in Equation (1), the time-varying copula parameter $\theta_c{}^t$ can also be expressed as a GLM function of the covariates x_i^t $(i = 1, 2, \ldots, m)$ via a proper link function $g_c(\cdot)$ as follows:

$$g_c\left(\theta_c^t\right) = \beta_0 + \sum_{i=1}^{m} \beta_i x_i^t \tag{7}$$

where $g_c(\cdot)$ relies on the limits of θ_c^t (Table 2); $\beta_i (i = 0, 1, 2, \ldots, m)$ are the GLM parameters. The copula-based GAMLSS with non-random sample selection is applied in this study so that the non-stationarities of marginal distributions can be considered [61,62]. The time-varying dependence parameter of copula is estimated by a penalized likelihood framework with integrated automatic multiple smoothing parameter selections [63]. It is hard to calculate the distance between the fitted and the empirical frequency so that the Probability Integral Transformation (PIT) is used for the goodness-of-fit test of time-varying copula [64,65]. The forward procedure is applied for the selection of $\theta_c{}^t$ based on GLM, while the BIC criteria is used to select the optimal copula function.

2.3.2. Nonstationary Most Likely Regional Composition (NS-MLRC) Method

The sketch of flood regional composition with cascade reservoirs is shown in Figure 2, in which the random variables X_1, X_2, …, X_n,Y_1, Y_2, …, Y_n and Z represent the natural inflow of reservoirs A_1, A_2, …, A_n, intermediate basins B_1, B_2, …, B_n, and downstream site C with the corresponding values x_1, x_2, …, x_n, y_1, y_2, …, y_n and z, respectively.

For the A_1-A_2 sub-partition, the inflow of downstream reservoir $A_2(x_2)$ consists of the inflow of reservoir $A_1(x_1)$ and the inflow of intermediate basin $B_1(y_1)$. The inflow of downstream reservoir A_2 is affected by the operation of upstream reservoir A_1. Then (x_1, y_1) is determined as the FRC of x_2, in which inflow x_1 can be turned into discharge based on the operation strategy of reservoir A_1. Thus, the discharge of reservoir A_2 is relevant to the FRC of A_2, i.e.,(x_1, y_1). For the cascade reservoir system, all the FRC $(x_1, y_1, y_2, \ldots, y_n, z)$ should satisfy the principle of the water balance equation [33]:

$$\begin{cases} x_1 + y_1 = x_2 \\ x_1 + y_1 + y_2 = x_3 \\ \ldots \\ x_1 + y_1 + y_2 + \ldots + y_n = z \end{cases} \tag{8}$$

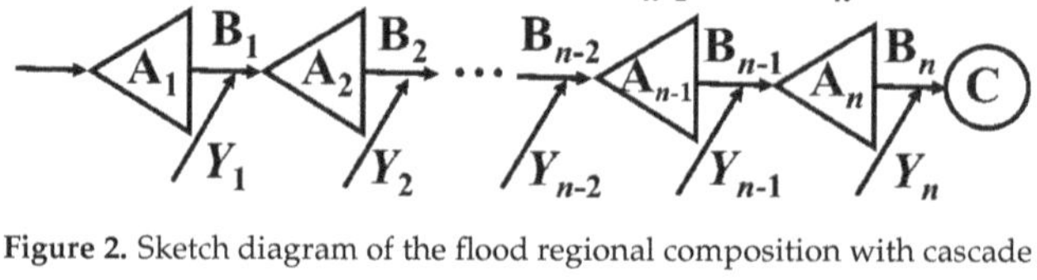

Figure 2. Sketch diagram of the flood regional composition with cascade reservoirs.

For the inherently stochastic mechanism of the flood generation, there exists multiple compositions of $(x_1, y_1, y_2, \ldots, y_n, z)$ in compliance with the water balance equation. The output of the MLRC method is the FRC with the largest occurrence probability. Nevertheless, under changing environments, the FRC derived by the MLRC method may vary over time. According to Sklar's theorem, the time-varying copula is expressed as follows

$$f(x_1^t, x_2^t, \ldots, x_{n-1}^t, x_n^t, z^t) = c(u_1^t, u_2^t, \ldots, u_{n-1}^t, u_n^t, v^t|\theta_c^t) \cdot \prod_{i=1}^{n} f_{X_i}(x_i{}^t|\mu_i{}^t, \sigma_i{}^t, \nu_i{}^t) \cdot f_Z(z^t|\mu_Z{}^t, \sigma_Z{}^t, \nu_Z{}^t) \tag{9}$$

where $c(\cdot)$ denotes the PDF of copula; $u_1^t, u_2^t, \ldots, u_{n-1}^t, u_n^t, v^t$ are the corresponding CDFs of $x_1^t, x_2^t, \ldots, x_{n-1}^t, x_n^t, z^t$, respectively; $f_{X_i}(x_i{}^t|\mu_i{}^t, \sigma_i{}^t, \nu_i{}^t)$ $(i = 1, 2, \ldots, n)$ and $f_Z(z^t|\mu_Z{}^t, \sigma_Z{}^t, \nu_Z{}^t)$ are the optimal time-varying marginal distributions of flood values occurring at reservoirs $A_1, A_2, \ldots, A_n$ and downstream site C, respectively.

The composition $(x_1^t, x_2^t, \ldots, x_{n-1}^t, x_n^t, z^t)$ is more likely to occur with the increase of PDF $f(x_1^t, x_2^t, \ldots, x_{n-1}^t, x_n^t, z^t)$ [66]. The NS-MLRC is obtained when $f(x_1^t, x_2^t, \ldots, x_{n-1}^t, x_n^t, z^t)$ is maximized by subjecting water balance constraint:

$$\max f(x_1^t, x_2^t, \ldots, x_{n-1}^t, x_n^t, z^t) = c(u_1^t, u_2^t, \ldots, u_{n-1}^t, u_n^t, v^t|\theta_c^t) \cdot \prod_{i=1}^{n} f_{X_i}(x_i{}^t|\mu_i{}^t, \sigma_i{}^t, \nu_i{}^t) \cdot f_Z(z^t|\mu_Z{}^t, \sigma_Z{}^t, \nu_Z{}^t)$$
$$s.t. \begin{cases} x_1^t + y_1^t = x_2^t \\ x_1^t + y_1^t + y_2^t = x_3^t \\ \cdots \\ x_1^t + y_1^t + y_2^t + \ldots + y_n^t = z^t = z_p \end{cases} \tag{10}$$

The joint PDF $f(x_1^t, x_2^t, \ldots, x_{n-1}^t, x_n^t, z^t)$ is maximized when its first-order derivative equals zero so that the following equation should be satisfied:

$$\begin{cases} \frac{\partial f(x_1^t, x_2^t, \ldots, x_{n-1}^t, x_n^t, z^t)}{\partial x_1^t} = 0 \\ \frac{\partial f(x_1^t, x_2^t, \ldots, x_{n-1}^t, x_n^t, z^t)}{\partial x_2^t} = 0 \\ \cdots \\ \frac{\partial f(x_1^t, x_2^t, \ldots, x_{n-1}^t, x_n^t, z^t)}{\partial z^t} = 0 \\ x_1^t + y_1^t = x_2^t \\ x_1^t + y_1^t + y_2^t = x_3^t \\ \cdots \\ x_1^t + y_1^t + y_2^t + \ldots + y_n^t = z^t = z_p \end{cases} \tag{11}$$

After deriving the composition $(x_1^t, x_2^t, \ldots, x_{n-1}^t, x_n^t, z^t)$ from Equation (11), the NS-FRC $(x_1^t, y_1^t, y_2^t, \ldots, y_n^t)$ can be obtained subsequently using NS-MLRC method. The Newton iteration algorithm is adopted to solve Equation (11) [33].

According to the flood prevention standard of cascade reservoirs, the design flood hydrographs of each sub-system are calculated by the peak and volume amplitude method [67,68] based on the results of the NS-MLRC method. River channel flood routing and reservoir flood control regulations are adopted to derive the design flood hydrograph at the downstream site C [34,69].

3. Study Area and Data

The Han River basin in China is located between 106–115° E and 30–35° N (see Figure 3), and has a total length of 1530 km. As one of the most important tributaries of the Yangtze River, The Han River rises in the southern of Qinling mountains, and flows from the northwest to the southeast. This mountainous region lies in the humid zone with a subtropical monsoonal climate. The annual average temperature is between 14 and 16 °C. The annual precipitation varies from 700 to 1100 mm, and about 70 to 80% of the annual precipitation occurs during the flood season (May to October), in which heavy rains in early summer and continuous rainfall in autumn often cause major floods.

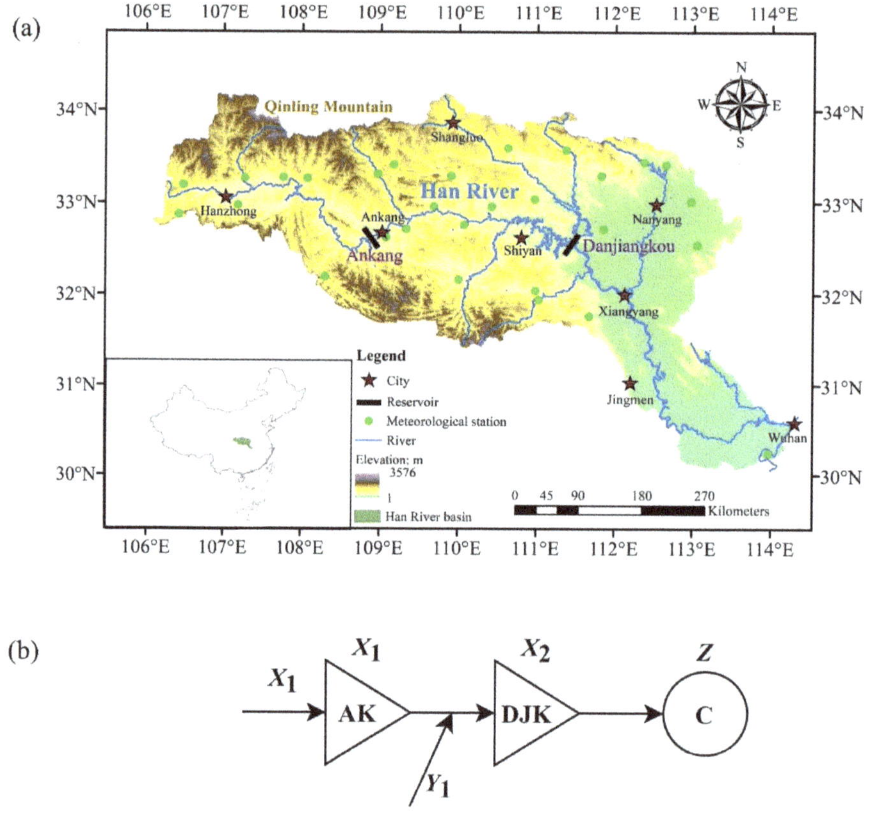

Figure 3. (**a**) Map of the Han River basin and (**b**) sketch diagram of the flood regional composition with AK and DJK cascade reservoirs.

3.1. Cascade Reservoirs

The Ankang (AK) and Danjiangkou (DJK) cascade reservoirs are located at the upper and middle reach of the Han River basin, respectively. The AK reservoir was built in 1992 and provides hydropower generation and flood control, while the DJK reservoir was built in 1973 and its primary functions are flood control, water supply, hydropower generation, and irrigation. These two reservoirs are selected for the case study because their inflow data is lengthy enough (over 60 years) and they have large storage capacities.

The basic information of AK and DJK cascade reservoirs is listed in Table 3. The characteristic parameter values and current flood control operation rules of the AK and DJK reservoirs are provided by the Changjiang (Yangtze River) Water Resources Commission (CWRC), Ministry of Water Resource [70]. More information about AK and DJK reservoirs can be found in the references [71,72].

Table 3. Characteristic parameter values of the cascade reservoirs in the Han River basin.

Characteristic Parameters	Ankang (AK) Reservoir	Danjiangkou (DJK) Reservoir
Normal pool water level (m)	330	170
Flood limited water level in summer (m)	325	160
Flood limited water level in autumn (m)	325	163.5
Design flood water level (m)	333.1	172.2
Total storage capacity (billion m^3)	3.34	33.91
Flood control capacity in summer (billion m^3)	0.36	14.1
Flood control capacity in autumn (billion m^3)	0.36	11.1
Installed hydropower capability (MW)	800	900
Flood control design standard (year)	1000	1000

3.2. Dataset

Four categories of data series were collected, including restored streamflow data, observed hydro-climate data, population growth data and GCMs outputs from CMIP5. The observed data series provide information up to 2020, while a projected dataset of the future period from 2021 to 2095 is also used in this study.

(1) Restored mean daily streamflow data from both the inflow of AK reservoir and the inflow of DJK reservoir were provided by the Hydrology Bureau of the Changjiang (Yangtze River) Water Resources Commission during 1954–2020. The restoration of streamflow data can be taken as a natural flow series that eliminates the regulation impact of reservoirs.

(2) Observed daily precipitation series from 27 stations during 1951–2020 were obtained from the National Climate Center of the China Meteorological Administration (source: http://data.cma.cn/ (accessed on 16 April 2021)).

(3) Given the unavailability of population data at the basin scale, the total registered population of all the prefecture-level cities amidst in the Han River basin was collected. These cities include Wuhan, Shiyan, Jingmen, Xiangyang in Hubei province, Hanzhong, Ankang and Shangluo in Shanxi province, and Nanyang in Henan province. The annual registered population data in Han River basin were obtained from the China Compendium of Statistics 1949–2008 [73], the website of the National Bureau of Statistics of China (source: http://www.stats.gov.cn/tjsj/ndsj/ (accessed on 1 July 2021)) and the websites of statistical bureaus of the provinces and cities mentioned above. For future projection, the logistic growth model that adapts the growth restriction resulting from limited natural resources is applied to predict the growth of population [74]. Based on the logistic growth model, the evolution of the population for the period 1950–2100 is illustrated in Figure 4.

(4) Future daily precipitation is simulated by GCM using climate change scenarios [75]. Our lab research team, Tian et al. (2021) used 10 different GCMs (see Table 4) and two representative concentration pathways (RCPs) of 4.5 and 8.5 from the IPCC Fifth Assessment to project climate change for the Han River basin, which are employed in this study [76–78]. The RCP 4.5 scenario represents a future of medium emission where climate policies limit [79]. Without climate change policies, RCP 8.5 scenario presumes that high emissions of greenhouse gases continue in the future.

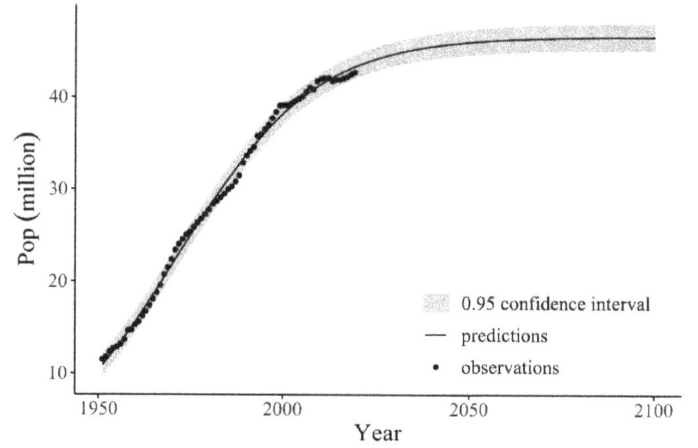

Figure 4. Evolution of the population in Han River basin during 1950–2100.

Table 4. Description of the 10 GCMs used in this study.

Model Name	Department	Country	Spatial Resolution (Number of Meridional Cells × Number of Latitudinal Lattices)
BCC-CSM1.1(m)	BCC	China	128 × 64
BNU-ESM	GCESS	China	128 × 64
CanESM2	CCCMA	Canada	128 × 64
CCSM4	NCAR	USA	288 × 192
CNRM-CM5	CNRM-CERFACS	Canada	256 × 128
CSIRO-Mk3.6.0	CSIRO-QCCCE	Australia	192 × 96
GFDL-ESM2G	NOAA-GFDL	USA	144 × 90
MRI-CGCM3	MRI	Japan	320 × 160
MPI-ESM-LR	MPI-M	Germany	192 × 96
NorESM1-M	NCC	Norway	144 × 96

The outputs of the GCMs not only involve the historical period before 2006 as a reference, but also cover 2021–2095 for future projection. A daily bias correction method is applied in this study for statistical downscaling [80]. Six statistics containing mean, standard deviation, 85th, 90th, 95th and 99th percentiles of future precipitation series are used to test the performance of the daily bias correction method [80]. Taking the BCC-CSM1-1 model as an example, the bias of raw and the corrected model for daily precipitation series during 1991–2005 are illustrated in Figure 5, where the horizontal and vertical coordinates follow the 27 meteorological stations and the bias about the six statistics, respectively. These findings indicate the great performance of the statistical downscaling method so that it can be employed for future projection. Figure 6 shows the projected future precipitation under RCP 4.5 and RCP 8.5 scenarios. The arithmetic mean values of the 10 GCMs are employed in this study for nonstationary flood frequency analysis.

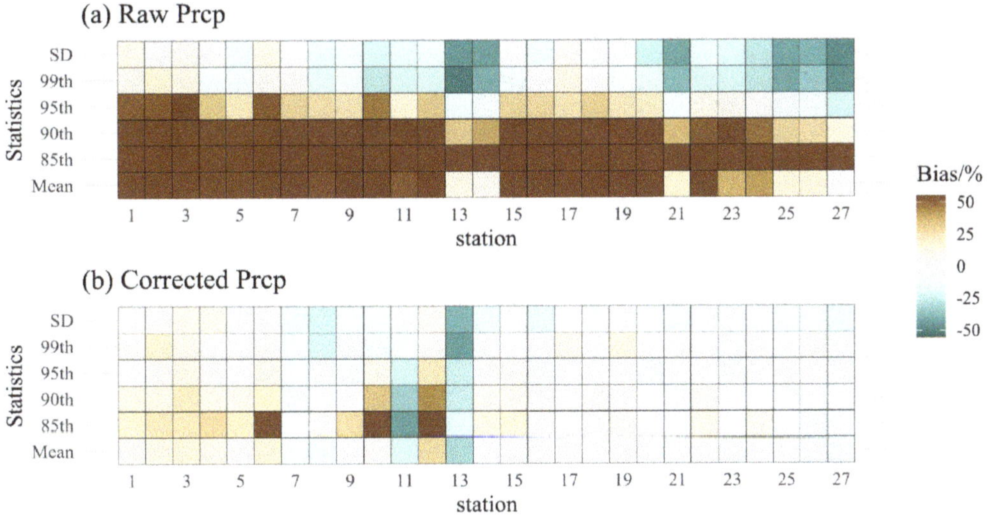

Figure 5. The bias of (**a**) raw and (**b**) corrected BCC-CSM1-1 outputs during 1991–2005.

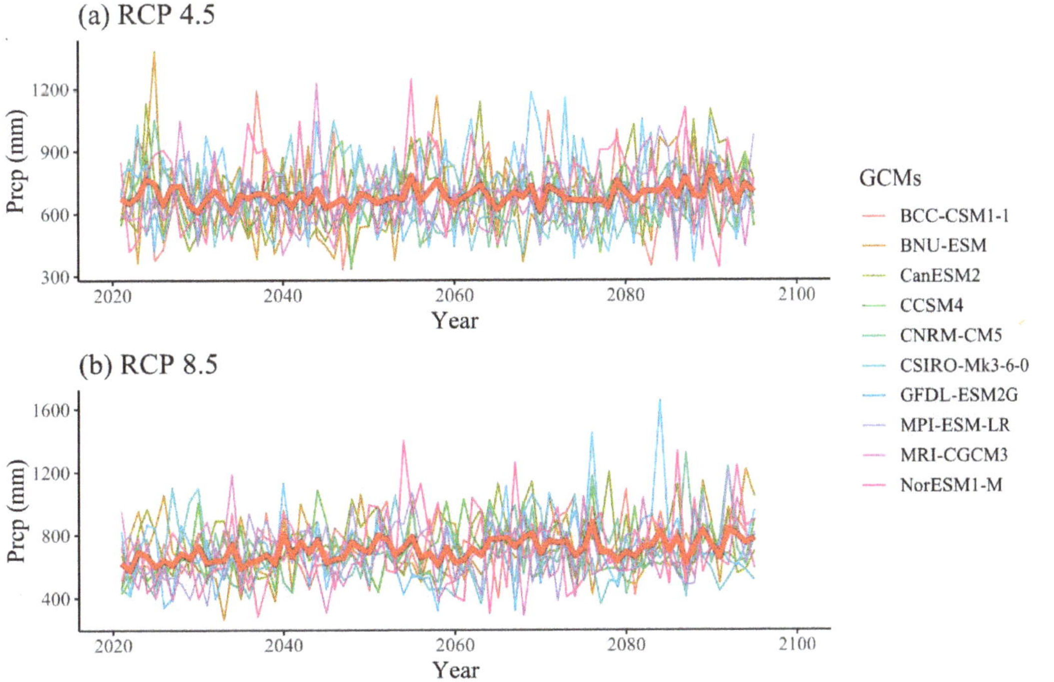

Figure 6. Projected annual precipitation under (**a**) RCP 4.5 and (**b**) RCP 8.5 scenarios for 2021–2095. The bold red lines are the arithmetic mean value of the 10 GCMs.

4. Result Analysis

4.1. Preliminary Analysis

According to the regulation characteristics of cascade reservoirs in the Han River basin, the annual maximum 15-day (denoted as W_{15}) flood volume series of the AK and DJK reservoir are selected for flood frequency analysis. Indicated by the fitted trend lines, the overall decreasing trends of W_{15} series in both AK and DJK reservoirs are shown in Figure 7. The significance of trends in the flood volume series and explanatory variables (i.e., Pop and Prcp) during 1954–2020 is analyzed by the Mann–Kendall test, and the Pettitt's test is employed to detect the change points in the W_{15} series, which is summarized in Table 5. Findings show decreasing trends of the W_{15} series in both the AK and DJK reservoirs and the detected change-points are located at 1985–1986 (p-value < 0.05) for two reservoirs. The nonparametric multiple change point analysis method detects the change point of the W_{15} series between AK and DJK reservoirs, which takes place in 1985 (p-value < 0.05). This preliminary analysis demonstrates that both the univariate series and the dependence structure between the W_{15} series are all nonstationary.

Table 5. Results of trend test, change-point detection and multiple change point analysis for the W_{15} series for the years 1954–2020 (* means the maximum of the absolute value of the vector U).

Reservoir	Mann-Kendall Test		Pettitt's Test			Nonparametric Multiple Change Point Analysis	
	S	p-Value	Change Point/Year	U *	p-Value	Change Point/Year	p-Value
AK	−555	0.0027	32/1985	564	0.0038	32/1985	<0.05
DJK	−613	0.0009	32/1985	596	0.0018		

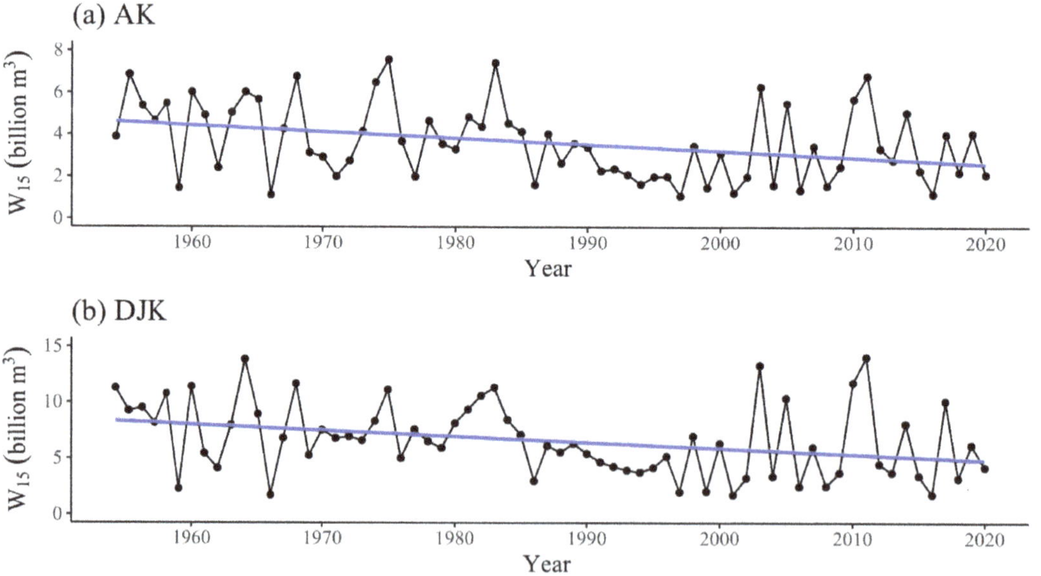

Figure 7. The W_{15} series with the fitted linear trend lines of (**a**) AK and (**b**) DJK reservoirs.

4.2. Nonstationary Design Flood Volumes

4.2.1. Univariate Nonstationary Flood Frequency Analysis

Based on the selection procedure mentioned in Section 2.2.2, we assume that the lower bounds of forward stepwise procedure for both μ and σ are constant (i.e., $\mu \sim 1$, $\sigma \sim 1$), while the upper bounds are (Pop + Prcp). Table 6 lists the distribution parameters and BIC values for W_{15} at the AK and DJK reservoirs, in which the time-varying BIC_{NS} and constant BIC_S are also calculated. According to the minimum BIC_{NS} values, the Log-normal distribution is the best one for the W_{15} series in both the AK and DJK reservoirs. To demonstrate the fitting process, the detailed forward procedure applied in Log-normal distribution parameter μ is displayed in Table 7. The worm plot and centile curves for the Log-normal distribution are plotted in Figure 8. In the worm plot, all scatter points are within the 95% confidence intervals, illustrating a good fit between the optimal distribution and empirical frequency series. In terms of centile curves, the percentages of observation points within the 5th, 25th, 50th, 75th and 95th intervals are 5.97%, 26.86%, 59.70%, 70.14% and 94.02% (5.97%, 28.35%, 53.73%, 71.64% and 94.02%) for AK (DJK), respectively. The above results indicate that the selected optimal Log-normal distribution is more adequate to model the nonstationarity of the W_{15} series, and to express the similar type of GLMs for both AK and DJK reservoirs. As presented in Table 6, the location parameter of Log-normal distribution for modelling the W_{15} series is linked to Prcp and Pop while the scale parameter of W_{15} is constant. Furthermore, the BIC_{NS} value is always smaller than the BIC_S value for the same distribution. Compared to the stationary distributions, the optimal nonstationary Log-normal distribution can satisfactorily capture the nonstationarity in flood frequency analysis for both the AK and DJK reservoirs.

Table 6. Distribution parameters and BIC values for W_{15} at the AK and DJK reservoirs.

Reservoir	Probability Distribution	Distribution Parameters			BIC_{NS}	BIC_S
		$\ln(\mu)$	$\ln(\sigma)$	ν		
AK	Gamma	$2.3979 + 0.2352 \times \text{Prcp} - -0.1615 \times \text{Pop}$	-0.9393		548.57	574.00
	Weibull	$2.6101 + 0.2199 \times \text{Prcp} - -0.1562 \times \text{Pop}$	0.9802		551.79	574.55
	Log-normal	$\exp(2.2262 + 0.2456 \times \text{Prcp} - -0.1544 \times \text{Pop})$	-0.9300		546.41	575.94
	Gumbel	$\exp(-0.2612 + 9.029 \times \text{Prcp} - -6.2479 \times \text{Pop})$	2.6700		575.45	596.75
	Pearson type III	$2.0914 + 0.2527 \times \text{Prcp} - -0.1451 \times \text{Pop}$	-0.9480	-0.7595	550.09	578.12
DJK	Gamma	$2.5707 + 0.2874 \times \text{Prcp} - -0.1442 \times \text{Pop}$	-1.0020		619.73	657.38
	Weibull	$2.6964 + 0.2749 \times \text{Prcp} - -0.1185 \times \text{Pop}$	1.0350		626.52	656.78
	Log-normal	$\exp(2.4567 + 0.2967 \times \text{Prcp} - -0.1508 \times \text{Pop})$	-0.9900		618.39	661.21
	Gumbel	$\exp(-54.34 + 19.15 \times \text{Pop})$	3.2450		646.41	678.77
	Pearson type III	$2.4442 + 0.2977 \times \text{Prcp} - -0.1510 \times \text{Pop}$	-0.9903	-0.1016	622.57	660.86

Table 7. The detailed forward procedure of Log-normal distribution for W_{15} at the DJK reservoir.

Steps		μ	σ	BIC
Step 1	Start	μ ~1	σ ~1	661.21
	+Prcp	μ ~Prcp	σ ~1	623.95
	+Pop	μ ~Pop	σ ~1	656.82
Step 2	Start	μ ~Prcp	σ ~1	623.95
	+Pop	μ ~Prcp + Pop	σ ~1	618.39
Step 3	Start	μ ~Prcp + Pop	σ ~1	618.39
	+Prcp	μ ~Prcp + Pop	σ ~Prcp	615.69
	+Pop	μ ~Prcp + Pop	σ ~Pop	622.47
Step 4	Start	μ ~Prcp + Pop	σ ~1	618.39
	−Pop	μ ~Prcp	σ ~1	623.95
	−Prcp	μ ~Pop	σ ~1	656.82
Optimal		μ ~ Prcp + Pop	σ ~1	618.39

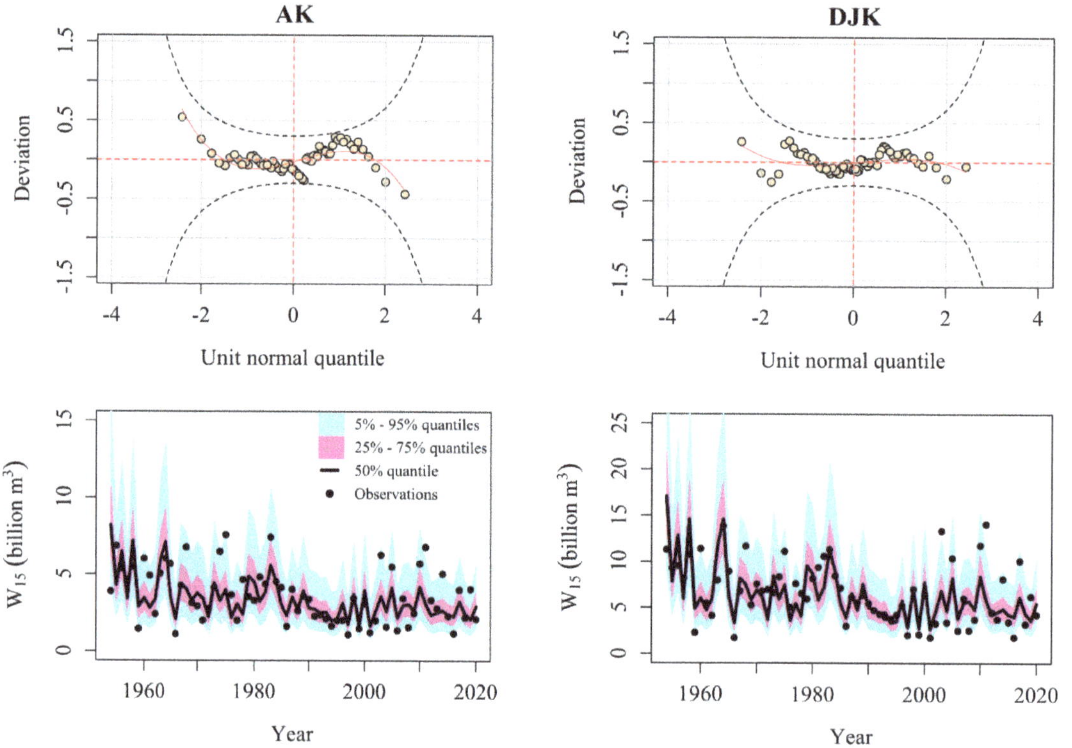

Figure 8. Diagnostic plots (worm plot and centile curves) for evaluating the goodness-of-fit of the optimal nonstationary Log-normal distribution for the W_{15} series in the AK and DJK reservoirs.

4.2.2. Design Flood Volumes in DJK Reservoir

The nonstationary design value is essential for handling changing environments in considering future flood risk management. Since the total storage capacity of the DJK reservoir is approximately 31 times larger than that of the AK reservoir (Table 3), the DJK reservoir is considered as the key flood prevention project in the Han River basin. If we assume that the design lifespan for the DJK reservoir is 100 years (from 1973 to 2072), then the years of 2021–2072 are chosen as future projections. The design flood volume under nonstationary conditions is calculated by the ER criteria. Three scenarios, i.e., stationary distribution condition (S1), nonstationary conditions based on RCP 4.5 (S2), or RCP 8.5 (S3), are compared. It is worth noting that the S1 scenario only takes cascade reservoir regulation into account while the nonstationary attributes of climate change and population growth are omitted. To produce an intuitive comparison among these scenarios, Figure 9 plots the stationary and nonstationary design flood volumes W_{15} for return period $m \in [2, 1000]$. It should be noted that the design flood volumes in the stationary reference scenario are estimated by the Pearson type III distribution and curve-fitting estimation method recommended by the Ministry of Water Resources in China [32]. It is found that the design flood volumes of S2 (S3) are lower than S1 when the return periods are less than 300 (200) years, respectively. The design flood volumes of S3 are always larger than S2. Compared with the nonstationary distribution scenarios (i.e., S2 and S3), the method used under the S1 scenario overestimates the flood volumes at a lower return period while the extreme event is underestimated at a higher return period. For the 1000-year return period, the design W_{15} in the DJK reservoir under S1, S2 and S3 scenarios are 20.041, 21.680 and 22.352 billion m^3, respectively. These results indicate that climate change and population growth have greater impact on the features of future W_{15} under the S2 and S3 scenarios so that the large uncertainty and flood hazard should be considered for future hydrological design and water resources management.

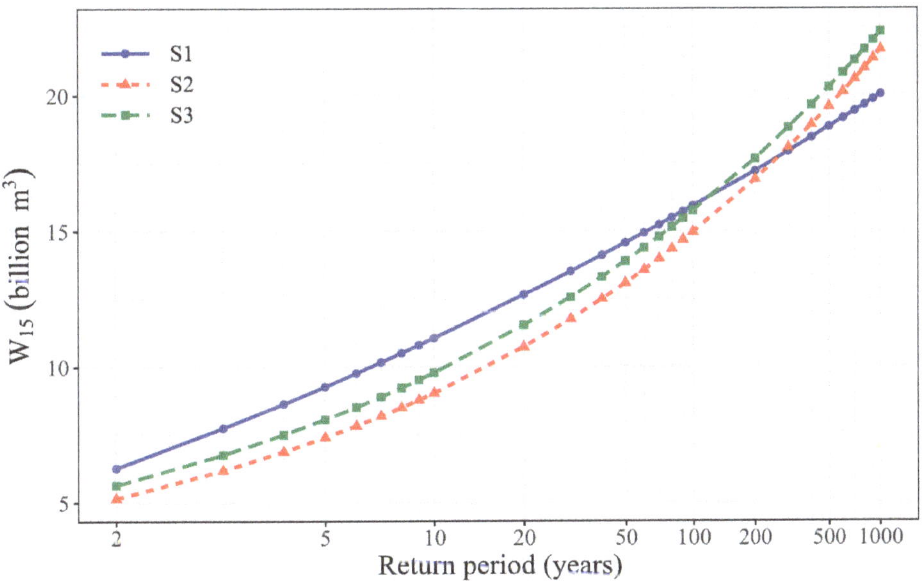

Figure 9. Comparison of design W15 with different return periods and scenarios for the DJK reservoir.

4.3. Design Flood Estimation at Downstream Site

4.3.1. Nonstationary Flood Regional Composition

After the estimation of the design flood volumes at the DJK reservoir, the flood regional composition method can be applied for analyzing the flood generation mechanism of the cascade reservoirs in the Han River basin. For nonstationary analysis of the dependence structure, the forward selection procedure is implemented to model the dependence parameter of three copulas, i.e., the Gumbel-Hougaard, Frank, and Clayton copula. The lower and upper bounds of forward stepwise procedure for θ_c are constant (i.e., θ_c ~1) and (Pop + Prcp), respectively. Table 8 summarizes the fitted dependence parameters and the goodness-of-fit results for the optimal copulas under three scenarios for nonstationary modeling of the W_{15} series in AK and DJK reservoirs. The p-KS(Z_1) and p-KS(Z_2) are p-values of the KS test for the two Rosenblatt's probabilities integral transformations Z_1 and Z_2, which should be uniformly and independently distributed on [0,1]. The p-Kendall is the p-value of the Kendall rank correlation test for Z_1 and Z_2. Table 8 shows that the p-KS(Z_1), p-KS(Z_2) and the p-Kendall values are larger than 0.05, which statistically supports the validity of the hypothetical models. The time-varying Gumbel-Hougaard copula is the optimal copula to model the dependence structure of the W_{15} series by comparing the BIC values. Obviously, the nonstationary spatial correlation of flood events can be sufficiently explained by the NS-MLRC method.

Table 8. Dependence parameters and goodness-of-fit results for candidate copulas in nonstationary modeling.

Copula	θ_c	*p*-KS(Z_1)	*p*-KS (Z_2)	*p*-Kendall	BIC_{NS}	BIC_S
Gumbel-Hougaard	$\exp(-0.3019 + 0.2642 \times \text{Pop}) + 1$	0.674	0.464	0.783	1092.48	1128.20
Frank	$5.515 + 1.144 \times \text{Pop}$	0.674	0.644	0.935	1095.81	1137.03
Clayton	0.9647	0.674	0.852	0.249	1096.95	1132.60

Figure 3b illustrates that the flood volumes of the DJK reservoir are decomposed to the outflow of the AK reservoir and the inflow of the AK-DJK inter-basin. It is noted that the return period of the design flood is 1000 years in both the AK and DJK reservoirs, as shown in Table 3. The Log-normal distributions with explanatory variables shown in Table 6 are applied as the marginal distributions of the time-varying copulas. Following the procedures described in Section 2.3.2, the NS-MLRC results of the cascade reservoirs can be estimated.

Based on the NS-MLRC method, the nonstationary FRC of the AK and DJK reservoirs with the maximum occurrence probabilities are derived. It should be noted that there are one hundred (from 1973 to 2072) FRCs derived by the NS-MLRC method, compared with the individual result calculated by the MLRC method. Table 9 compares the FRC results under three scenarios during design lifespan with the same flood prevention standard according to ER. It is shown that the designed W_{15} in the AK reservoir (X_1) of S2 and S3 is greater than that of nonstationary cases, while the design W_{15} at the (Y_1) are overestimated when compared with S1.

Table 9. The FRC results derived by the MLRC (S1 scenario) and NS-MLRC (S2 and S3 scenarios) methods during design lifespan with the same flood prevention standards based on ER. X_1, Y_1 and X_2 represent the designed flood volume W_{15} at the AK reservoir, inter-basin, and the DJK reservoir, respectively.

Method	Scenario	Volumes (Billion m^3)	X_1	Y_1	X_2
The MLRC	S1		10.605	9.435	20.041
The NS-MLRC	S2	Minimum	10.760	6.779	21.680
		Median	12.990	8.693	
		Mean	12.960	8.724	
		Maximum	14.900	10.921	
	S3	Minimum	11.120	6.944	22.352
		Median	13.210	9.144	
		Mean	13.120	9.232	
		Maximum	15.410	11.321	

4.3.2. Design Flood Hydrographs at Downstream Site

The estimation of design floods at downstream sites can be obtained using design inflow hydrographs and cascade reservoir regulations based on the derived FRC results. Different FRCs may differ in terms of their occurrence probabilities [35], and the FRCs with maximum occurrence probabilities among the lifespan of projects are analyzed. To generate the design flood at the downstream site C (Figure 3), the current flood control regulation and the commonly used Muskingum model provided by the CWRC are used. After flood control operation, 1000-year design flood hydrographs for the downstream sites under three scenarios with or without reservoir regulation are plotted in Figure 10. Table 10 lists the results of the 1000-year design floods under three scenarios with or without reservoir regulation.

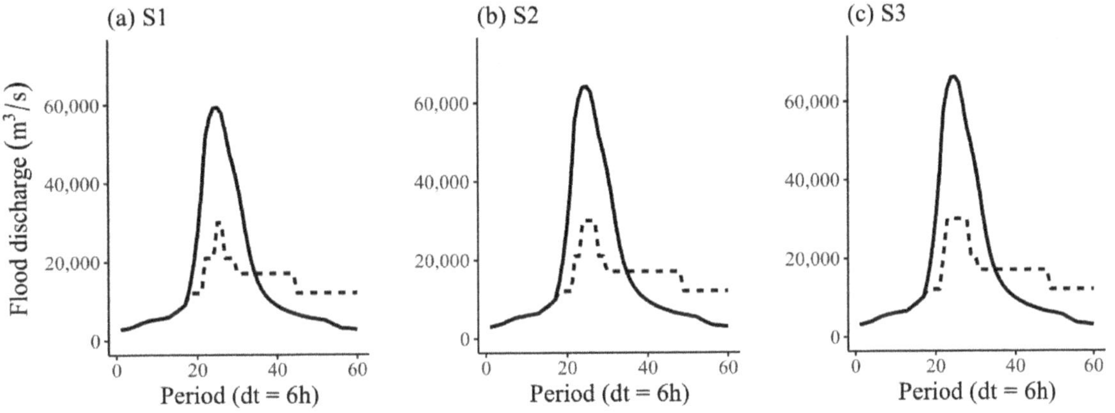

Figure 10. Design flood hydrographs of 1000-year return period at downstream under three scenarios with (dash line) or without (solid line) cascade reservoir regulation.

Table 10. Comparison of 1000-year design floods under three scenarios with or without reservoir regulation.

Scenario	Variable	Reservoir Regulation	Ankang Reservoir		Danjiangkou Reservoir	
S1	Q_{max} (m^3/s)	no	31,393		59,323	
		yes	28,093	−10.51%	30,000	−49.43%
	W_1 (billion m^3)	no	2.663		5.031	
		yes	2.515	−5.53%	2.398	−52.35%
	W_3 (billion m^3)	no	6.497		12.277	
		yes	6.141	−5.48%	5.594	−54.43%
S2	Q_{max} (m^3/s)	no	31,846		64,175	
		yes	30,236	−5.06%	30,000	−53.25%
	W_1 (billion m^3)	no	2.701		5.443	
		yes	2.550	−5.60%	2.592	−52.38%
	W_3 (billion m^3)	no	6.591		13.282	
		yes	6.200	−5.92%	5.875	−55.76%
S3	Q_{max} (m^3/s)	no	32,919		66,164	
		yes	31,171	−5.31%	30,000	−54.66%
	W_1 (billion m^3)	no	2.792		5.612	
		yes	2.632	−5.71%	2.592	−53.81%
	W_3 (billion m^3)	no	6.813		13.69	
		yes	6.526	−4.21%	6.350	−53.62%

Figure 10 shows that the 1000-year design flood hydrographs vary considerably with less peaks and gentler flood processes at downstream sites after cascade reservoir regulation. As shown in Table 10, the design floods of the cascade reservoirs decrease significantly on account of the regulation of the AK and DJK reservoirs. Taking Q_{max} as an example, under the S1, S2, and S3 scenarios, the 1000-year design Q_{max} of AK (DJK) at downstream sites decreased by 10.97% (49.43%), 5.06% (53.25%) and 5.31% (54.66%), respectively due to cascade reservoir regulation. Furthermore, the W_1 and W_3 flood volumes are less than half of these without cascade reservoir regulation.

5. Discussion

5.1. Nonstationary Characteristics

Flood events differ according to climate, reservoir regulation, water consumption, and land use, which are the potential driving forces that link to flood design. When traditional hydrologic design criteria are extended for accommodating nonstationary conditions, the future covariates should be obtained on account of the hydraulic lifespan. However, it is difficult to predict the changes of some variables, such as deforestation, so that other land use indices are not considered in this study. Based on the case study in the Han River, the influence of climate change, population growth and cascade reservoir regulation on the design flood at the downstream site are discussed as follows.

Many studies have identified how the changing pattern brings diverse control on a flow process [28,33], which can be classified as either a direct and/or indirect impact. Climate change and population growth will alter the regional hydrological characteristics of the basin, and affect the flood data series as well as the flood design (Figure 8) [43]. Compared to these prolonged effects, the cascade reservoir regulation clips the flood peak and decreases the flood volumes in a transitory and swift way (Figure 10). Among the multiple driving patterns, the cascade reservoir regulation plays a dominant role in affecting the design floods at downstream sites. Furthermore, the cascade reservoir flood control strategies can be re-regulated for adapting the slow-to-change non-stationarity (such as climate change and population growth) in future work.

5.2. The Worst FRC during Reservoir Lifespan

In practical operations, the worst FRC during reservoir lifespan is more noteworthy in comparison with the most likely FRC. The reservoir water level is at the flood limited water level during flood season to provide sufficient storage for possible floods [81]. When a flood event occurs, the highest water level during the period of reservoir flood regulation can quantificationally indicate the flood hazard, such as dam-break risk, downstream inundation risk and so on. In this study, the highest water level of the DJK reservoir during the whole flood regulation process is applied as an indicator to specifically represent the hazard of flood events.

Although the MLRC method has a strong statistical basis and generates a single regional composition, it is unable to evaluate the composition of the worst regional flood. The NS-MLRC method can include varying factors such as precipitation and population growth. In this study, variant FRCs derived by the NS-MLRC method are used to amplify the typical flood hydrograph, forming one hundred (from 1973 to 2072) FRC types. After cascade reservoir regulation, the comparison of the highest water levels derived from both the most likely FRCs and the worst FRCs at the DJK reservoir during its lifespan is demonstrated in Table 11. It indicates that the highest water level estimated by the most likely FRC is lower than that of the worst FRC, which means the worst FRC under the S2 and S3 scenarios is more adverse than the most likely FRC. In practice, the worst FRC in S3 is worthy of noting for future flood risk management.

Table 11. Comparison of the highest water level derived from both the most likely FRC and the worst FRC during reservoir lifespan at the DJK reservoir for different scenarios.

Flood Regional Composition (FRC)	S1	S2	S3
The most likely FRC	168.27 m	168.97 m	168.96 m
The worst FRC	-	169.02 m	169.37 m

6. Conclusions

There is an increasing need to develop an effective design flood estimation framework to deal with nonstationary data series caused by climate change and anthropogenic activities. In this study, the univariate flood frequency analysis, nonstationary hydrologic design criteria and the NS-MLRC method were adopted to derive the nonstationary design flood volumes in the Han River. The design flood hydrographs at the downstream site were estimated after cascade reservoir regulation. The main conclusions are summarized as follows:

(1) The proposed NS-MLRC method can be effectively implemented as an extension of the MLRC method for explaining the nonstationary spatial correlation of the flood events.

The multiple nonstationary driving forces, i.e., climate change and population growth, can be captured and precisely quantified by the proposed design flood estimation framework.

(2) The slow-to-change impacts of climate change and population growth are presented in design flood volumes according to the nonstationary flood frequency analysis method and ER criteria. The long-lasting driving factors imply the larger risks of the flood hazard. The 1000-year design W_{15} of the DJK reservoir under the stationary distribution scenario (S1), RCP 4.5-based (S2), and the RCP 8.5-based (S3) nonstationary scenario are 20.041, 21.680 and 22.352 billion m^3, respectively.

(3) The swift effects of cascade reservoirs are reflected in design flood hydrographs with lower peaks and less volumes based on the NS-MLRC method and flood control operation rules. For instance, the 1000-year design Q_{max} of the AK (DJK) downstream site under the stationary distribution scenario (S1), RCP 4.5-based (S2), and RCP 8.5-based (S3) nonstationary scenario decrease by 10.97% (49.43%), 5.06% (53.25%) and 5.31% (54.66%), respectively due to cascade reservoir regulation.

As the dominant element affecting flood design, the current cascade reservoir operation strategies can be improved to accommodate the slow-to-change nonstationarity in further research.

Author Contributions: Y.X.: Conceptualization, Data curation, Methodology, Visualization. S.G., L.X.: Methodology, Supervision. J.T.: Resources, Data Curation. F.X.: Methodology, Validation. All authors have read and agreed to the published version of the manuscript.

Funding: This study was financially supported by the National Natural Science Foundation of China (51879192), China Three Gorges Corporation (0799254), and the Research Council of Norway (FRINATEK Project No. 274310).

Data Availability Statement: Data sharing is not applicable to this article as no new data were created or analyzed in this study.

Acknowledgments: The numerical calculations in this paper have been done on the supercomputing system in the Supercomputing Center of Wuhan University.

Conflicts of Interest: The authors declare no competing interests. The corresponding author is responsible for submitting a competing interest statement on behalf of all authors of the paper.

References

1. Milly, P.C.D.; Betancourt, J.; Falkenmark, M.; Hirsch, R.M.; Kundzewicz, Z.W.; Lettenmaier, D.P.; Stouffer, R.J. Stationarity Is Dead: Whither Water Management? *Science* **2008**, *319*, 573–574. [CrossRef]
2. Benyahya, L.; Gachon, P.; St-Hilaire, A.; Laprise, R. Frequency Analysis of Seasonal Extreme Precipitation in Southern Quebec (Canada): An Evaluation of Regional Climate Model Simulation with Respect to Two Gridded Datasets. *Hydrol. Res.* **2013**, *45*, 115–133. [CrossRef]
3. Onyutha, C.; Willems, P. Uncertainty in Calibrating Generalised Pareto Distribution to Rainfall Extremes in Lake Victoria Basin. *Hydrol. Res.* **2014**, *46*, 356–376. [CrossRef]
4. Qu, C.; Li, J.; Yan, L.; Yan, P.; Cheng, F.; Lu, D. Non-Stationary Flood Frequency Analysis Using Cubic B-Spline-Based Gamlss Model. *Water* **2020**, *12*, 1867. [CrossRef]
5. Zhang, T.; Su, X.; Feng, K. The Development of a Novel Nonstationary Meteorological and Hydrological Drought Index Using the Climatic and Anthropogenic Indices as Covariates. *Sci. Total Environ.* **2021**, *786*, 147385. [CrossRef]
6. Javelle, P.; Ouarda, T.B.M.J.; Lang, M.; Bobée, B.; Galéa, G.; Grésillon, J.-M. Development of Regional Flood-Duration–Frequency Curves Based on the Index-Flood Method. *J. Hydrol.* **2002**, *258*, 249–259. [CrossRef]
7. Brandimarte, L.; Di Baldassarre, G. Uncertainty in Design Flood Profiles Derived by Hydraulic Modelling. *Hydrol. Res.* **2012**, *43*, 753–761. [CrossRef]
8. Jiang, C.; Xiong, L.; Xu, C.-Y.; Guo, S. Bivariate Frequency Analysis of Nonstationary Low-Flow Series Based on the Time-Varying Copula. *Hydrol. Process.* **2015**, *29*, 1521–1534. [CrossRef]
9. López, J.; Francés, F. Non-Stationary Flood Frequency Analysis in Continental Spanish Rivers, Using Climate and Reservoir Indices as External Covariates. *Hydrol. Earth Syst. Sci.* **2013**, *17*, 3189–3203. [CrossRef]
10. Yan, L.; Li, L.; Yan, P.; He, H.; Li, J.; Lu, D. Nonstationary Flood Hazard Analysis in Response to Climate Change and Population Growth. *Water* **2019**, *11*, 1811. [CrossRef]
11. Debele, S.E.; Bogdanowicz, E.; Strupczewski, W.G. Around and about an Application of the GAMLSS Package to Non-Stationary Flood Frequency Analysis. *Acta Geophys.* **2017**, *65*, 885–892. [CrossRef]

12. Hui, R.; Herman, J.; Lund, J.; Madani, K. Adaptive Water Infrastructure Planning for Nonstationary Hydrology. *Adv. Water Resour.* **2018**, *118*, 83–94. [CrossRef]
13. Villarini, G.; Smith, J.A.; Serinaldi, F.; Bales, J.; Bates, P.D.; Krajewski, W.F. Flood Frequency Analysis for Nonstationary Annual Peak Records in an Urban Drainage Basin. *Adv. Water Resour.* **2009**, *32*, 1255–1266. [CrossRef]
14. Salvadore, E.; Bronders, J.; Batelaan, O. Hydrological Modelling of Urbanized Catchments: A Review and Future Directions. *J. Hydrol.* **2015**, *529*, 62–81. [CrossRef]
15. Wang, H.; Mei, C.; Liu, J.; Shao, W. A New Strategy for Integrated Urban Water Management in China: Sponge City. *Sci. China Technol. Sci.* **2018**, *61*, 317–329. [CrossRef]
16. Sto Domingo, N.D.; Refsgaard, A.; Mark, O.; Paludan, B. Flood Analysis in Mixed-Urban Areas Reflecting Interactions with the Complete Water Cycle through Coupled Hydrologic-Hydraulic Modelling. *Water Sci. Technol. J. Int. Assoc. Water Pollut. Res.* **2010**, *62*, 1386–1392. [CrossRef]
17. Xu, Y.; Xu, J.; Ding, J.; Chen, Y.; Yin, Y.; Zhang, X. Impacts of Urbanization on Hydrology in the Yangtze River Delta, China. *Water Sci. Technol. J. Int. Assoc. Water Pollut. Res.* **2010**, *62*, 1221–1229. [CrossRef]
18. Zhao, G.; Gao, H.; Naz, B.S.; Kao, S.-C.; Voisin, N. Integrating a Reservoir Regulation Scheme into a Spatially Distributed Hydrological Model. *Adv. Water Resour.* **2016**, *98*, 16–31. [CrossRef]
19. He, S.; Guo, S.; Chen, K.; Deng, L.; Liao, Z.; Xiong, F.; Yin, J. Optimal Impoundment Operation for Cascade Reservoirs Coupling Parallel Dynamic Programming with Importance Sampling and Successive Approximation. *Adv. Water Resour.* **2019**, *131*, 103375. [CrossRef]
20. Zhang, Q.; Chen, Y.; Tao, X.; Xu, C.-Y.; Chen, X. A Spatial Assessment of Hydrologic Alteration Caused by Dam Construction in the Middle and Lower Yellow River, China. *Hydrol. Process.* **2008**, *22*, 3829–3843. [CrossRef]
21. Wang, W.; Li, H.-Y.; Leung, L.R.; Yigzaw, W.; Zhao, J.; Lu, H.; Deng, Z.; Demisie, Y.; Blöschl, G. Nonlinear Filtering Effects of Reservoirs on Flood Frequency Curves at the Regional Scale. *Water Resour. Res.* **2017**, *53*, 8277–8292. [CrossRef]
22. Strupczewski, W.G.; Singh, V.P.; Feluch, W. Non-Stationary Approach to at-Site Flood Frequency Modelling I. Maximum Likelihood Estimation. *J. Hydrol.* **2001**, *248*, 123–142. [CrossRef]
23. Strupczewski, W.G.; Kaczmarek, Z. Non-Stationary Approach to at-Site Flood Frequency Modelling II. Weighted Least Squares Estimation. *J. Hydrol.* **2001**, *248*, 143–151. [CrossRef]
24. Strupczewski, W.G.; Singh, V.P.; Mitosek, H.T. Non-Stationary Approach to at-Site Flood Frequency Modelling. III. Flood Analysis of Polish Rivers. *J. Hydrol.* **2001**, *248*, 152–167. [CrossRef]
25. Zhang, Q.; Gu, X.; Singh, V.P.; Xiao, M.; Chen, X. Evaluation of Flood Frequency under Non-Stationarity Resulting from Climate Indices and Reservoir Indices in the East River Basin, China. *J. Hydrol.* **2015**, *527*, 565–575. [CrossRef]
26. Wen, Q.; Sun, P.; Zhang, Q.; Li, H. Nonstationary Ecological Instream Flow and Relevant Causes in the Guai River Basin, China. *Water* **2021**, *13*, 484. [CrossRef]
27. Stasinopoulos, D.M.; Rigby, R.A. Generalized Additive Models for Location Scale and Shape (GAMLSS) in r. *J. Stat. Softw.* **2007**, *23*. [CrossRef]
28. Du, T.; Xiong, L.; Xu, C.-Y.; Gippel, C.J.; Guo, S.; Liu, P. Return Period and Risk Analysis of Nonstationary Low-Flow Series under Climate Change. *J. Hydrol.* **2015**, *527*, 234–250. [CrossRef]
29. Mondal, A.; Mujumdar, P.P. Return Levels of Hydrologic Droughts under Climate Change. *Adv. Water Resour.* **2015**, *75*, 67–79. [CrossRef]
30. Su, C.; Chen, X. Assessing the Effects of Reservoirs on Extreme Flows Using Nonstationary Flood Frequency Models with the Modified Reservoir Index as a Covariate. *Adv. Water Resour.* **2019**, *124*, 29–40. [CrossRef]
31. Xiong, B.; Xiong, L.; Xia, J.; Xu, C.-Y.; Jiang, C.; Du, T. Assessing the Impacts of Reservoirs on Downstream Flood Frequency by Coupling the Effect of Scheduling-Related Multivariate Rainfall with an Indicator of Reservoir Effects. *Hydrol. Earth Syst. Sci.* **2019**, *23*, 4453–4470. [CrossRef]
32. Ministry of Water Resources. *Regulation for Calculating Design Flood of Water Resources and Hydropower Projects*; China Water & Power Press: Beijing, China, 2006.
33. Guo, S.; Muhammad, R.; Liu, Z.; Xiong, F.; Yin, J. Design Flood Estimation Methods for Cascade Reservoirs Based on Copulas. *Water* **2018**, *10*, 560. [CrossRef]
34. Xiong, F.; Guo, S.; Yin, J.; Tian, J.; Rizwan, M. Comparative Study of Flood Regional Composition Methods for Design Flood Estimation in Cascade Reservoir System. *J. Hydrol.* **2020**, *590*, 125530. [CrossRef]
35. Xiong, F.; Guo, S.; Liu, P.; Xu, C.-Y.; Zhong, Y.; Yin, J.; He, S. A General Framework of Design Flood Estimation for Cascade Reservoirs in Operation Period. *J. Hydrol.* **2019**, *577*, 124003. [CrossRef]
36. Sarhadi, A.; Burn, D.H.; Concepción Ausín, M.; Wiper, M.P. Time-Varying Nonstationary Multivariate Risk Analysis Using a Dynamic Bayesian Copula. *Water Resour. Res.* **2016**, *52*, 2327–2349. [CrossRef]
37. Jiang, C.; Xiong, L.; Yan, L.; Dong, J.; Xu, C.-Y. Multivariate Hydrologic Design Methods under Nonstationary Conditions and Application to Engineering Practice. *Hydrol. Earth Syst. Sci.* **2019**, *23*, 1683–1704. [CrossRef]
38. Feng, Y.; Shi, P.; Qu, S.; Mou, S.; Chen, C.; Dong, F. Nonstationary Flood Coincidence Risk Analysis Using Time-Varying Copula Functions. *Sci. Rep.* **2020**, *10*, 3395. [CrossRef]
39. Liang, Z.; Hu, Y.; Huang, H.; Wang, J.; Li, B. Study on the Estimation of Design Value under Non-Stationary Environment. *South-North Water Transf. Water Sci. Technol.* **2016**, *14*, 50–53. [CrossRef] [PubMed]

40. Wang, J.; Li, B.; Wang, H. Concept of Equivalent Reliability for Estimating the Design Flood under Non-Stationary Conditions. *Water Resour. Manag.* **2018**, *32*, 997–1011. [CrossRef]
41. Koutrouvelis, I.A.; Canavos, G.C. Estimation in the Pearson Type 3 Distribution. *Water Resour. Res.* **1999**, *35*, 2693–2704. [CrossRef]
42. Villarini, G.; Serinaldi, F.; Smith, J.A.; Krajewski, W.F. On the Stationarity of Annual Flood Peaks in the Continental United States during the 20th Century: Stationarity of Annual Flood Peaks. *Water Resour. Res.* **2009**, *45*. [CrossRef]
43. Yan, L.; Xiong, L.; Guo, S.; Xu, C.-Y.; Xia, J.; Du, T. Comparison of Four Nonstationary Hydrologic Design Methods for Changing Environment. *J. Hydrol.* **2017**, *551*, 132–150. [CrossRef]
44. Xiong, B.; Xiong, L.; Chen, J.; Xu, C.-Y.; Li, L. Multiple Causes of Nonstationarity in the Weihe Annual Low-Flow Series. *Hydrol. Earth Syst. Sci.* **2018**, *22*, 1525–1542. [CrossRef]
45. Mann, H.B. Nonparametric Tests against Trend. *Econometrica* **1945**, *13*, 245–259. [CrossRef]
46. Kendall, M.G. *Rank Correlation Methods*; Griffin: Oxford, UK, 1948.
47. Yue, S.; Pilon, P.; Cavadias, G. Power of the Mann–Kendall and Spearman's Rho Tests for Detecting Monotonic Trends in Hydrological Series. *J. Hydrol.* **2002**, *259*, 254–271. [CrossRef]
48. Pettitt, A.N. A Non-Parametric Approach to the Change-Point Problem. *J. R. Stat. Soc. Ser. C Appl. Stat.* **1979**, *28*, 126–135. [CrossRef]
49. Matteson, D.S.; James, N.A. A Nonparametric Approach for Multiple Change Point Analysis of Multivariate Data. *J. Am. Stat. Assoc.* **2014**, *109*, 334–345. [CrossRef]
50. Katz, R.W.; Parlange, M.B.; Naveau, P. Statistics of Extremes in Hydrology. *Adv. Water Resour.* **2002**, *25*, 1287–1304. [CrossRef]
51. Liu, D.; Guo, S.; Lian, Y.; Xiong, L.; Chen, X. Climate-Informed Low-Flow Frequency Analysis Using Nonstationary Modelling. *Hydrol. Process.* **2015**, *29*, 2112–2124. [CrossRef]
52. Dobson, A.J.; Barnett, A.G. *An Introduction to Generalized Linear Models*, 4th ed.; Chapman and Hall/CRC: Boca Raton, FL, USA, 2018; ISBN 978-1-315-18278-0.
53. Cole, T.J.; Green, P.J. Smoothing Reference Centile Curves: The LMS Method and Penalized Likelihood. *Stat. Med.* **1992**, *11*, 1305–1319. [CrossRef]
54. Wang, C.; Zeng, B.; Shao, J. Application of Bootstrap Method in Kolmogorov-Smirnov Test. In Proceedings of the 2011 International Conference on Quality, Reliability, Risk, Maintenance, and Safety Engineering, Xi'an, China, 17–19 June 2011; pp. 287–291.
55. Van Buuren, S.; Fredriks, M. Worm Plot: A Simple Diagnostic Device for Modelling Growth Reference Curves. *Stat. Med.* **2001**, *20*, 1259–1277. [CrossRef]
56. Lima, C.H.R.; Lall, U.; Troy, T.J.; Devineni, N. A Climate Informed Model for Nonstationary Flood Risk Prediction: Application to Negro River at Manaus, Amazonia. *J. Hydrol.* **2015**, *522*, 594–602. [CrossRef]
57. Schwarz, G. Estimating the Dimension of a Model. *Ann. Stat.* **1978**, *6*, 461–464. [CrossRef]
58. Akaike, H. A New Look at the Statistical Model Identification. *IEEE Trans. Autom. Control* **1974**, *19*, 716–723. [CrossRef]
59. Read, L.K.; Vogel, R.M. Reliability, Return Periods, and Risk under Nonstationarity. *Water Resour. Res.* **2015**, *51*, 6381–6398. [CrossRef]
60. Nelsen, R. An Introduction to Copulas. *Technometrics* **2000**, *42*, 317. [CrossRef]
61. Marra, G.; Radice, R. Bivariate Copula Additive Models for Location, Scale and Shape. *Comput. Stat. Data Anal.* **2017**, *112*, 99–113. [CrossRef]
62. Wojtyś, M.; Marra, G.; Radice, R. Copula Based Generalized Additive Models for Location, Scale and Shape with Non-Random Sample Selection. *Comput. Stat. Data Anal.* **2018**, *127*, 1–14. [CrossRef]
63. Marra, G.; Radice, R.; Bärnighausen, T.; Wood, S.N.; McGovern, M.E. A Simultaneous Equation Approach to Estimating Hiv Prevalence with Nonignorable Missing Responses. *J. Am. Stat. Assoc.* **2017**, *112*, 484–496. [CrossRef]
64. Rosenblatt, M. Remarks on a Multivariate Transformation. *Ann. Math. Stat.* **1952**, *23*, 470–472. [CrossRef]
65. Genest, C.; Rémillard, B.; Beaudoin, D. Goodness-of-Fit Tests for Copulas: A Review and a Power Study. *Insur. Math. Econ.* **2009**, *44*, 199–213. [CrossRef]
66. Salvadori, G.; De Michele, C.; Durante, F. On the Return Period and Design in a Multivariate Framework. *Hydrol. Earth Syst. Sci.* **2011**, *15*, 3293–3305. [CrossRef]
67. Zhong, Y.; Guo, S.; Liu, Z.; Wang, Y.; Yin, J. Quantifying Differences between Reservoir Inflows and Dam Site Floods Using Frequency and Risk Analysis Methods. *Stoch. Environ. Res. Risk Assess.* **2017**, *6*, 419–433. [CrossRef]
68. Yin, J.; Guo, S.; Liu, Z.; Yang, G.; Zhong, Y.; Liu, D. Uncertainty Analysis of Bivariate Design Flood Estimation and Its Impacts on Reservoir Routing. *Water Resour. Manag.* **2018**, *32*, 1795–1809. [CrossRef]
69. Franchini, M.; Bernini, A.; Barbetta, S.; Moramarco, T. Forecasting Discharges at the Downstream End of a River Reach through Two Simple Muskingum Based Procedures. *J. Hydrol.* **2011**, *399*, 335–352. [CrossRef]
70. *Ministry of Water Resources The Designed Operation Rules of Danjiangkou Reservoir for Water Diversion*; Water Resources and Hydropower Press: Beijing, China, 2016.
71. Yang, G.; Guo, S.; Li, L.; Hong, X.; Wang, L. Multi-Objective Operating Rules for Danjiangkou Reservoir under Climate Change. *Water Resour. Manag.* **2016**, *30*, 1183–1202. [CrossRef]
72. He, S.; Guo, S.; Yang, G.; Chen, K.; Liu, D.; Zhou, Y. Optimizing Operation Rules of Cascade Reservoirs for Adapting Climate Change. *Water Resour. Manag.* **2020**, *34*, 101–120. [CrossRef]

73. Department of Comprehensive Statistics of the National Bureau of Statistics of China. *China Compendium of Statistics 1949–2008*; Chinese Statistics Press: Beijing, China, 2010.
74. Tsoularis, A.; Wallace, J. Analysis of Logistic Growth Models. *Math. Biosci.* **2002**, *179*, 21–55. [CrossRef]
75. IPCC. *Climate Change 2014: Impacts, Adaptation, and Vulnerability: Working Group II Contribution to the Fifth Assessment Report of the Intergovernmental Panel on Climate Change*; Intergovernmental Panel on Climate Change, Ed.; Cambridge University Press: New York, NY, USA, 2014; ISBN 978-1-107-64165-5.
76. Tian, J.; Guo, S.; Deng, L.; Yin, J.; Pan, Z.; He, S.; Li, Q. Adaptive Optimal Allocation of Water Resources Response to Future Water Availability and Water Demand in the Han River Basin, China. *Sci. Rep.* **2021**, *11*, 7879. [CrossRef]
77. Moss, R.H.; Edmonds, J.A.; Hibbard, K.A.; Manning, M.R.; Rose, S.K.; van Vuuren, D.P.; Carter, T.R.; Emori, S.; Kainuma, M.; Kram, T.; et al. The next Generation of Scenarios for Climate Change Research and Assessment. *Nature* **2010**, *463*, 747–756. [CrossRef] [PubMed]
78. Van Vuuren, D.P.; Edmonds, J.A.; Kainuma, M.; Riahi, K.; Weyant, J. A Special Issue on the RCPs. *Clim. Chang.* **2011**, *109*, 1–4. [CrossRef]
79. Thomson, A.M.; Calvin, K.V.; Smith, S.J.; Kyle, G.P.; Volke, A.; Patel, P.; Delgado-Arias, S.; Bond-Lamberty, B.; Wise, M.A.; Clarke, L.E.; et al. RCP4.5: A Pathway for Stabilization of Radiative Forcing by 2100. *Clim. Chang.* **2011**, *109*, 77. [CrossRef]
80. Chen, J.; Brissette, F.P.; Chaumont, D.; Braun, M. Performance and Uncertainty Evaluation of Empirical Downscaling Methods in Quantifying the Climate Change Impacts on Hydrology over Two North American River Basins. *J. Hydrol.* **2013**, *479*, 200–214. [CrossRef]
81. Liu, P.; Li, L.; Guo, S.; Xiong, L.; Zhang, W.; Zhang, J.; Xu, C.-Y. Optimal Design of Seasonal Flood Limited Water Levels and Its Application for the Three Gorges Reservoir. *J. Hydrol.* **2015**, *527*, 1045–1053. [CrossRef]

water

Article

The Impact of the Uncertain Input Data of Multi-Purpose Reservoir Volumes under Hydrological Extremes

Stanislav Paseka * and **Daniel Marton**

Institute of Landscape Water Management, Faculty of Civil Engineering, Brno University of Technology, 602 00 Brno, Czech Republic; marton.d@fce.vutbr.cz
* Correspondence: paseka.s@fce.vutbr.cz

Abstract: The topic of uncertainties in water management tasks is a very extensive and highly discussed one. It is generally based on the theory that uncertainties comprise epistemic uncertainty and aleatoric uncertainty. This work deals with the comprehensive determination of the functional water volumes of a reservoir during extreme hydrological events under conditions of aleatoric uncertainty described as input data uncertainties. In this case, the input data uncertainties were constructed using the Monte Carlo method and applied to the data employed in the water management solution of the reservoir: (i) average monthly water inflows, (ii) hydrographs, (iii) bathygraphic curves and (iv) water losses by evaporation and dam seepage. To determine the storage volume of the reservoir, a simulation-optimization model of the reservoir was developed, which uses the balance equation of the reservoir to determine its optimal storage volume. For the second hydrological extreme, a simulation model for the transformation of flood discharges was developed, which works on the principle of the first order of the reservoir differential equation. By linking the two models, it is possible to comprehensively determine the functional volumes of the reservoir in terms of input data uncertainties. The practical application of the models was applied to a case study of the Vír reservoir in the Czech Republic, which fulfils the purpose of water storage and flood protection. The obtained results were analyzed in detail to verify whether the reservoir is sufficiently resistant to current hydrological extremes and also to suggest a redistribution of functional volumes of the reservoir under conditions of measurement uncertainty.

Keywords: multi-purpose reservoir; functional volume; uncertainties; Monte Carlo method; hydrological extremes; simulation-optimization model; optimal storage volume; simulation model; retention volume; transformation of flood discharges

Citation: Paseka, S.; Marton, D. The Impact of the Uncertain Input Data of Multi-Purpose Reservoir Volumes under Hydrological Extremes. *Water* **2021**, *13*, 1389. https://doi.org/10.3390/w13101389

Academic Editors: Rajesh R. Shrestha and Mohammad Reza Najafi

Received: 30 March 2021
Accepted: 12 May 2021
Published: 16 May 2021

1. Introduction

According to the latest evaluated data on the state of weather and climate in the world by the World Meteorological Organization [1], warming is continuing to increase, which has been observable for several decades. According to the Intergovernmental Panel on Climate Change (IPCC) [2], one of the five reasons for concern (RFCs) that illustrate the consequences of global warming and summarize key impacts and risks across sectors and regions is extreme weather events. These include, for example, an increase in the number of heat waves and an increase in the periodicity and intensity of droughts and floods. Currently, due to global warming, water scarcity is growing very rapidly worldwide and an increasing number of drought-affected areas are emerging. Problems with the security of water resources are beginning to be evident, even in areas where the population has not been very aware of drought. On the other hand, there is also a more frequent incidence of floods worldwide.

The people of Central Europe are also beginning to feel these problems more strongly [3]. Drought periods have appeared in Central Europe in recent years, in 1949, 1961 and 1963, from 1991 to 1994 and in 2003 [4], and they have been repeated to a greater extent since about 2011 [4], persisting until now. In the climate of Central Europe, the reduction in average precipitation is not recorded, but there are changes in the distribution of precipitation. In other words, the periods of drought are becoming more prolonged, alternating with more intense precipitation, which can cause floods from torrential rains. The above is also intensified by the extensive regional floods that Central Europe has faced in recent years; for instance, the regional floods in 1997, 2002, 2006, 2009, 2010 and 2013 [5].

The values of annual river flow trends in Central Europe have a negative tendency, especially in the spring and summer months [6]. In the future, a further decrease in flows is probable, especially in low-water periods [7], and the probability of lower flow occurrence will increase. In addition, the prospects for the coming years are not very optimistic given the frequency and length of droughts and the occurrence of floods, even if estimates from climate models are not completely fulfilled. This is confirmed by the IPCC Fifth Assessment Report [8]. According to this report, climate change is expected to increase drought risks and water scarcity in urban areas with a very high degree of reliability. Indicators of a medium degree of reliability point to higher flood risks at the regional level [8]. In addition, the longer the period without the occurrence of major regional floods, the more likely it becomes that this event will occur.

From the point of view of hydrological extremes, a large degree of uncertainty affects water management applications, whether it is uncertainty stemming from climate change or from the measurement of input data. The overall concept of uncertainty can currently be perceived from several perspectives, and there are many applications of uncertainty. Uncertainty can be classified [9] into two categories, namely aleatoric and epistemic uncertainty. The application of the aleatoric and epistemic uncertainty typology is complex in technical tasks but it can be determined by the author when creating a model based on many factors, knowledge of the problem and the decision-making process [9]. Uncertainty arising from measurements falls into the group of aleatoric uncertainty. This category of uncertainties is tied to a certain probability distribution and shows the variability associated with the system or the environment. Therefore, it can be described using stochastic simulations [10].

No major watercourses flow into the Czech Republic and the only basic source of water is precipitation that falls on its territory. Under these hydroclimatic conditions, water resources management must be focused primarily on increasing the retention capacity of water in the landscape and its subsequent infiltration into underground sources, but also on strengthening surface water resources. One of the appropriate adaptation measures for the future change of climatic conditions is water reservoirs, either in the sense of appropriate management of existing reservoirs or the construction of new reservoirs. This is confirmed by the Strategy on Adaptation to Climate Change in the Czech Republic [11], which represents a national adaptation strategy including water management. One of the recommended strategies in the document [11] is the optimization of the function of existing reservoirs and water management systems with regard to the more intensive occurrence of hydrological extremes.

In the Czech Republic, water reservoirs have been designed according to historical or derived hydrological time series, and individual volumes in reservoirs were designed separately. Water reservoir handling codes are outdated and do not take into account the uncertainties arising from the processing of input data and the uncertainty of future climate change. These uncertainties can jeopardize the reliability of water supply. It is therefore appropriate to undertake thorough analyses of reservoir volumes and revisions of handling codes and the Czech Technical Standards [12].

The aim of this study was to present a complex solution for the design of functional volumes of a multi-purpose reservoir under conditions of uncertainty of input data measurement. To meet the main objective, it was first necessary to meet two sub-objectives:

(a) The first sub-objective was to develop a simulation-optimization model of the reservoir to determine the optimal storage volume of the reservoir under conditions of input data uncertainty (UNCE_RESERVOIR). The reservoir model is based on the balance equation of the reservoir and involves optimization using the grid method with the required temporal reliability.

(b) The second sub-objective was to develop a simulation model for the transformation of uncertain flood discharges to determine the retention volume of the reservoir under conditions of input data uncertainty (TRANSFORM_WAVE). The model is based on the first order of the reservoir differential equation.

(c) The main objective was to link the two models and analyze what effect the optimized reservoir storage volume will have on the transformation effect of the reservoir.

The main model integration comprises the development of a comprehensive solution for the functional volumes of the reservoir in terms of input data uncertainties applied to the input data. This innovative approach responds to changes in future climatic conditions and contributes to reducing the risk of water supply disturbances during low-water periods with a safe design for the size of the retention volume in extreme floods. The combination of both models creates a procedure and, subsequently, a tool, which can be applied to the design of any new reservoirs or to the redistribution of volumes of existing multi-purpose reservoirs, not only in Central Europe but also across the world. Adequate input data is an important condition. The procedure and models were applied and tested on the real Vír reservoir in the basin of the Morava River in the Czech Republic.

2. Background

The uncertainties themselves, from the point of view of current knowledge, were first described in the work "Risk, Uncertainty, and Profit" [13]. Measurement uncertainties as we know them today became common practice in calibration laboratories only in 1990 with the Western European Calibration Association publication "WECC 19/90" [14], which outlined, for the first time, the general principles of uncertainty and defined further procedures for introducing uncertainty theory into metrological practice. This was followed by other regulations and guidelines [15,16], according to which the theory of measurement uncertainties developed two classifications: so-called type A uncertainty and type B uncertainty. This classification is based on the method of obtaining the given uncertainty. Reference [17] deals with the distribution and promotion of measurement uncertainties using simulation with the stochastic Monte Carlo method.

Worldwide, research on uncertainties in hydrology and water management is extensive, including the uncertainty of measurement of hydrological quantities, uncertainties in hydrological applications and the influence of measurement uncertainties on multi-purpose reservoirs. In hydrology itself, estimates of uncertainties in the form of uncertain and incomplete data or ignorance of systems were first described by the GLUE method [18,19]. Uncertainties of flow measurements or possible approaches to fill in missing flow series have been discussed [20–24]. Uncertainties applied to predict water flow in streams and floods using the Monte Carlo method have been mentioned in other studies [25,26]. Hydrological applications involving the investigation of the effects of hydrological uncertainties arising from measurements on the volume of water in reservoirs have been discussed in two articles [27,28]. The most recent publications have examined the risks of uncertainty and its effect on the storage volume of reservoirs using Monte Carlo simulations [29,30]. Based on uncertain future flood inflows to a reservoir, an analysis of the probable risks for a reservoir was performed in [31] and an analysis of flood control risks with an uncertain prognosis of water inflows was undertaken through Monte Carlo simulation of the reservoir system in [32]. A simultaneous solution for storage and retention volume under uncertainty conditions was investigated in [33] for the largest multi-purpose water reservoir in Vietnam using the simulation model MIKE 11. Other articles that evaluate new approaches to decision-making methods for solving problems involved in multi-purpose reservoirs under conditions of uncertainties are [34–36].

In general, the models for determining the storage volumes of reservoirs are based on the balance between the inflow of water into the reservoir and the outflow of water from the reservoir, including abstractions. The model approaches include simulation [37,38] and an unconventionally statistical approach [39]. For the transformation of flood discharges or the determination of the retention volume of the reservoir, the approaches of the models generally involve simulation and can be solved with numerical Runge–Kutta differential methods [40–42], Klemeš graphical methods based on a differential equation [43,44] or mathematical and physical modeling.

Further similar research dealing with the volumes of multi-purpose reservoirs under uncertain conditions can be found in articles [45–51]. The results obtained generally show that current reservoirs and optimal operating rules for the water supply system cannot deal with the problem of performance degradation under uncertainties associated with inflow conditions and water demand growth. The search for the trade-off between water supply and flood protection leads to individual approaches and to the improvement of reservoirs' optimal operation models. Articles [45–51] further develop models at tropical monsoon climate locations or in areas with significant annual periods of dry and wet seasons. Current climate conditions in Central Europe are changing. The frequencies of drought and floods episodes are increasing and the problem focus on the trade-off between water supply and flood protection can only be solved when the issues are treated separately. Under these climate change conditions, newly joined water supply and flood modeling including uncertain input data will be an issue for water science in the Czech Republic.

In the Czech Republic, the influence of random errors in hydrological data for the value of the reservoir storage volume was first investigated in 1984 [52]. This report introduced the principle of random errors into time series using the Monte Carlo method and their subsequent application to reservoir storage volume calculations was defined. This method was developed in [53–57] where a detailed analysis of the influence of the uncertainties of the input data on the simulation model of the storage volume of the reservoir was undertaken. In articles [52–57], comprehensive models were constructed to address the issues described above in sub-objectives (a) and (b). These models were first tested in isolation, as described in [58,59], and these articles present the development of their integration.

The research in [54–57] has shown that the uncertainties of the input hydrological data can: (i) negatively affect or underestimate the size of the storage volume of a reservoir, (ii) realistically cause unexpected operational failure of a reservoir, (iii) result in high economic damage and (iv) lead under certain conditions to reservoirs being erroneously classified into the more significant classes of waterworks according to the Czech Technical Standard [12] based on storage volumes. This standard classifies water reservoirs based on their strategic importance. The classification is evaluated by the time reliability R_T, which is defined as the ratio of months without water failures, represented as water deficit, and the number of all months for a given time series [60–62]. Category A, with the highest priority, is for $R_T \geq 99.5\%$ and category D, with the lowest priority, is for $R_T \geq 95.0\%$. A water deficit is identified when the storage volume of the reservoir is in an unsatisfactory state.

3. Case Study

A case study was undertaken on the Vír I multi-purpose reservoir located on the Svratka River in the basin of the Morava River in the Czech Republic. The reservoir is located 150 km east of the capital city of Prague (see Figure 1) and supplies water to Brno, the second largest city. The concrete gravity dam falls under the administration of the Morava River Basin State Enterprise. The long-term inflow Q_A is 3.28 m^3 s^{-1}. The required water outflow from the reservoir O_R is 2.5 m^3 s^{-1}. The storage volume V_Z is 44.056 million m^3 at an elevation of 464.45 m above sea level, the volume of inactive storage is 3.80 million m^3, the retention volume V_R is 8.337 million m^3 and the total volume of the reservoir V_{TOT} is 56.193 million m^3. The reservoir has two bottom outlets with diameters of 1800 mm. At the maximum water level, the flow rate of the bottom outlets is maximally 2×40 m^3 s^{-1}. The emergency spillway of the reservoir is an uncontrolled crest structure at an elevation of 467.05 m above sea level, with a total length of 60.5 m (5×12.1 m) and a capacity of 180.5 m^3 s^{-1} at the level of the uncontrollable retention volume. According to the Czech Technical Standard [12], the reservoir falls into category A, the highest priority, i.e., with a time reliability $R_T \geq 99.5\%$. The main purposes of the reservoir are storage of drinking water and flood protection.

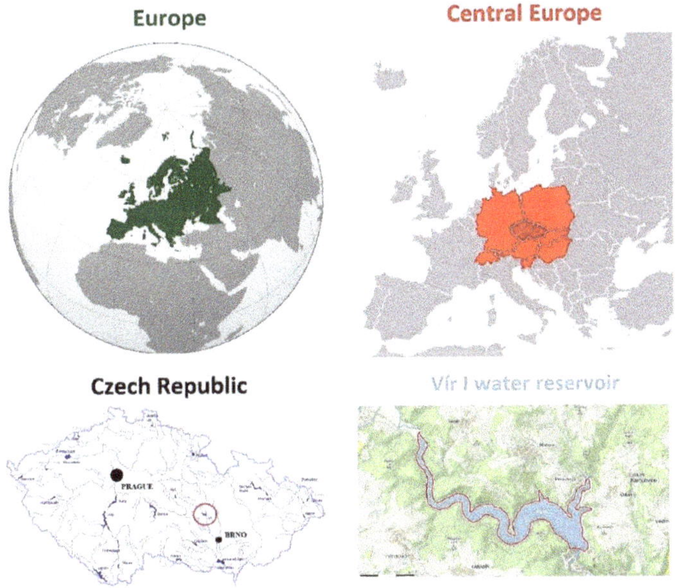

Figure 1. Location of the Vír I water reservoir.

Hydrological data regarding the inflow of water into the reservoir used for the determination of the storage volume of the reservoir was in the form of historical monthly flows for the last 66 years and an updated flood discharge from 2008 was used to calculate the retention volume of the reservoir. Specifically, this was a flood discharge of Q_{1000} according to the classical regression without seasonality differentiation and with a time step of 360 min. Hydrological data were acquired from the Czech Hydrometeorological Institute.

The calculation of the storage volume of the reservoir aimed to take into account the losses of water from the reservoir, specifically evaporation from the water surface and seepage through the body of the dam. The amount of the evaporation was determined by the annual evaporation, i.e., 613 mm, and the seepage size through the dam body was $0.15\ \mathrm{l\ s^{-1}}$ per $1000\ \mathrm{m^2}$ [63].

To calculate the retention volume of a reservoir, it is necessary to determine its control maximum water level (CML) according to [63,64]. This CML value corresponds to the level of the dam crest, i.e., 470.45 m above sea level.

4. Methodology

4.1. Problem Formulation

The operation of multi-purpose reservoirs is complicated due to the conflict between different objectives. In recent years, it has become desirable to maximize both storage volume and protective volume, as multi-purpose reservoirs may not fully respond to current and future climatic conditions. The solution is to find a suitable ratio between the size of the storage and the retention volume of water in the reservoir in conditions of uncertainty. In other words, we need to comprehensively determine the functional volumes of the reservoir. For this purpose, the models presented below apply input data uncertainties to their solution through the Monte Carlo method. The connection of both models for the determination of the functional volumes of the reservoir, including an example of the introduction of input uncertainties, is shown in the flow chart in Figure 2.

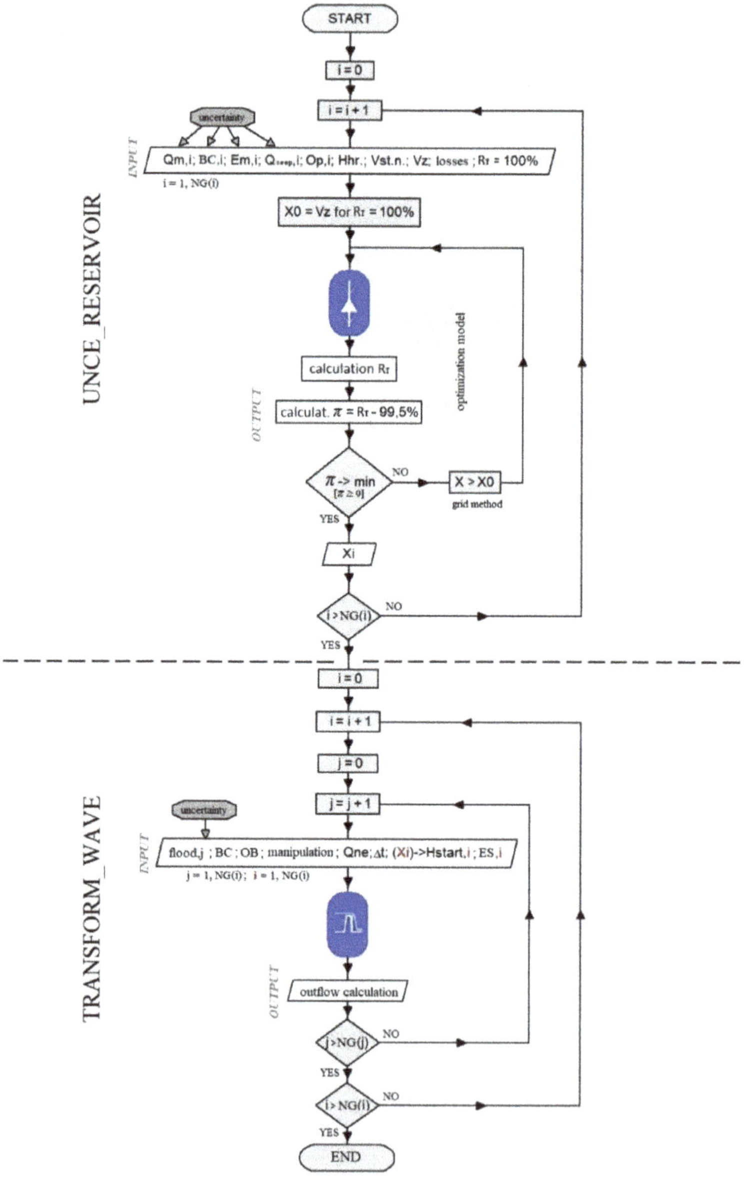

Figure 2. Flow chart of a complex solution for the functional volumes of a reservoir.

4.2. UNCE_RESERVOIR—Simulation-Optimization Model of the Reservoir for Determining the Storage Volume of the Reservoir

The developed simulation-optimization model determines the optimal storage volume of the reservoir V_Z in conditions of uncertainty. To determine the optimal storage volume of the reservoir $V_Z = f(O_R, R_T)$, which is a function of the required outflow O_R and predetermined temporal reliability $R_T < 100\%$, repeated calculations determining the temporal reliability $R_T = f(O_R, V_Z)$ at the predetermined required outflow O_R and storage volume V_Z were used.

The parameter sought was therefore the storage volume V_Z. The criterion was temporal reliability according to the temporal reliability R_T and the water management solution allowed water failures in the reservoir according to the categorization of the reservoir. The initial condition was a full reservoir at the beginning of the test period (a value of 0 on the left side of Equation (1) characterizes the full storage volume of the reservoir) and the boundary condition was a series of water inflows into the reservoir at the appropriate time steps. For each time step, a balance was made between the required outflow O_R and the historical inflow of water into the reservoir Q. In addition, the limiting condition $\sum(O_R - Q)$ was tested, i.e., whether or not the reservoir was emptied at the end of each month (the $V_{Z,MAX}$ value on the right side of Equation (1) characterizes the empty storage volume of the reservoir). If it was emptied, a failure of the water outflow from the reservoir was judged to exist. This meant that in all months when the outflow of water O_i was less than the required outflow O_R, the reservoir failed to supply enough water to the distribution system. The total sum of all failure months according to Equation (2) was recorded and the temporal reliability R_T was calculated (see Equation (3)).

The basis of the simulation model of the subtask $R_T = f(O_R, V_Z)$ is the modified balance equation of the reservoir in the sum form converted into the following inequality (Equation (1)) [65]:

$$0 \leq \sum_{i=0}^{k}(O_i - Q_i)\Delta t \leq V_{Z,MAX}, \quad (1)$$

where O_i is the water outflow from the reservoir (m^3 s^{-1}) in a given month for $i = 1, \ldots, k$; Q_i is the inflow of water into the reservoir (m^3 s^{-1}) in a given month for $i = 1, \ldots, k$ and Δt is the time step of the calculation of one month.

The classification of the reservoir storage volume failure for the calculation of temporal reliability is expressed by Equation (2) [53]:

$$Z_{t,i} = \begin{cases} Z_{t,i} = 1, & O_i > O_R \\ Z_{t,i} = 0, & O_i < O_R \end{cases}, \quad (2)$$

where $Z_{t,i} = 1$ describes the state of the V_Z reservoir in the fault-free (satisfactory) time step of the calculation and $Z_{t,i} = 0$ describes the state of the V_Z reservoir in the faulty (unsatisfactory) time step of the calculation.

The degree of temporal reliability of the improved outflow O_R as a result of the outflow control is the probability that the actual outflow of water from the reservoir will not fall below the value of the improved outflow O_R. In this case, the temporal reliability is applied according to the temporal reliability R_T, which can be calculated from the values $Z_{t,i}$ according to Equation (2) [66]:

$$R_T = \frac{\left(\sum_{i=1}^{k} Z_{t,i}\right) - 0.3}{k + 0.4} \cdot 100, \ [\%], \quad (3)$$

where $\sum_{i=1}^{k} Z_{t,i}$ is the sum of the records of the faulty and fault-free months and k is the number of all months of the input time series.

First, the value of the parameter (storage volume) is selected and a new variant of reservoir operation is repeatedly simulated according to Equation (1), and then the monitored criterion π (decrease in temporal reliability R_T) is evaluated according to Equations (2) and (3). The solution is a variant in which the criterion coincides with the required value. In this variant, the selected parameter becomes the result of the solution. The task leads to an optimization in which the solved parameter is unknown and the criterion is the difference between the calculated and the required temporal reliability, which is minimized. The reservoir model uses a simple optimization method called the grid method, where parameters with a fixed step are selected at allowable intervals. The calculation also includes, among other things, water losses from the reservoir, specifically water losses by evaporation from the water surface and seepage of the dam body. The principle for introducing water losses from the reservoir into the solution is that the loss flows are counted using repeated simulation.

4.3. *TRANSFORM_WAVE—Reservoir Simulation Model for Determining the Retention Volume of the Reservoir*

For the second hydrological extreme, flooding, a simulation model was developed for the transformation of uncertainty-affected flood discharges. In the model, the transformation of the flood discharge is performed by mathematical modification of the original Klemeš graphical method [43]. The Klemeš graphical method (see Figure 3) is based on the first order of the reservoir differential equation and expresses the relationship between the inflow and outflow of water from the reservoir as a function of time and the volume of water retained in the reservoir.

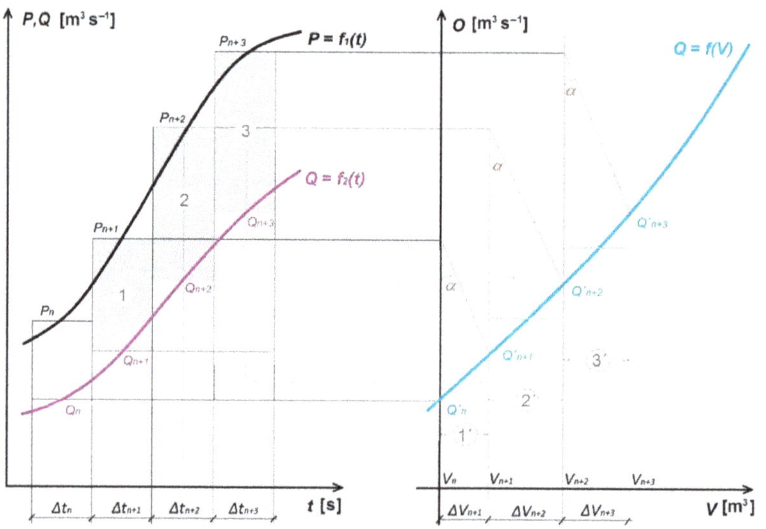

Figure 3. Principle of the Klemeš graphical method [43].

The following simplifications were introduced to construct the model: (i) Although the inflow and outflow values change continuously over a period of time during the process, they were considered to be constant in a given time interval. The inflow (marked as *P* and *Q* in Figure 3) and outflow in each interval are thus represented by a single instantaneous value; (ii) The Klemeš method was set to enter the calculation only when the emergency spillway is exceeded. If the water is below the level of the emergency spillway, then the calculation only balances the volume of water inflow and outflow to the previous reservoir volume.

The main data for the Klemeš method are the hydrograph of the flood, the parameters of the bottom outlets, the emergency spillway and the line of flooded volumes for the construction of the so-called transformation curve, which can be seen in the right part of Figure 3 and is marked as Q'. The transformation curve expresses the total outflow of water from the reservoir depending on its filling. It characterizes the potential outflow of water from the reservoir and is a function of the volume of the reservoir. To determine it, according to [67], the capacity of the bottom outlets O_B should be calculated according to Equation (4) and the emergency spillway capacity Q_{ES} according to Equation (5):

$$O_B = \mu\, S\, \sqrt{2\, g\, h_w}, \tag{4}$$

$$Q_{ES} = m\, b\, \sqrt{2\, g}\, h_{es}^{\frac{3}{2}}, \tag{5}$$

where μ is the outflow coefficient (-), S is the area of the outlet opening (bottom outlets) (m^2), g is the gravity acceleration (m s^{-2}), h_w is the height of the water above the bottom outlets (m), m is the overflow coefficient (-), b is the width of the emergency spillway (m) and h_{es} is the height of the water above the spillway (m).

Furthermore, it is necessary to use the Klemeš method to construct the reduction angle α for the transformation, which is based on the size of the flow and the time interval and characterizes the proportionality between the created area surrounded by inflow and outflow in a given time interval and the length of the horizontal line of the transformation curve of the given time interval. The graphical construction by the Klemeš method takes place gradually in individual time intervals, first on the ascending and then on the descending branch of the flood discharge, through the mentioned transformation curve and the reduction angle α.

4.4. Monte Carlo Method for Applying Input Uncertainties to the Reservoir Simulation Model

The following assumptions were introduced to create an algorithm that generates random series with a burden in the form of uncertainties. The general input value X resulting from the measurement was considered as a random (stochastic) quantity. This assumption makes it possible to generate new X_i values around the input value X resulting from the measurement completely randomly and independently of each other. The quantity X_i is therefore random and independent of the previous and following values. The randomly generated quantities X_i are the result of a number of mutually independent phenomena, which makes it possible to describe the input value with a corresponding normal probability distribution $N(\mu(X),\sigma(X))$. The introduction of a normal probability distribution makes it possible to enter the vicinity of the resulting value of a random variable using the mean value $\mu(X)$ as the measured value and the standard deviations $\sigma(X)$ as the standard uncertainty. Only the standard measurement uncertainty of type B $u_{B,X}$ was considered in the calculations. Finally, a simplification was introduced, whereby the standard uncertainty of measurement $u_{B,X}$ is deployed using the relative value of the coefficient of variation $C_v(X)$ (see Equation (10)) and the resulting standard deviation $\sigma(X)$ is then calculated according to Equation (9).

The essence of the random series generator is repeated use of the Monte Carlo method. Subsequently, for each mean value $\mu_t(X)$, distribution curves $F_t(X)$ of the normal standardized probability distribution $N(\mu(X),\sigma(X))$ are created for t = 1, 2, ... , NE, where NE is the total number of elements (e.g., the total number of average monthly inflows or the total number of points from the flooded volume line). Using a pseudo-random number generator, generating random numbers from the interval $\xi \in \langle 0,1 \rangle$, random waveforms of a number of elements X_t are repeatedly generated, which are referred to as random positions of $NX_{t,i}$ values, in the interval of specified uncertainty for i = 1, 2, ... , NG, where NG is the total number of repetitions (generations). The general principle for generating random positions of input parameters can be found in previous studies [52,53]. This described procedure for generating random elements can be applied to all quantities entering the water management solution for the storage and retention volume of the reservoir, except the bathygraphic curves of the reservoir. In this case, a compilation of two independent Monte Carlo generators was required. Each generator constructs a random point position (e.g., water level height) with a second random point position added to it (e.g., the volume of water in the reservoir). Together, the random positions of two points then create a random point coordinate (e.g., a random point coordinate of a flooded volume line). A series of random points then form random lines of flooded volumes burdened with uncertainties. A symbolic depiction of the introduction of considered quantities burdened with uncertainties is shown Figure 4.

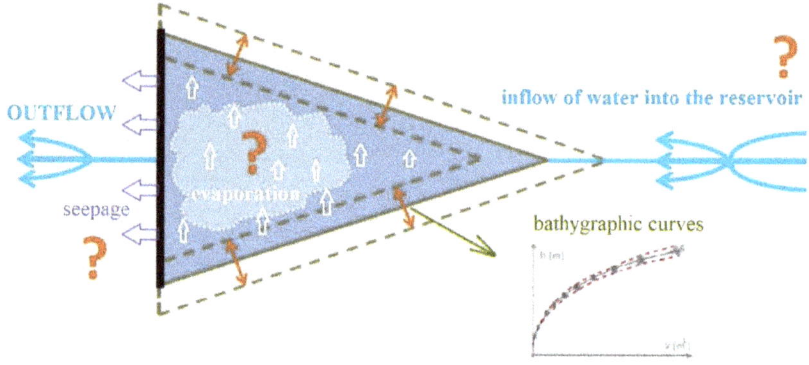

Figure 4. Symbolic depiction of the introduction of considered quantities burdened with uncertainties.

4.5. Methods for Evaluation

The generated input uncertainties in the data for the calculation of the water management solution of the reservoir provide spectra of storage and retention volume sizes. For a suitable presentation of the achieved results, the calculations were statistically evaluated and quantiles and the overshoot probability curve were also used.

4.5.1. Mean Value

The mean value is the value of the first general moment, denoted as $\mu(X)$, and is expressed in the following form (Equation (6)):

$$\mu(x) = \int_x^{\cdot} x f_x(x) dx. \tag{6}$$

The mean value belongs to the so-called position characteristics and its value is the x-coordinate of the center of gravity of the probability density. The method of moments involves an estimate of the mean value, as expressed by Equation (7):

$$\mu(x) \approx \overline{x} = \frac{\sum_{i=1}^{n} x_i}{n}, \tag{7}$$

where $\overline{x}$ is the sample mean or mean value, x_i are elements of random selection and n is the number of elements of random selection.

4.5.2. Variance and Standard Deviation

The standard deviation is expressed as the square root of the variance $D(X)$. The variance rate of the random variable X regarding the diameter μX is given by the second central moment, or the variance $D(X)$, which is expressed in the following form (Equation (8)):

$$D(x) = \int_x^{\cdot} (x - \mu(x))^2 f_x(x) dx. \tag{8}$$

The standard deviation is also based on the second central moment and is denoted by $\sigma(x)$. Using the method of moments, the standard deviation is expressed by Equation (9):

$$\sigma(x) = \sqrt{D(x)} = \sqrt{\frac{\sum_{i=1}^{n} (x_i - \mu(x))^2}{(n-1)}}. \tag{9}$$

4.5.3. Coefficient of Variation

Similar to the variance and standard deviation, the coefficient of variation is based on the second central moment. The coefficient of variation is denoted by $C_v(x)$ and is expressed as the ratio of the standard deviation and the mean value (Equation (10)):

$$C_v(x) = \frac{\sigma(x)}{\mu(x)} = \sqrt{\frac{\sum_{i=1}^{n} (k_i - \mu(x))^2}{(n-1)}}, \; for \; k_i = \frac{x_i}{\mu(x)}. \tag{10}$$

4.5.4. Coefficient of Variation

The overshoot probability function or overshoot probability curve determines the probability that a random variable will be greater than or equal to the value of A. The overshoot probability curve is a decreasing function and takes values from one to zero. It can be obtained by integration from the probability density on the right. The shape of the overshoot probability is given in Equation (11):

$$P(A) = P[x \geq A] = \int_{b}^{A} f(x)dx. \tag{11}$$

4.5.5. Quantile

The quantile indicates the measure of the position of the probability distribution of a random variable. In other words, quantiles describe the points at which the distribution function of a random variable passes through a given value. In the case of a continuous distribution having the distribution function $F_x(x)$, the p-quantile x_p is a value of a random variable X for which values less than x_p occur only with probability p, i.e., for which the distribution function $F_x(x_p)$ is equal to the probability p (Equation (12)):

$$P(X < x_p) = F_x(x_p) = p. \tag{12}$$

When presenting the results of quantiles in the overshoot probability function, 5, 10, 15, 20 and 25% quantiles correspond to the 95, 90, 85, 80 and 75% quantiles of the distribution function.

5. Results and Discussion

5.1. Storage Volume Modeling

It was first necessary to determine the inputs into the UNCE_RESERVOIR model for the updated length of the historical series of water inflows into the Vír reservoir up to 2018. In other words, it was necessary to determine the required water outflow from the O_R reservoir for temporal reliability $R_T \geq 99.5\%$ and existing V_Z. The calculation was performed without input uncertainties, including consideration of water losses from the reservoir. As a result, the O_R had to be reduced to achieve a satisfactory $R_T = f(O_R, V_Z)$, as shown in Table 1.

Table 1. Temporal reliability R_T results for changing input O_R for the updated data regarding water inflow into the reservoir.

O_R (m^3 s^{-1})	>>>	R_T (%)
2.5		98.776
2.4		99.028
2.3		99.533
2.31		99.404

In the next step, V_Z was calculated deterministically and without input uncertainties, including the consideration of water losses from the reservoir for the calculated O_R based on Table 1 and for $R_T \geq 99.5\%$, i.e., the calculation of $V_Z = f(O_R, R_T = 99.5\%)$. The resulting optimized V_Z should be close to the real reservoir volume. Based on the comparison of the optimized V_Z with the real V_Z, the O_R was slightly changed. These values are given in Table 2 along with the O_R value for the further calculations that follow.

Table 2. O_R results for varying input O_R and $R_T \geq 99.5\%$ for the updated data regarding water inflow into the reservoir.

O_R (m^3 s^{-1})	>>>	V_Z (m^3)
2.3		43,657,000
2.31		44,371,700
2.305		44,069,000

After determining the exact value of O_R needed to meet the significance of the reservoir with regard to the updated line of water inflow into the reservoir and the resulting V_Z approaching the real V_Z, the optimized storage volumes V_Z of the reservoir for the whole range of water inflow into the reservoir and with input uncertainties were calculated and evaluated. Input uncertainties from the measurements were applied: (i) constantly for all inputs with sizes u_B = ±1, ±2, ±3, ±5 and ±7% and (ii) differently for entered values according to the probable size of uncertainty for each input; specifically, ±3% for the inflow of water into the reservoir, ±5% for bathygraphic curves, ±4% for evaporation and ±3% for seepage through the reservoir body. The number of repetitions (generations) *NG* was always set to 300 repetitions.

Figure 5 shows the calculated filling and emptying curves. These curves are shown for all considered input uncertainties and Table 3 shows the calculated mean values of the optimized stock volumes, including standard deviations. Furthermore, these results were evaluated by adding the upper and lower limits through a coefficient of expansion k = 2, which corresponds to a probability density coverage of approximately 95%. The last line also shows the 95% quantile of the optimized storage volume of the tested reservoir.

Table 3. Results of the analysis of optimal storage volumes V_Z of the tested reservoir.

(m^3)	u_B = ±0%	u_B = ±1%	u_B = ±2%	u_B = ±3%	u_B = ±5%	u_B = ±7%	u_B Different
$\mu(Vz)$	44,069,000	44,098,652	44,121,960	44,154,544	44,010,168	44,078,572	44,148,504
$\pm 2\sigma(Vz)$	0	545,346	1,137,567	1,627,581	2,574,596	3,958,923	1,621,724
$Vz_{bottom\ 2\sigma(Vz)}$	44,069,000	43,553,306	42,984,393	42,526,963	41,435,572	40,119,649	42,526,780
$Vz^{upper\ 2\sigma(Vz)}$	44,069,000	44,643,998	45,259,527	45,782,125	46,584,764	48,037,495	45,770,228
$95\%_{quant.\ Vz}$	44,069,000	44,673,900	45,114,800	45,496,700	46,569,400	47,560,700	45,628,200

The reservoir filling and emptying courses shown in Figure 5 effectively demonstrate the increase in the variance of possible solutions with increasing input uncertainties, which is also evident in Table 3. Different input uncertainty settings according to the probable uncertainty affected the reservoir filling and emptying courses as an approximate uniform input uncertainty setting of u_B = ±3%.

For the probable sizes of the input data uncertainty (different), the resulting optimized storage volume of the reservoir for 95% probability coverage was 44.149 million m^3 ± 1.622 million m^3, i.e., the result lay in the interval {42.527 million m^3; 45.771 million m^3}. To be on the safe side, it is desirable to work with the resulting upper interval, i.e., a higher storage volume, in a stochastic solution. Expressed by the 95% quantile, the resulting optimized volume was 45.628 million m^3; compared to the upper quantile, this value of V_Z is 0.31% lower. The relatively safe and therefore recommended final value of the storage volume from this analysis was 45.770 million m^3. Based on the updated input series of water inflows into the reservoir and the introduction of input uncertainties, including consideration of water losses from the reservoir, it is therefore recommended that the existing storage volume of the Vír I reservoir be increased by up to 3.9%, specifically by 1.71 million m^3.

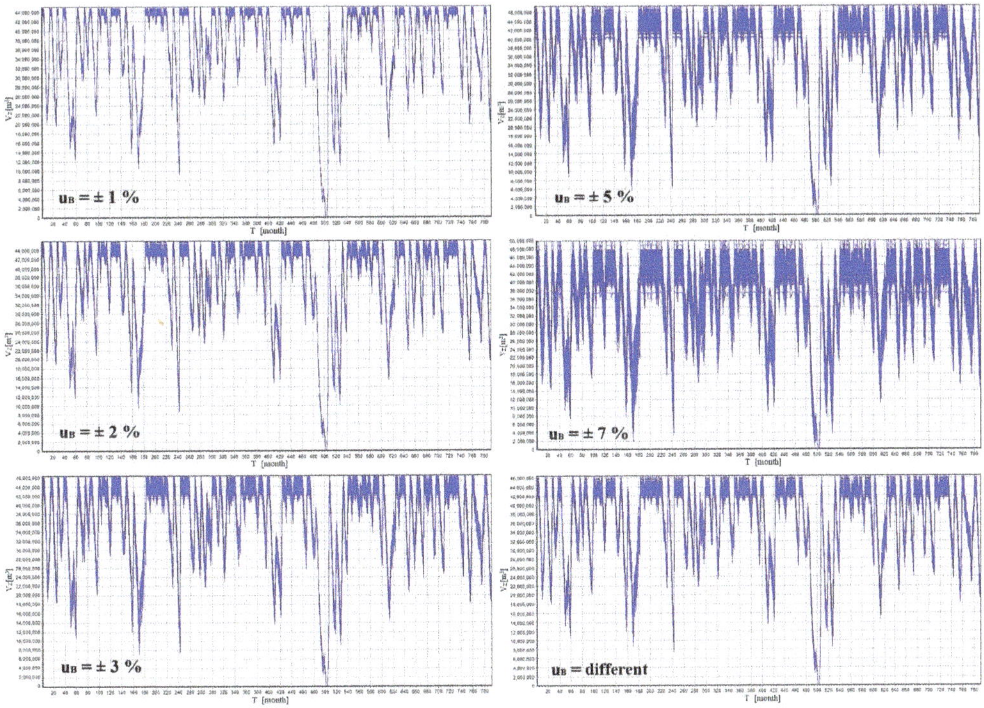

Figure 5. Courses of filling and emptying of the optimized V_Z of the tested reservoir for constant u_B = ±1, ±2, ±3, ±5 and ±7% and for different u_B.

The results from Table 3 are also presented in the form of a bar chart in Figure 6, where the resulting optimized stock volumes for all tested input uncertainties are plotted. The lower and upper limits for ±2 standard deviations of the storage volumes are marked in yellow and the solution using the 95% quantiles in red. The final bar in blue depicts the results for V_Z with the probable different input uncertainties.

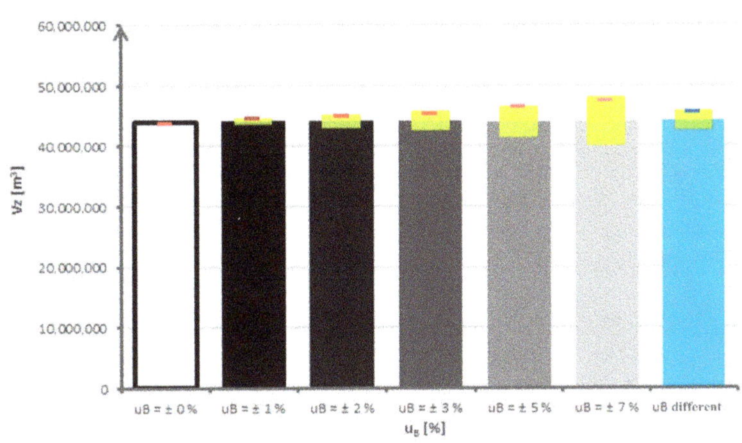

Figure 6. Bar chart of the resulting optimized storage volumes $\mu(V_Z)$ of the Vír I reservoir for $\pm 2\sigma(V_Z)$ and 95% quantiles of the tested input uncertainties u_B.

Furthermore, increases in the values of the variances of the resulting optimized storage volumes $\pm 2\sigma(V_Z)$ compared to the existing V_Z depending on the input uncertainties u_B were expressed as percentages. The results are shown in Table 4.

Table 4. Percentage increases (%) of the resulting variances $\pm 2\sigma(V_Z)$ of the optimized reservoir volume depending on the input uncertainties u_B for the tested reservoir and another reservoir.

	$u_B = \pm 1\%$	$u_B = \pm 2\%$	$u_B = \pm 3\%$	$u_B = \pm 5\%$	$u_B = \pm 7\%$	u_B **Different**
Vír	1.24	2.58	3.69	5.84	8.99	3.68
Vranov	8.04	8.35	9.14	10.89	13.42	9.20

Table 4 shows that, for the tested reservoir Vír (line two), the results in the form of $\pm 2\sigma(V_Z)$ demonstrate a steady increase depending on u_B. Testing on another reservoir, Vranov (line three), showed that there are not always such steady increases in results but that these depend on the growing uncertainties of the input data. Thus, different reservoirs can react completely differently to input uncertainties.

The results for the sets of optimal reservoir storage volumes obtained from the reservoir filling and emptying process depicted in Figure 5 are shown in Figure 7 in the form of an empirical overshoot probability curve. Each empirical curve consists of 300 final values for optimal storage volumes V_Z, which were sorted from minimum to maximum. Each overshoot probability curve corresponds to one type of input uncertainty setting. Furthermore, the 95% quantile (i.e., 5%) used for evaluation is marked on each curve.

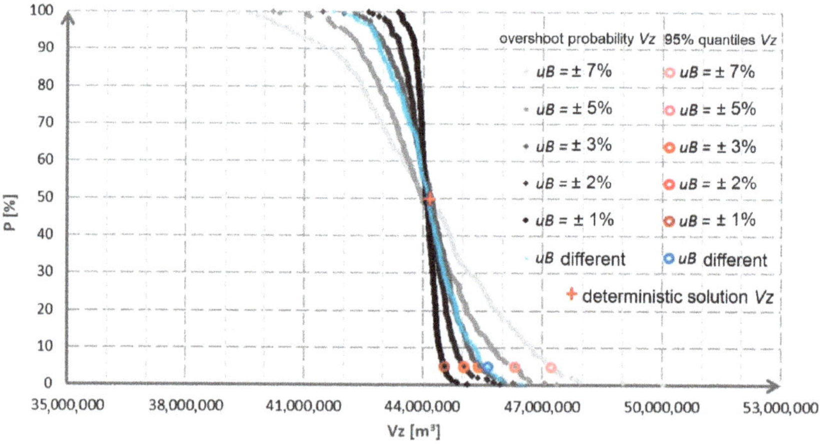

Figure 7. Overshoot probability curves for the resulting optimized storage volumes of the tested reservoir for different sizes of input uncertainties.

In Figure 7 we can see that, for the probability $P = 50\%$, the optimized values of V_Z for all courses take on values just above 44 million m^3, which is similar to the deterministic solution $V_Z = 44.056$ million m^3. This confirms the correctness of the random number generator. The courses of the individual curves are relatively symmetrical. The resulting mean values of the optimized storage volumes $\mu(V_Z)$ in Table 3 increase in comparison with the current value of the storage volumes by percentage values from 0.10% (for $u_B = \pm 5\%$) to + 0.22% (for $u_B = \pm 3\%$). The average of deviations for all input uncertainties was about + 0.09%. The obtained results for the stochastic optimized storage volumes reach only slightly higher values than in the deterministic solution. These factors again confirm the correctness of the random number generator and the appropriateness of using the Monte Carlo method.

5.2. Retention Volume Modeling

Further calculations were performed to determine how increasing the storage volume of the reservoir can affect the transformation of the flood discharge and the change in the retention volume of the reservoir. For this calculation, several variants of the design of the storage volume of the reservoir were considered. The developed simulation optimization model UNCE_RESERVOIR (described in Section 4.2) makes it possible to store all calculated optimal storage volumes according to the number of set repetitions and then to calculate the retention volume of the reservoir for any number of repetitions for each of these volumes using the second simulation model, TRANSFORM_WAVE (described in Section 4.3). To simplify and shorten the calculation time, only a few optimal storage volumes were selected to obtain a single complex solution of functional volumes out of 300 possible solutions with input uncertainties entered differently (according to the probable size of uncertainty at each input), i.e., the blue course of the probability curve in Figure 7. Specifically, the optimal storage volumes resulting from 75, 80, 85, 90 and 95% quantiles (i.e., 25, 20, 15, 10 and 5% in the overshoot probability function) and the upper limit of the resulting expanded uncertainty $+2\sigma$ were selected from this calculation setting (see Figure 8).

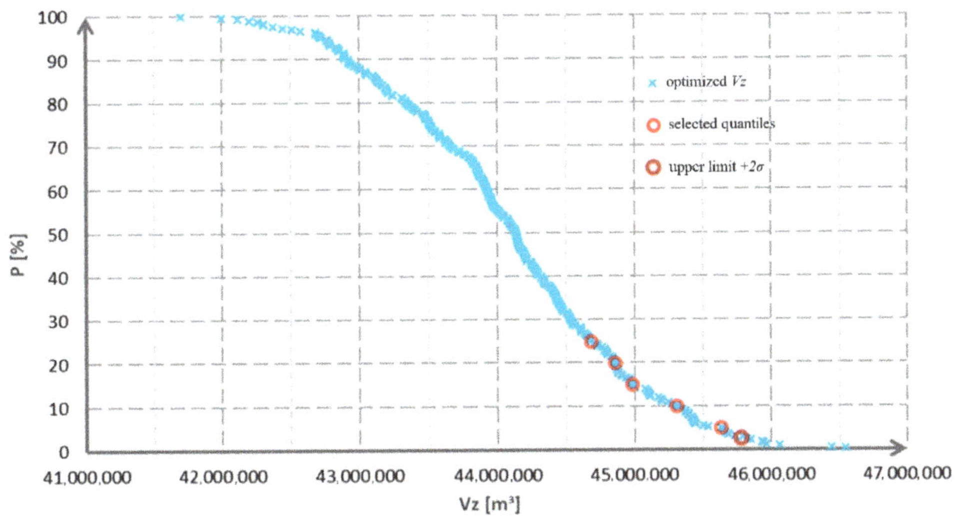

Figure 8. Optimal V_Z values selected from the overshoot probability curve of the obtained optimized storage volumes of the Vír I reservoir.

For the selected optimal V_Z values of the reservoir, Table 5 shows the specific V_Z values of the optimal storage volumes in column three and the corresponding water heights in the reservoir h_{V_Z} in column two. It is clear that, with the increasing quantile (downwards in Table 5), the optimal value of V_Z increases and therefore the height of water in the reservoir h_{V_Z} also increases. Since the emergency spillway of the reservoir is always considered at the same height h_{VRC} (column four) in accordance with the chosen location, the increase of controllable retention V_{RC} (column five) and h_{VRC} is at the expense of the uncontrollable retention volume h_{VRU} (column six) and V_{RU} (column seven).

Table 5. The calculated volumes of water in the reservoir and the corresponding heights of water in the reservoir for selected optimal V_Z values (m^3), including the peak size of the transformed flood discharges Q_{1000}.

	h_{V_Z} (m)	Optimal V_Z (m^3)	h_{VRC} (m)	V_{RC} (m^3)	h_{VRU} (m)	V_{RU} (m^3)	V_R (m^3)	V_{TOTAL} (m^3)	Q_{PEAK} ($m^3\ s^{-1}$)	Height to CML (m)
Current state	63.00	44,056,000	65.60	5,286,000	67.00	3,051,000	8,337,000	56,193,000	-	2.00
Calculation for the current state	63.00	44,056,000	65.60	5,286,000	67.80 ± 0.64	5,002,000 + 1,554,000	10,288,000 11,842,000	58,144,000 59,698,000	172.11 ± 62.54	1.20 0.56
75% quantile V_Z	63.32	44,682,300	65.60	4,659,700	67.80 ± 0.61	4,999,000 + 1,499,000	9,658,700 11,157,700	58,141,000 59,640,000	177.91 ± 61.36	1.20 0.58
80% quantile V_Z	63.41	44,858,400	65.60	4,483,600	67.80 ± 0.60	4,990,000 + 1,471,000	9,473,600 10,944,600	58,132,000 59,603,000	179.68 ± 61.56	1.20 0.60
85% quantile V_Z	63.47	44,984,800	65.60	4,357,200	67.78 ± 0.60	4,950,000 + 1,458,000	9,307,200 10,765,200	58,092,000 59,550,000	181.12 ± 61.54	1.22 0.62
90% quantile V_Z	63.64	45,310,700	65.60	4,031,300	67.71 ± 0.55	4,770,000 + 1,336,000	8,801,300 10,137,300	57,912,000 59,248,000	185.86 ± 61.43	1.29 0.75
95% quantile V_Z	63.79	45,628,200	65.60	3,713,800	67.70 ± 0.51	4,755,000 + 1,239,000	8,468,800 9,707,800	57,897,000 59,136,000	188.41 ± 60.41	1.30 0.79
Upper limit V_Z (+2σ)	63.90	45,770,228	65.60	3,571,772	67.67 ± 0.49	4,676,000 + 1,181,000	8,247,772 9,428,772	57,818,000 58,999,000	190.87 ± 59.59	1.33 0.85

To calculate the retention volume, a flood discharge Q_{1000} with an input standard uncertainty $u_B = \pm 10\%$ was selected, which was chosen as the minimum following [68], in which the reliability classes of hydrological data are given, including the probable variance of errors. The calculation was performed in the variant without pre-drainage of water from the reservoir. The number of uncertainties generated for the flood discharge was chosen to be 300 repetitions. The starting water level in the reservoir at the beginning of the flood discharge transformation solution was always the full storage volume V_Z, i.e., the newly calculated height h_{Vz}. Based on the above information on bottom outlets and the emergency spillway along with Equations (4) and (5), the outlet coefficient of the bottom outlets or the outflow coefficient μ and the overflow coefficient m were determined. Specifically, for the tested reservoir, the outflow coefficient $\mu = 0.435$ (-) and overflow coefficient $m = 0.407$ (-). In the variant without pre-drainage, the level is kept at the level of the full storage volume and, as soon as this level is exceeded, the bottom outlets are opened to a harmless flow Q_{NE}. When the emergency spillway is exceeded, the bottom outlets are smoothly closed and, after the flooding, they are smoothly opened to a water height of 0.5 m above the emergency spillway. The level after the flood is again kept at the level of the full storage volume.

Figure 9 shows the results of individual transformations for selected optimal V_Z values. The generated courses of these flood discharges are shown here in red, the results of the transformed waves in blue and the courses of the water heights in the reservoir during the transformations in green.

In Figure 9, the transformations of the generated flood discharges have similar courses and there are also similar courses in the heights reached, with the only difference being that the starting level is from the selected optimal V_Z. The input uncertainty has a clear effect on flood discharges but also on the results of transformed floods and water heights in the reservoir.

5.3. *Summary of Results*

It is worth noting the creation of several bundles of transformed flood discharges (blue lines), which are probably caused by the size of the time step. When the emergency spillway is exceeded in another time step, the difference is at least one time step, i.e., 6 h, which is a consequence of the formation of bundles and jumps in the peaks of the transformed floods. If the time step were finer, these bundles would not be formed and the courses of the transformed floods, including peaks, would be spread more smoothly.

Comparing the results for the starting height of 63 m from the existing storage volume, a slight decrease in the peak of the water heights in the reservoir is evident. This decrease is also confirmed in Table 5, specifically in column six, i.e., the height h_{VRU}, the height of the uncontrollable retention volume or the achieved peak of the height of the water in the reservoir during the transformations of flood discharges. These values include the expanded uncertainty of $\pm 2\sigma$. In Table 5, column seven then shows the results of the uncontrollable retention volume V_{RU}, including the upper limit of the expanded uncertainty $+2\sigma$. Furthermore, column eight shows the total volume of the retention volume V_R, including the upper limit of the expanded uncertainty $+2\sigma$ (bold), and the total volume of the reservoir V_{TOTAL} is shown in column nine. Finally, the last columns, 10 and 11, show the values of Q_{1000} flood peaks, including the expanded uncertainty $\pm 2\sigma$, and the height between the uncontrollable retention volume (maximum water level in the reservoir) and the control maximum level CML (dam height), including the upper limit of the expanded uncertainty $+2\sigma$ (bold).

In Table 5, we can see the results from the transformations of the generated flood discharges for selected optimal values V_Z. Specifically, the first row shows the values for the current state of the reservoir according to [63] and the second row contains the results of the transformation of the flood discharge Q_{1000} for the current state of V_Z. The following rows show the results for selected quantiles of optimal V_Z values.

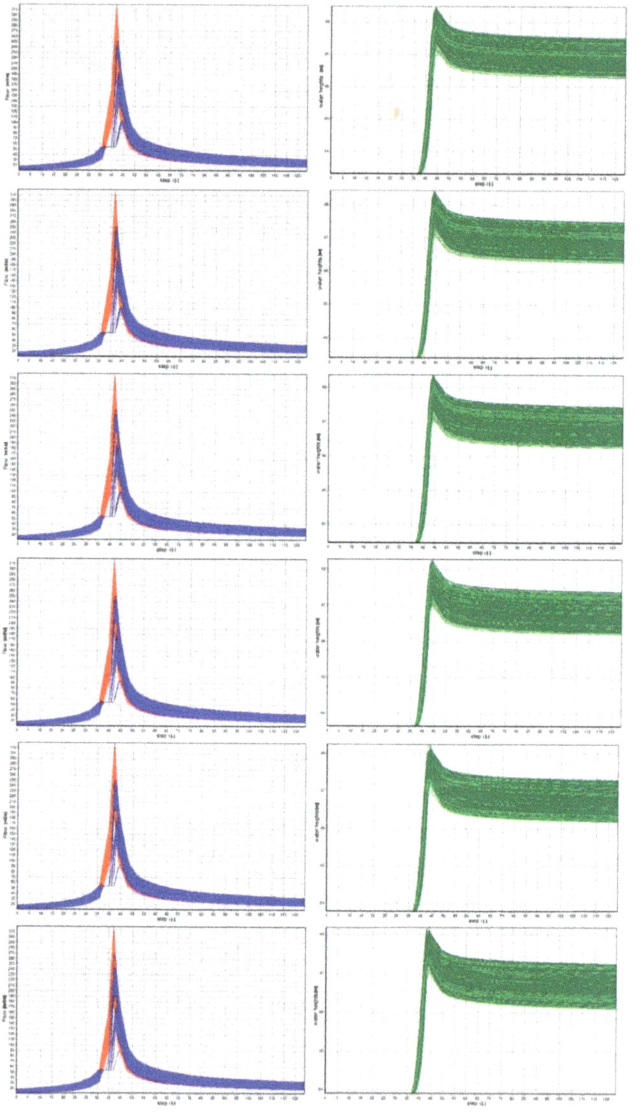

Figure 9. Results of the transformation of generated Q_{1000} waves and water heights in the reservoir for the input $u_B = \pm 10\%$ for the 75, 80, 85, 90, 95% quantiles and for the upper limit $+2\sigma$ of the optimal V_Z.

The results in Table 5 show that, with a higher storage volume at a constant emergency spillway height (h_{VRC} level), the following decrease: (i) the controllable retention volume V_{RC}, (ii) the height of the uncontrollable retention volume h_{VRU}, (iii) the uncontrollable retention volume of the reservoir V_{RN}, (iv) the retention volume of the reservoir V_R and (v) the total reservoir volume V_{TOTAL}. The decrease in h_{VRU} has the effect of increasing the difference in size between the h_{VRU} and CML (column 11), which is desirable for the solution. On the other hand, the peak flow Q_{PEAK} increases. It should be noted that, although the values of volumes and heights decrease (columns five to nine), they are significantly higher (lower for column 11) than the actual values of the existing tested reservoir.

This is because a given higher value V_Z (starting level) corresponds to a larger volume from the line of flooded volumes and, therefore, more water is captured in the initial flood step and subsequent steps than at the starting height of 63 m. As a result, the resulting retention volume and total reservoir volume decrease with increasing height V_Z. In contrast, the difference between the h_{VRU} and CML increases, which is a key parameter when designing or changing the functional volumes of a reservoir. This suggests a design for the safest solution in terms of optimal V_Z, i.e., either the 95% quantile of V_Z or the upper limit of V_Z (+2σ). In these designs, the increases in the retention volume V_R by 1.37 million m^3 and 1.09 million m^3 would correspond to increases of 16.5% and 13.1% compared to the actual V_R. However, we must not forget that with this choice the peak flow of the transformed flood discharge increases. For example, for the transformation from the current state, the peak flow of the upper limit of the expanded uncertainty (+2σ) is 234.65 m^3 s^{-1} and, for the transformation from the 95% quantile of V_Z, the peak flow is 248.82 m^3 s^{-1}, which is another key parameter in the design of functional volumes of a reservoir.

Finally, the calculated retention volumes of water in the reservoir for selected optimal V_Z values were also evaluated using quantiles, as was the case with the storage volume of the reservoir. Figure 10 shows the results of water height peaks in the reservoir during transformations of generated flood discharges burdened with input uncertainties $u_B = \pm 10\%$ in the form of the probabilities of exceeding these peaks for selected optimal storage volumes (black and white shades) and for the current reservoir storage volume (brown shade).

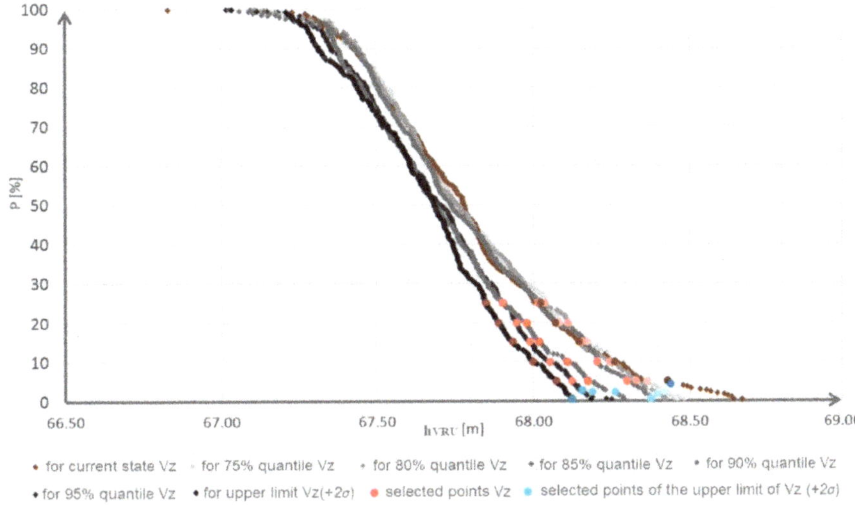

Figure 10. Lines indicating the probabilities of exceeding the calculated water height peaks in the reservoir during the transformation of the uncertain flood discharge Q_{1000} for selected optimized storage volumes.

In Figure 10 we can see again that at a higher starting height (higher percentage of V_Z quantiles) lower peaks of the water level in the reservoir are achieved, i.e., the h_{VRU} level or the maximum water level in the reservoir does not rise so high. At the same time, selected quantiles and the upper limit +2σ are marked on the plotted curves, similarly to the selection of optimized V_Z. The complete results for the water height peaks in the reservoir with possible functional volumes are summarized in detail in Table 6, which is designed in the same form as Table 5. This table shows the results of flood discharge transformations for selected quantiles and the upper limit +2σ. Peak flow Q_{PEAK} is expressed for the given variants in the form +2σ.

Table 6. The calculated volumes of water in the reservoir and their corresponding heights for selected optimal V_Z values (m^3), including the sizes of peaks of transformed flood discharges Q_{1000} for selected quantiles and the upper limit $+2\sigma$.

	h_{V_Z} (m)	*Optimal* V_Z (m^3)	h_{VRC} (m)	V_{RC} (m^3)	**Selected Quantiles and** V_{RU} **(+2σ)**	h_{VRU} (m)	V_{RU} (m^3)	V_R (m^3)	V_{TOTAL} (m^3)	*Upper limit* (+2σ) Q_{PEAK} ($m^3\ s^{-1}$)	*Height to CML* (m)
For the current state	63.00	44,056,000	65.60	5,286,000	75% quan. V_{RU}	68.02	5,533,000	10,819,000	58,675,000	234.65	0.98
					80% quan. V_{RU}	68.07	5,655,000	10,941,000	58,797,000		0.93
					85% quan. V_{RU}	68.15	5,857,000	11,143,000	58,999,000		0.85
					90% quan. V_{RU}	68.26	6,118,000	11,404,000	59,260,000		0.74
					95% quan. V_{RU}	68.43	6,532,000	11,818,000	59,674,000		0.57
					up. l. V_{RU} (+2σ)	68.44	6,556,000	11,842,000	59,698,000		0.56
75% quantile V_Z	63.32	44,682,300	65.60	4,659,700	75% quan. V_{RU}	68.04	5,582,000	10,241,700	58,724,000	239.27	0.96
					80% quan. V_{RU}	68.08	5,679,000	10,338,700	58,821,000		0.92
					85% quan. V_{RU}	68.17	5,898,000	10,557,700	59,040,000		0.83
					90% quan. V_{RU}	68.26	6,118,000	10,777,700	59,260,000		0.74
					95% quan. V_{RU}	68.36	6,362,000	11,021,700	59,504,000		0.64
					up. l. V_{RU} (+2σ)	68.42	6,498,000	11,157,700	59,640,000		0.58
80% quantile V_Z	63.41	44,858,400	65.60	4,483,600	75% quan. V_{RU}	68.03	5,557,000	10,040,600	58,699,000	241.24	0.97
					80% quan. V_{RU}	68.11	5,752,000	10,235,600	58,894,000		0.89
					85% quan. V_{RU}	68.15	5,857,000	10,340,600	58,999,000		0.85
					90% quan. V_{RU}	68.24	6,069,000	10,552,600	59,211,000		0.76
					95% quan. V_{RU}	68.33	6,288,000	10,771,600	59,430,000		0.67
					up. l. V_{RU} (+2σ)	68.40	6,461,000	10,944,600	59,603,000		0.60
85% quantile V_Z	63.47	44,984,800	65.60	4,357,200	75% quan. V_{RU}	68.02	5,533,000	9,890,200	58,675,000	242.66	0.98
					80% quan. V_{RU}	68.08	5,679,000	10,036,200	58,821,000		0.92
					85% quan. V_{RU}	68.16	5,874,000	10,231,200	59,016,000		0.84
					90% quan. V_{RU}	68.20	5,972,000	10,329,200	59,114,000		0.80
					95% quan. V_{RU}	68.30	6,215,000	10,572,200	59,357,000		0.70
					up. l. V_{RU} (+2σ)	68.38	6,408,000	10,765,200	59,550,000		0.62
90% quantile V_Z	63.64	45,310,700	65.60	4,031,300	75% quan. V_{RU}	67.90	5,241,000	9,272,300	58,383,000	246.29	1.10
					80% quan. V_{RU}	67.98	5,436,000	9,467,300	58,578,000		1.02
					85% quan. V_{RU}	68.02	5,533,000	9,564,300	58,675,000		0.98
					90% quan. V_{RU}	68.11	5,752,000	9,783,300	58,894,000		0.89
					95% quan. V_{RU}	68.17	5,898,000	9,929,300	59,040,000		0.83
					up. l. V_{RU} (+2σ)	68.25	6,106,000	10,137,300	59,248,000		0.75
95% quantile V_Z	63.79	45,628,200	65.60	3,713,800	75% quan. V_{RU}	67.90	5,241,000	8,954,800	58,383,000	248.82	1.10
					80% quan. V_{RU}	67.95	5,362,000	9,075,800	58,504,000		1.05
					85% quan. V_{RU}	67.99	5,460,000	9,173,800	58,602,000		1.01
					90% quan. V_{RU}	68.05	5,606,000	9,319,800	58,748,000		0.95
					95% quan. V_{RU}	68.12	5,777,000	9,490,800	58,919,000		0.88
					up. l. V_{RU} (+2σ)	68.21	5,994,000	9,707,800	59,136,000		0.79

Table 6. *Cont.*

	h_{V_Z} (m)	*Optimal* V_Z (m^3)	h_{VRC} (m)	V_{RC} (m^3)	Selected Quantiles and V_{RU} (+2σ)	h_{VRU} (m)	V_{RU} (m^3)	V_R (m^3)	V_{TOTAL} (m^3)	*Upper limit* (+2σ) Q_{PEAK} (m^3 s^{-1})	*Height to CML* (m)
Upper limit V_Z (+2σ)	63.90	45,770,228	65.60	3,571,772	75% quan. V_{RU}	67.85	5,118,000	8,689,772	58,260,000	250.46	1.15
					80% quan. V_{RU}	67.89	5,216,000	8,787,772	58,358,000		1.11
					85% quan. V_{RU}	67.93	5,313,000	8,884,772	58,455,000		1.07
					90% quan. V_{RU}	68.00	5,484,000	9,055,772	58,626,000		1.00
					95% quan. V_{RU}	68.07	5,655,000	9,226,772	58,797,000		0.93
					up. l. V_{RU} (+2σ)	68.15	5,857,000	9,428,772	58,999,000		0.85

Table 6 shows that the most suitable solution is at the bottom of the table. It consists in a significant increase in V_Z at the expense of the size of the flood peak because, with such a solution, higher water height peaks in the reservoir during the transformation are achieved but, at the same time, there is a greater height between the h_{VRU} and CML. In the case of large floods, which the Q_{1000} undoubtedly is, the safety of the dam itself must be considered, i.e., elimination of the overflow of the CML.

6. Conclusions and Recommendations

This article applied the uncertainties resulting from the measurements that enter into water management solutions to a case study of the Vír reservoir, Czech Republic. For this reservoir, the following were developed and tested: (i) a simulation-optimization model that determines the optimal storage volume of the reservoir and (ii) a simulation model for the transformation of uncertain flood discharges that determines the retention volume of the reservoir. The obtained results lead to the following key conclusions:

- Input uncertainty significantly affects the results of V_Z and V_R calculations.
- To be on the safe side, it is appropriate to increase the values of either V_Z or V_R in accordance with the calculated uncertainties. Specifically, the input uncertainties discussed here highlighted the need to increase the existing V_Z of the tested reservoir by up to 1.71 million m^3 (3.9%) and the existing V_R by up to 1.37 million m^3 (16.5%).
- For a comprehensive determination of functional volumes, calculations of the transformation of the updated flood discharge burdened with uncertainty for selected optimal values of V_Z were performed. These led to the determination of how an increase in V_Z can affect the transformation of the flood discharge and the change in the V_R of the reservoir.
- Based on the above, Table 6 was created with solution options for V_Z and V_R under conditions of uncertainty, including possible flood peaks and water height peaks in the reservoir.
- The developed simulation-optimization (i) and simulation (ii) models of the reservoir, the methods used and the introduction of uncertainties on the input data proved their functionality in solving the functional volumes of the water in the reservoir.
- Uniqueness can be observed in the connection between the solutions of the functional volumes of the reservoir for input data under conditions of uncertainty.
- The source codes of both models are written in such a way as to maintain generality and thus can be quickly used to test other existing or planned reservoirs anywhere in the world, if suitable data are available.

The introduction of uncertainties into the input data and their subsequent analysis proved the influence of extreme values on the final solution. It can be assumed, that even for other reservoirs with different input uncertainties, such uncertainties will have an impact on the existing functional volumes of these reservoirs. This only confirms the importance of this issue and highlights the need for detailed analyses of waterworks.

The models were applied to an updated historical series dating up to the present, i.e., in a period that can be considered a period with ongoing climate change. In addition, the assumption of uncertainty in the measurement of input data was incorporated into the solution.

From the point of view of measurement uncertainty and climate change uncertainty, the achieved results are relevant to the first stage of water management analyses, which involves capturing possible changes in the development of hydroclimatic extremes. In other words, the volumes of reservoirs are assessed with regard to their current state but with the assumption that they must incorporate the already ongoing process of climate change.

In the second stage of the solution, inputs can be inserted into the models to capture the future uncertainty of climate change thanks to the general notation of the source codes of the models. Instead of historical series of water inflows into the reservoir, simulations of non-stationary hydrological flow series, hydrological time series with negative trends of flow development or directly hydrological simulations describing climate change based on climate model scenarios can be used.

Likewise, flood discharges can be entered in the form of real extreme floods, artificial flood hydrographs or predicted floods affected by climate change.

Although the present analysis was performed for only one reservoir and the results cannot be generalized at present, the created software tool as a whole allows the analysis of existing reservoirs, or the dimensioning of new reservoirs, during the ongoing process of climate change under conditions of extreme fluctuation and for non-stationary hydrological data around the world.

Author Contributions: Conceptualization, S.P. and D.M.; methodology, S.P. and D.M.; software, S.P. and D.M.; validation, S.P.; formal analysis, S.P.; investigation, S.P.; resources, S.P.; data curation, S.P. and D.M.; writing—original draft preparation, S.P.; writing—review and editing, S.P. and D.M.; visualization, S.P.; supervision, D.M.; project administration, S.P.; funding acquisition, S.P. and D.M. All authors have read and agreed to the published version of the manuscript.

Funding: This research was funded by "Effective management of water resources in South Moravia", research project number [FAST-S-20-6346].

Institutional Review Board Statement: Not applicable.

Informed Consent Statement: Not applicable.

Data Availability Statement: Not applicable.

Conflicts of Interest: The authors declare no conflict of interest.

References

1. WMO. *Statement on the State of the Global Climate in 2019*; World Meteorological Organization: Geneva, Switzerland, 2020; ISBN 978-92-62-11248-5.
2. IPCC. *Global Warming of 1.5 C. An IPCC Special Report on the Impacts of Global Warming of 1.5 C above Pre-Industrial Levels and Related Global Greenhouse Gas Emission Pathways, in the Context of Strengthening the Global Response to the Threat of Climate Change, Sustainable Development, and Efforts to Eradicate Poverty*; MassonDelmotte, V., Zhai, P., Portner, H.O., Roberts, D., Skea, J., Shukla, P.R., Pirani, A., Moufouma-Okia, W., Péan, C., Pidcock, R., et al., Eds.; IPCC: Geneva, Switzerland, 2018. Available online: http://ipcc.ch/report/sr15/ (accessed on 20 February 2021).
3. Trnka, M.; Vizina, A.; Hanel, M.; Balek, J.; Hlavinka, P.; Semerádová, D.; Chuchma, F.; Dumbrovský, M.; Daňhelka, J.; Dubrovský, M.; et al. *Pozorované Změny a Výhled pro Vodní Bilanci a Potřebu Vody v Zemědělské Krajině České Republiky. Vodohospodářská Konference Vodní Nádrže 2017*; Povodí Moravy: Brno, Czech Republic, 2017; ISBN 978-80-905368-5-2.
4. Zahradníček, P.; Trnka, M.; Brázdil, R.; Možný, M.; Štěpánek, P.; Hlavinka, P.; Žalud, Z.; Malý, A.; Semerádová, D.; Dobrovolný, P.; et al. The extreme drought episode of August 2011–May 2012 in the Czech Republic. *Int. J. Clim.* **2015**, *35*, 3335–3352. [CrossRef]
5. Duchan, D.; Dráb, A.; Říha, J. Flood Protection in the Czech Republic. In *Management of Water Quality and Quantity*; Springer Nature: Cham, Switzerland, 2019; pp. 333–363; ISBN 978-3-030-18358-5.
6. Stahl, K.; Hisdal, H.; Hannaford, J.; Tallaksen, L.M.; van Lanen, H.A.J.; Sauquet, E.; Demuth, S.; Fendekova, M.; Jódar, J. Streamflow trends in Europe: Evidence from a dataset of near-natural catchments. *Hydrol. Earth. Syst. Sci.* **2010**, *14*, 2367–2382. [CrossRef]
7. Hanel, M.; Kašpárek, L.; Mrkvičková, M.; Horáček, S.; Vizina, A.; Novický, O.; Fridrichová, R. *Odhad Dopadů Klimatické Změny na Hydrologickou Bilanci v ČR a Možná Adaptační Opatření*; Výzkumný Ústav Vodohospodářský T. G. Masaryka, v.v.i: Praha, Czech Republic, 2011; ISBN 978-80-87402-22-1.
8. IPCC. *Summary for Policymakers, Climate Change 2014, Synthesis Report. Contribution of Working Groups I, II and III to the Fifth Assessment Report of the Intergovernmental Panel on Climate Change*; Core Writing Team, Pachauri, R.K., Meyer, L.A., Eds.; IPCC: Geneva, Switzerland, 2014.
9. Der Kiureghian, A.; Ditlevsen, O. Aleatory or epistemic? Does it matter? *Struct. Saf.* **2009**, *31*, 105–112. [CrossRef]
10. Dantan, J.; Gayton, N.; Qureshi, A.; Lemaire, M.; Etienne, A. Tolerance Analysis Approach based on the Classification of Uncertainty (Aleatory/Epistemic). *Procedia CIRP* **2013**, *10*, 287–293. [CrossRef]
11. Czech Government Document: Strategie Přizpůsobení se Změně Klimatu v Podmínkách ČR. Ministerstvo Životního Prostředí. 2015. Available online: http://www.mzp.cz/C1257458002F0DC7/cz/zmena_klimatu_adaptacni_strategie/$FILE/OEOK-Adaptacni_strategie-20151029.pdf (accessed on 1 October 2020).
12. Czech Technical Standard ČSN 75,2405 Reservoir Storage Capacity Analysis. Available online: https://csnonline.agentura-cas.cz/Detailnormy.aspx?k=501441 (accessed on 3 October 2020).
13. Knight, F.H. *Risk, Uncertainty, and Profit*; Boston, Hart, Schaffner & Marx; Houghton Mifflin Company: Boston, MA, USA, 1921.

14. WECC doc. 19–1990:"Guidelines for Expression of the Uncertainty in Calibrations". 1990. Available online: http://www.qcalibration.com/image/uncertainty.pdf (accessed on 1 October 2020).
15. International Organization for Standardization. *Guide to the Expression of Uncertainty in Measurement*; International Organization for Standardization: Geneva, Switzerland, 1993.
16. *Document: Expression of the Uncertainty in Measurement in Calibration*. EAL Task Force, EA 4/02. 1999. Available online: https://www.isobudgets.com/pdf/uncertainty-guides/european-co-operation-for-accreditation-ea-4-02-m-1999-expression-of-the-uncertainty-of-measurement-in-calibration.pdf (accessed on 3 October 2020).
17. Document: ISO GUM Suppl. 1 (DGUIDE 99998). *Guide to the Expression of Uncertainty in Measurement (GUM)—Supplement 1: Numerical Methods for the Propagation of Distributions*; International Organization for Standardization: Geneva, Switzerland, 2004; Available online: http://geste.mecanica.ufrgs.br/medterm/ISO_GUM_sup1.pdf (accessed on 3 October 2020).
18. Beven, K.J.; Binley, A.M. The future of distributed models: Model calibration and uncertainty prediction. *Hydrol. Processes* **1992**, *6*, 279–298. [CrossRef]
19. Beven, K. Towards integrated environmental models of everywhere: Uncertainty, data and modelling as a learning process. *Hydrol. Earth Syst. Sci.* **2007**, *11*, 460–467. [CrossRef]
20. Baldassarre, G.D.; Montanari, A. Uncertainty in river discharge observations: A quantitative analysis. *Hydrol. Earth Syst. Sci.* **2009**, *13*, 913. [CrossRef]
21. Shrestha, R.R.; Simonovic, S.P. Fuzzy set theory based methodology for the analysis of measurement uncertainty in river discharge and stage. *Can. J. Civ. Eng.* **2010**, *37*, 429–439. [CrossRef]
22. Tomkins, K.M. Uncertainty in streamflow rating curves: Methods, controls and consequences. *Hydrol. Process.* **2012**, *28*, 464–481. [CrossRef]
23. Westerberg, I.K.; McMillan, H.K. Uncertainty in hydrological signatures. *Hydrol. Earth Syst. Sci.* **2015**, *19*, 3951–3968. [CrossRef]
24. Westerberg, I.K.; McMillan, H.K. Uncertainty in hydrological signatures for gauged and ungauged catchments. *Water Resour. Res.* **2016**, *52*. [CrossRef]
25. Whitehead, P.; Hornberger, G.; Black, R. Effects of parameter uncertainty in a flow routing model/Les effets de l'incertitude des paramètres dans un modèle du calcul du cheminement. *Hydrol. Sci. Bull.* **2009**, *24*, 445–464. [CrossRef]
26. Akbari, G.H.; Nezhad, A.H.; Barati, R. Developing a model for analysis of uncertainties in prediction of floods. *J. Adv. Res.* **2012**, *3*, 73–79. [CrossRef]
27. Winter, T.C. Uncertainties in estimating the water balance of lakes. *JAWRA J. Am. Water Resour. Assoc.* **1981**, *17*, 82–115. [CrossRef]
28. LaBaugh, J.W.; Winter, T.C. The impact of uncertainties in hydrologic measurement on phosphorus budgets and empirical models for two Colorado reservoirs. *Limnol. Oceanogr.* **1984**, *29*, 322–339. [CrossRef]
29. Campos, J.; Filho, F.S.; Lima, H. Risks and uncertainties in reservoir yield in highly variable intermittent rivers: Case of the Castanhão Reservoir in semi-arid Brazil. *Hydrol. Sci. J.* **2014**, *59*, 1184–1195. [CrossRef]
30. Kuria, F.W.; Vogel, R.M. A global water supply reservoir yield model with uncertainty analysis. *Environ. Res. Lett.* **2014**, *9*, 095006. [CrossRef]
31. Li, B.; Liang, Z.; Zhang, J.; Chen, X.; Jiang, X.; Wang, J.; Hu, Y. Risk analysis of reservoir flood routing calculation based on inflow forecast uncertainty. *Water* **2016**, *8*, 486. [CrossRef]
32. Chen, J.; Zhong, P.A.; Wang, M.L.; Zhu, F.L.; Wan, X.Y.; Zhang, Y. A risk-based model for real-time flood control operation of a cascade reservoir system under emergency conditions. *Water* **2018**, *10*, 167. [CrossRef]
33. Le Ngo, L.; Madsen, H.; Rosbjerg, D. Simulation and optimisation modelling approach for operation of the Hoa Binh reservoir, Vietnam. *J. Hydrol.* **2007**, *336*, 269–281. [CrossRef]
34. Paseka, S.; Kapelan, Z.; Marton, D. Multi-Objective Optimization of Resilient Design of the Multipurpose Reservoir in Conditions of Uncertain Climate Change. *Water* **2018**, *10*, 1110. [CrossRef]
35. Ren, K.; Huang, S.; Huang, Q.; Wang, H.; Leng, G.; Wu, Y. Defining the robust operating rule for multi-purpose water reservoirs under deep uncertainties. *J. Hydrol.* **2019**, *578*, 124134. [CrossRef]
36. Meysami, R.; Niksokhan, M.H. Evaluating robustness of waste load allocation under climate change using multi-objective decision making. *J. Hydrol.* **2020**, *588*, 125091. [CrossRef]
37. Savić, D.A.; Bicik, J.; Morley, M.S. A DSS generator for multiobjective optimisation of spreadsheet-based models. *Environ. Model. Soft.* **2011**, *26*, 551–561. [CrossRef]
38. Da Silva, R.T.; Sánchez-Román, R.; Teixeira, M.B.; Franzotti, C.L.; Folegatti, M.V. Software for calculation of reservoir active capacity with the sequent-peak algorithm. *Eng. Agrícola* **2013**, *33*, 501–510. [CrossRef]
39. Fletcher, S.; Ponnambalam, K. Estimation of reservoir yield and storage distribution using moments analysis. *J. Hydrol.* **1996**, *182*, 259–275. [CrossRef]
40. Liang, Q. Flood Simulation Using a Well-Balanced Shallow Flow Model. *J. Hydraul. Eng.* **2010**, *136*, 669–675. [CrossRef]
41. Yuan, B.; Sun, J.; Yuan, D.-K.; Tao, J.-H. Numerical simulation of shallow-water flooding using a two-dimensional finite volume model. *J. Hydrodyn.* **2013**, *25*, 520–527. [CrossRef]
42. Vatankhah, A.R. Evaluation of Explicit Numerical Solution Methods of the Muskingum Model. *J. Hydrol. Eng.* **2014**, *19*, 06014001. [CrossRef]
43. Klemeš, V. A simplified solution of the flood-routing problem. *Vodohospod. Časopis* **1960**, *8*, 317–326.
44. Klemeš, V. Dilettantism in hydrology: Transition or destiny? *Water Resour. Res.* **1986**, *22*, 177–188. [CrossRef]

45. Hsu, N.S.; Wei, C.C. A multipurpose reservoir real-time operation model for flood control during typhoon invasion. *J. Hydrol.* **2007**, *336*, 282–293. [CrossRef]
46. Tu, M.-Y.; Hsu, N.-S.; Tsai, F.T.-C.; Yeh, W.W.-G. Optimization of Hedging Rules for Reservoir Operations. *J. Water Resour. Plan. Manag.* **2008**, *134*, 3–13. [CrossRef]
47. Shiau, J.T. Analytical optimal hedging with explicit incorporation of reservoir release and carryover storage targets. *Water Resour. Res.* **2011**, *47*. [CrossRef]
48. Chaleeraktrakoon, C.; Chinsomboon, Y. Dynamic rule curves for flood control of a multipurpose dam. *HydroResearch* **2015**, *9*, 133–144. [CrossRef]
49. Ding, W.; Zhang, C.; Peng, Y.; Zeng, R.; Zhou, H.; Cai, X. An analytical framework for flood water conservation considering forecast uncertainty and acceptable risk. *Water Resour. Res.* **2015**, *51*, 4702–4726. [CrossRef]
50. Lin, N.M.; Rutten, M. Optimal Operation of a Network of Multi-purpose Reservoir: A Review. *Procedia Eng.* **2016**, *154*, 1376–1384. [CrossRef]
51. Ding, W.; Zhang, C.; Cai, X.; Li, Y.; Zhou, H. Multiobjective hedging rules for flood water conservation. *Water Resour. Res.* **2017**, *53*, 1963–1981. [CrossRef]
52. Starý, M. *Zpráva o Výsledcích Řešení při Spolupráci na Normalizačním Rozborovém Úkolu HDP VH 83/6 RÚ*; VUT FAST v Brně: Brno, Czech Republic, 1984.
53. Marton, D.; Starý, M.; Menšík, P. The Influence of Uncertainties in the Calculation of Mean Monthly Discharges on Reservoir Storage. *J. Hydrol. Hydromech.* **2011**, *59*, 228–237. [CrossRef]
54. Marton, D.; Starý, M.; Menšík, P. Analysis of the influence of input data uncertainties on determining the reliability of reservoir storage capacity. *J. Hydrol. Hydrom.* **2015**, *63*, 287–294. [CrossRef]
55. Marton, D.; Starý, M.; Menšík, P.; Paseka, S. Hydrological Reliability Assessment of Water Management Solution of Reservoir Storage Capacity in Conditions of Uncertainty. In *Drought: Research and Science-Policy Interfacing*; CRC Press Taylor & Francis Group: Leiden, The Netherlands, 2015; pp. 377–382. [CrossRef]
56. Paseka, S.; Marton, D.; Menšík, P. Uncertainties of reservoir storage capacity during low water period. In *Proceedings of the SGEM International Multidisciplinary Geoconference: Hydrology and Water Resources*; STEF92 Technology Ltd.: Sofia, Bulgaria, 2016; pp. 789–796. ISBN 978-619-7105-61-2.
57. Marton, D.; Paseka, S. Uncertainty Impact on Water Management Analysis of Open Water Reservoir. *Environments* **2017**, *4*, 10. [CrossRef]
58. Paseka, S.; Marton, D. Optimal Assessment of Reservoir Active Storage Capacity under Uncertainty. In *Proceedings of the SGEM International Multidisciplinary Geoconference: Water Resources, Sofia, Bulgaria, 12 July 2019*; STEF92 Technology Ltd.: Sofia, Bulgaria, 2019; pp. 427–434. ISBN 978-619-7105-61-2.
59. Paseka, S.; Marton, D. Assessing the Impact of Flood Wave Uncertainty to Reservoir Flood Storage Capacity. In *Proceedings of the SGEM International Multidisciplinary Geoconference: Water Resources, Sofia, Bulgaria, 12 July 2019*; STEF92 Technology Ltd.: Sofia, Bulgaria, 2019; pp. 49–56. ISBN 978-619-7105-61-2.
60. Kritskiy, S.N.; Menkel, M.F. *Water Management Computations*; GIMIZ: Leningrad, Russia, 1952.
61. Klemeš, V. Reliability estimates for a storage reservoir with seasonal input. *J. Hydrol.* **1967**, *7*, 198–216. [CrossRef]
62. Hashimoto, T.; Stedinger, J.R.; Loucks, D.P. Reliability, resiliency, and vulnerability criteria for water resource system performance evaluation. *Water Resour. Res.* **1982**, *18*, 14–20. [CrossRef]
63. *Document Water Reservoir: Manipulační Řád pro Vodní Dílo Vír na Řece Svratce v km 114,900*; Povodí Moravy, s. p.: Brno, Czech Republic, 2011.
64. Czech Technical Standard ČSN 75 2935 The Safety Assessment of Hydraulic Structures during Floods. Available online: https://csnonline.agentura-cas.cz/Detailnormy.aspx?k=94534 (accessed on 25 October 2020).
65. Starý, M. *Reservoir and Reservoir System (MODUL 01)*; Education Tutorial, Faculty of Civil Engineering, Brno University of Technology: Brno, Czech Republic, 2006.
66. Starý, M. *Hydrology (MODUL 03)*; Education Tutorial, Faculty of Civil Engineering, Brno University of Technology: Brno, Czech Republic, 2005.
67. Jandora, J.; Šulc, J. *Hydraulics (MODUL 01)*; Education Tutorial, Faculty of Civil Engineering, Brno University of Technology: Brno, Czech Republic, 2006.
68. Czech Technical Standard ČSN 75,1400 Hydrological Data of Surface Waters. Available online: http://www.technicke-normy-csn.cz/751400-csn-75-1400_4_32709.html (accessed on 15 February 2021).

water

MDPI

Article

Investigation of Attributes for Identifying Homogeneous Flood Regions for Regional Flood Frequency Analysis in Canada

Ziyang Zhang [1,2] **and Tricia A. Stadnyk** [1,3,*]

1 Department of Civil Engineering, University of Manitoba, Winnipeg, MB R3T 5V6, Canada; umzha494@myumanitoba.ca
2 Matrix Solutions Inc., 3001-6865 Century Ave, Mississauga, ON L5N 7K2, Canada
3 Department of Geography, University of Calgary, Calgary, AB T2N 1N4, Canada
* Correspondence: tricia.stadnyk@ucalgary.ca

Received: 22 July 2020; Accepted: 12 September 2020; Published: 15 September 2020

Abstract: The identification of homogeneous flood regions is essential for regional flood frequency analysis. Despite the type of regionalization framework considered (e.g., region of influence or hierarchical clustering), selecting flood-related attributes to reflect flood generating mechanisms is required to discriminate flood regimes among catchments. To understand how different attributes perform across Canada for identifying homogeneous regions, this study examines five distinctive attributes (i.e., geographical proximity, flood seasonality, physiographic variables, monthly precipitation pattern, and monthly temperature pattern) for their ability to identify homogeneous regions at 186 gauging sites with their annual maximum flow data. We propose a novel region revision procedure to complement the well-known region of influence and L-Moments techniques that automates the identification of homogeneous regions across continental domains. Results are presented spatially for Canada to assess patterning of homogeneous regions. Memberships of two selected regions are investigated to provide insight into membership characteristics. Sites in eastern Canada are highly likely to identify homogeneous flood regions, while the western prairie and mountainous regions are not. Overall, it is revealed that the success of identifying homogeneous regions depends on local hydrological complexities, whether the considered attribute(s) reflect primary flooding mechanism(s), and on whether catchment sites are clustered in a small geographic region. Formation of effective pooling groups affords the extension of record lengths across the Canadian domain (where gauges typically have <50 years of record), facilitating more comprehensive analysis of higher return period flood needs for climate change assessment.

Keywords: regional flood frequency analysis; flood-related attribute; region of influence; flood region revision process; Canadian annual maximum flow

---

1. Introduction

Designing future infrastructure for flood resiliency is necessary and crucial for emerging design standards. Flood frequency analysis (FFA) is often used to estimate flood quantiles for river infrastructure design to prevent structural failure or inadequacy during extreme flood events. Given its importance, a growing number of countries have carried out nation-wide study for advanced methods of FFA to improve design flood estimation [1–3]. Outcomes from these studies can be generalized into published guidelines, which are beneficial for domestic end-users in terms of simplicity and consistency and for reducing the element of subjectivity within the design process [4,5].

In Canada, flooding has been recognized as the most frequent and costliest of natural disasters over the past 100 years, claiming considerable economic and social losses for cities, urban clusters, and agricultural land use [6]. Consistent with the recent assessment of 23 unsolved problems in hydrology led by the International Association for Hydrological Sciences (IAHS), improved analysis of the magnitude of extreme (flood and drought) events and the variability in these events is a critical area of research in hydrology [7]. General requirements for design floods at primary crossings or for floodplain delineation in Canada necessitate a one in 50 year or 100 year event [8,9]. The 2T rule of traditional at-site FFA recommends at least 100 years flood data [4], while only 1.05% of flow gauging sites (67/6379) in the Canadian HYdrologic DATabase (HYDAT) satisfy this criterion [10]. As a result, FFA based on regional information, or regional flood frequency analysis (RFFA), is a particularly important method for Canadian design flood estimation.

The identification of homogeneous flood regions is of paramount importance for RFFA, as it marks the first step in the FFA process, forming the period of record for analysis [11,12]. The homogeneous region is a collection of hydrologically similar catchments so that flood information, such as annual maximum flow, can be reasonably and effectively transferred within the defined flood region using transformation methods, such as index-flood [13–17]. Many studies [18–20] have focused on investigating different regionalization frameworks and techniques; for example, in Canada, the statistical hydrology research group in Québec (Groupe de Recherche en Hydrologie Statistique, GREHYS) [11] compared four different regionalization techniques, including region of influence (ROI), canonical correlation analysis, hierarchical cluster analysis, and L-Moments statistics, for delineating homogeneous flood regions in Québec and Ontario. For the same study area, Wazneh et al. [21] endorsed catchment regionalization based on statistical depth function over ROI and canonical correlation approaches because of robust region identification process and improved accuracy of pooled estimation. Zadeh and Burn [22] delineated 1114 Canadian gauging sites into six super hydrological regions based on flood seasonality statistics, drainage area, and mean annual precipitation. The concept of delineated super regions was later adopted to calibrate a nonparametric model for ungauged pooled estimation [23,24]. Though regionalization techniques may differ, the selection of arbitrary, flood-related attributes is required for all regionalization techniques in order to effectively discriminate between flood behaviors among catchments sites. Geographical contiguity has been frequently used as an attribute because hydrological variability tends to be smaller within smaller geographical regions [4,19,25]. For large catchments with fewer gauging stations, however, cohesive flood behavior associated with geographic contiguity is often reduced. This is often the case for rural and northern Canada and regions with highly dynamic flood response, such as the Prairie Pothole region [26,27].

Other widely used flood attributes include physiographic, climatic, and statistical types. Each type of attribute can effectively measure flood similarity, and thus be used to identify homogeneous flood regions. Burn [28] considered coincident annual peak flood values as the prime flood-related variable for 41 sites in southern Manitoba, Canada. Catchment geographical distribution and local empirical knowledge were also embedded in the regionalization process. Three homogeneous regions were formulated, acknowledged by a statistical homogeneity test. In the same study area, Burn [29] derived a composite attribute to group catchments. The attribute comprised coefficient of variation (CV) of floods, mean annual flow divided by drainage area (QDA), and latitude and longitude of the gauging station. The results also showed that CV and QDA were relatively more effective than geographical proximity for forming homogeneous regions, with the CV attribute being more informative than the QDA. The type or the method of forming a composite attribute to describe multiple aspects of flood characteristics is often considered more informative for dividing catchments into distinct flood regions [19]. Weighting each variable within the composite attribute, however, introduces the element of subjectivity. Additionally, variable selection is often identified mostly on localized physiographic and climatic knowledge rather than analytical reasoning [11,29,30].

Recent studies of RFFA in Canada tend to focus on deriving a robust quantile regression model for ungauged frameworks [23,24,31–34]. Among these studies, the following variables were frequently considered in their quantile regression model as flood influential attributes: latitude and longitude of gauging stations, CV, QDA, mean annual precipitation, and basin slope.

The geographic extent of Canada means that water resources engineering practice is generally governed at the provincial level and the boundary, as opposed to federal jurisdiction, which is more common in other countries [35,36]. As a result, methods of RFFA have been inconsistently applied among government agencies, academic communities, and industrial partners [8,11,37]. To tackle this problem, the Natural Sciences Engineering Research Council of Canada (NSERC) funded FloodNet Strategic Network project unified researchers across Canada to develop nation-wide flood forecasting and water resources management strategies. An important mandate was to research standardized FFA methods and techniques tailored for Canadian hydrological environments [38]. Within this network, Sandink et al. and Zahmatkesh at al. [23,24] examined FFA using a quantile regression model that considered ungauged catchments across Canada. Zhang et al. [39] demonstrated the generalized extreme value (GEV) distribution fits Canadian annual maximum flow data considerably better than other well-known distributions, including generalized logistic, Pearson type III, and log Pearson type III distributions. Others [40–44] focused on developing regionalization techniques using peaks-over-threshold (POT) flood data, which is advantageous for gauging sites where annual maximum flood records are limited. Little attention has been paid to the examination of different flood-related attributes and their characteristics for identifying homogeneous flood regions.

Here, we consider five distinct categories of frequently used attributes (i.e., geographical proximity, physiographic variables, flood seasonality, monthly precipitation pattern, and monthly temperature pattern) and investigate their relevance in identifying homogeneous flood regions for RRFA applications on a continental, Canada-wide scale. Their abilities to identify homogenous regions are investigated across major hydrological sub-regions of Canada. Regional hydrological characteristics are used as input to analyze homogeneous region identification results. To increase efficiency of our analysis and minimize the element of subjectivity, a novel automated regionalization process that combines the well-known ROI [45,46] approach with a proposed automatic region revision algorithm (ARRA) is introduced and demonstrated for applicability to continental domains. Memberships of two regions are selected as a case study to provide insight into membership characteristics. Findings of this study are deemed to be an important contribution toward the Canadian statistical flood estimation guideline under the FloodNet project.

2. Materials and Methods

2.1. Rationale for Attribute Selection

Geographical proximity is selected based on the rationale that catchments closer to each other generally encompass similar hydrological and physiographical characteristics, and, therefore, catchments with smaller geographical proximity are likely to exhibit a similar flood regime and to form a homogeneous region. The presence of large spatial variability in flood characteristics might question the use of geographical proximity, therefore, directly using physiographic variables that exert key influence on the dominant flood generating mechanisms provides another way to group sites with similar flood behavior. Geographical proximity and physiographic variables are the most common flood-related attributes for catchment regionalization and thus are included in this study.

As previously noted, flood seasonality has the advantage of convenience in attribute extraction. In addition, it has been previously applied and was found to be beneficial for flood studies in Canada for catchment classification [42,47,48] and in the formation of homogeneous regions [49,50].

Monthly precipitation and temperature patterns consider monthly average precipitation and temperature for the location of the catchment site. These values are provided by Environment and Climate Change Canada (ECCC) [51], computed for each catchment site in this study using historical monthly climate grids for North America [52,53].

Flood generating mechanisms in Canada are generally dominated by either rainfall (pluvial), snowmelt (nival), and rain-on-snow (mixed) events [42,54]. The monthly patterning of precipitation and temperature are considered to contain key information concerning the dominant flood generation process. For example, precipitation accumulation during winter months dominates the magnitude of the spring melt event. Large precipitation values in summer and fall suggest rainfall-driven peak floods. Temperature values in the melt season influence the timing and the magnitude of spring peak floods. Therefore, we explore these attributes given their potential usefulness in mapping regional flood characteristics.

2.2. Datasets

Annual maxima flood samples are taken from the Canadian Reference Hydrometric Basin Network (RHBN). Developed by Water Survey of Canada, the RHBN constitutes 223 gauging sites in total at the start of this study, and is only a small subset of the Canadian hydrometric gauging network (6379 gauging with flow record sites in total) [10]. RHBN sites are identified as near pristine catchments, high quality flow measurements, with an absence of anthropogenic control [55,56]. These merits make their flood data ideal for RFFA. In addition to the 223 RHBN sites, only 186 sites have corresponding physiographic variables available, supported by ECCC [51]. Therefore, we consider a total of 186 gauging sites in this study, generating a total of 186 annual maximum flood samples. Although RHBN stations generally have flow records that are greater than 20 years in length, some sites are seasonally operated, which means that not all calendar years are able to derive the annual maximum flood. The average station record length among our samples is 48 years, with a maximum of 103 years and a minimum of eight years. More than 80% of samples have station record lengths greater than 30 years.

The geographical distribution of the 186 sites is presented in Figure 1, with corresponding record length distributions of the 186 sites presented in Figure 2 (the x-axis corresponding to the longitude, from west to east, noted by province or territory. Figures 1 and 2 indicate that most study sites in British Columbia and the Atlantic provinces have relatively higher record lengths compared to other regions. The prairie provinces, particularly Saskatchewan and Manitoba, have relatively fewer stations and relatively shorter record lengths. The three northern territories have the fewest number of gauging sites and an average record length of 40 years.

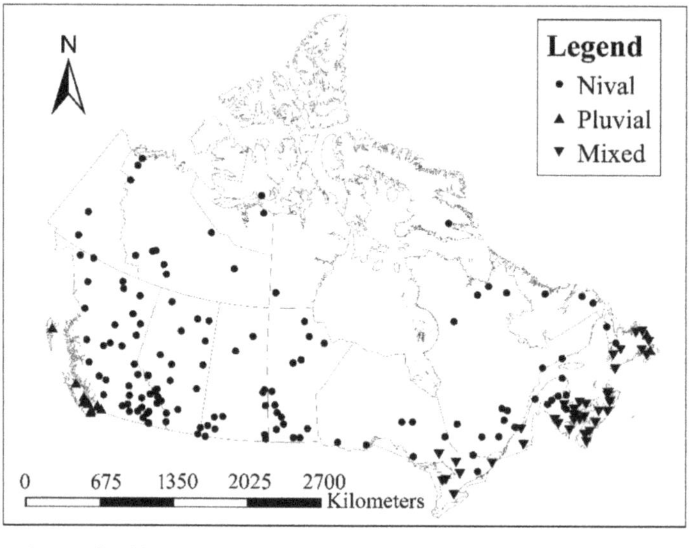

Figure 1. Geographical location of 186 study sites identifying primary cause of flood response.

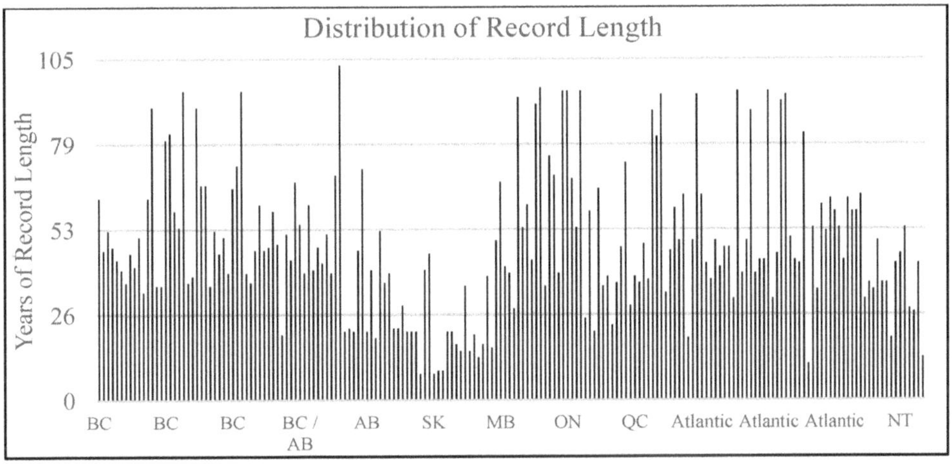

Figure 2. Distribution of record length for the 186 flood samples. Sites in the ten provinces are plotted in order of longitude from west (left) to east (right). Sites in the three territories are plotted to the right most of the figure, with three sites in northeast Québec embedded within the Atlantic provinces.

2.3. *Defining Attribute Similarity Distance*

2.3.1. Geographical Proximity

The latitude and the longitude of the gauging stations are used to calculate the geographical distance between two catchments. The similarity distance between catchment m and n is defined as:

$$d_{mn} = \left[(Lat_m - Lat_n)^2 + (Lon_m - Lon_n)^2 \right]^{0.5} \tag{1}$$

where Lat_m and Lat_n are the latitude and the longitude coordinates for the gauging site of catchment m. We use geographical coordinates for the above equation, which can cause minor discrepancies in the calculation or the comparison of one-degree longitude approaching the polar region.

2.3.2. Physiographic Variables

The selection of physiographic variables is based on the stepwise regression method, which has been used to select flood-related attributes in previous studies [57–59]. The stepwise regression method is an automatic procedure used to select explanatory variables based on the development of a multilinear regression model. Candidate variables are iteratively added and removed based on the use of statistical t-test until the predictive power of the regression model is optimized. In this study, 66 sets of different physiographic variables at each site are obtained from ECCC [51]. Because different variables have different units and scale, variables are normalized by their standard deviation prior to the regression. The dependent variable for the stepwise regression considers the median value of each flood sample, which corresponds to a 2-year return period flood. The median value is considered a robust indicator of flood characteristics and is meant to reduce impact from outlier flood values [4,34]. Consequently, the stepwise method recognizes the following variables as sufficiently explanatory of flood characteristics: (1) catchment area, (2) waterbody area in the catchment, (3) standard deviation of elevation across the catchment, (4) average annual air temperature for the catchment, and (5) average annual precipitation for the catchment. Variables (2) and (3) are derived from the ECCC National Hydrology Network database. Variables (4) and (5) are computed based on 10 km historical gridded climate data representing a 30 year period of record from 1981 to 2010. Data provided by ECCC are computed using historical monthly climate grids for North America [52,53].

The similarity distance between catchment m and n is calculated based on a weighted Euclidean distance formula defined as:

$$d_{mn} = \left[\sum_{j=1}^{k} w_j\left(x_{mj-}x_{nj}\right)^2\right]^{0.5} \tag{2}$$

where k is the number of physiographic variables, w_j is the weighting factor for the physiographic variable j, and x_{mj} is the standardized value for the physiographic variable j of catchment m. w_j controls the relative importance of variable j. Here, weights of 0.4 were assigned to the basin area and 0.15 to the remaining four variables. These weights corresponded to variable coefficients in the computed stepwise model, rounding to the nearest 0.05 digit.

2.3.3. Flood Seasonality

Similarity between catchments is measured using a unit polar coordinate system. A catchment is presented as a point in the polar coordinate space and can be positioned by angular and radial values. The angular value reflects the average date of flood occurrence, whereas the radial value reflects the variability in the occurrence date of floods. A larger radial value indicates smaller variability in occurrence date; a radial value of one indicates no variability in occurrence date, implying that all floods occur on the same day of each year.

Based on Burn [49], for a single flooding event, the date of occurrence of the event is transformed from Julian day to an angular value, where Julian day one is 1 January and day 365 is 31 December, using:

$$\theta_i = (JulianDate)_i\left(\frac{2\pi}{365}\right) \tag{3}$$

For a given catchment flood sample composed of k flooding events, its Cartesian coordinates $\overline{x}$ and $\overline{y}$ in the unit circle are calculated as:

$$\overline{x} = \frac{1}{k}\sum_{i=1}^{k} cos(\theta_i) \tag{4}$$

$$\overline{y} = \frac{1}{k}\sum_{i=1}^{k} sin(\theta_i) \tag{5}$$

Therefore, the similarity distance between catchments m and n is calculated as:

$$d_{mn} = \left[(\overline{x}_m - \overline{x}_n)^2 + \left(\overline{y}_m - \overline{y}_n\right)^2\right]^{0.5} \tag{6}$$

Followed by the Durocher et al. [42] classification, sites used in this study are further classified into nival, pluvial, and mixed regimes based on their flood seasonality statistics and localized geographic and climatic environments (i.e., classifications noted on Figures 1 and 3, respectively). Nival sites are subject to regular flood occurrence dates for the spring snowmelt period. These sites are generally located in cold regions of Canada such as continental interior, mountainous British Columbia, and northern Canada. A smaller number of sites are exclusively pluvial-driven with average annual flood occurrence from November to February. These sites are in the warmest regions of Canada, which are coastal southwest British Columbia and Vancouver Island. A substantial number of study sites are classified as mixed response. These sites experience warm to mild winters and are predominately located in southeastern Ontario, southern Québec, and the Atlantic provinces. Peak floods for these sites can be either spring snowmelt, rain-on-snow, or single heavy rainfall events. Their wide range of regularity in the flood seasonality space provides an effective indication of annual peak floods driven by multiple flood responses.

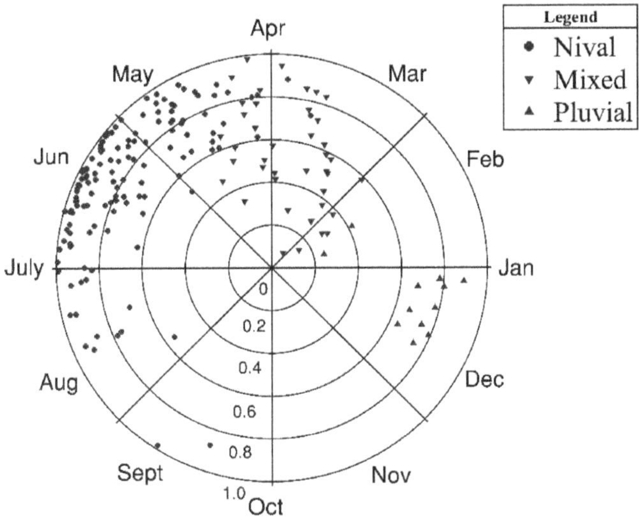

Figure 3. The 186 study sites plotted in flood seasonality space.

2.3.4. Monthly Precipitation Pattern

Similarity measures based on precipitation patterns are attributed to the values of monthly average precipitation from January to December for each catchment site. The correlation coefficient is selected to assess the similarity measure between two catchments. In contrast to Euclidean distance, the correlation coefficient is considered more effective when characterizing the pattern of two datasets, as it measures the degree of linearity of the datasets, while the Euclidean distance measures the distance between two points in a matric space. The correlation coefficient between catchment n and m is described as:

$$r_{nm} = \frac{\sum_{i=1}^{12}(x_{ni} - \overline{x_n})(x_{mi} - \overline{x_m})}{\sqrt{\sum_{i=1}^{12}(x_{ni} - \overline{x_n})^2}\sqrt{\sum_{i=1}^{12}(x_{mi} - \overline{x_m})^2}} \tag{7}$$

where x_{ni} is the monthly average precipitation value for month i of catchment n, and $\overline{x_n}$ is the average of the 12 monthly average precipitation values for catchment n expressed as:

$$\overline{x_n} = \frac{1}{12}\sum_{i=1}^{12} x_{ni} \tag{8}$$

r_{nm} ranges from −1 to 1, with values exactly equal to 1 (−1) indicating a perfect positive (negative) linear relationship between two datasets, and values exactly equal to 0 indicating no linear relationship. For the similarity measure of catchment m and n, r_{nm} closer to 1 indicates a stronger positive linear relationship between catchment m and n, therefore, the similarity distance based on the correlation coefficient is computed as:

$$d_{nm} = 1 - r_{nm} \tag{9}$$

2.3.5. Monthly Temperature Pattern

In common with the similarity measure for precipitation patterning, temperature patterning is computed from monthly average temperature for each catchment. Monthly average temperature data for catchment n and m are then input into Equations (7) and (8); Equation (9) is used to calculate the similarity distance between the two catchments.

2.4. Region of Influence Approach

The ROI approach [45,46] is used given its flexibility of identifying flood regions for each study site. The ROI defines target sites as having a unique flood region. The addition of other sites to the region proceeds in order of the shortest similarity distance to the greatest. Determining the number of sites in a region requires a trade-off between the size of the region and the quality of the region. A larger region benefits flood estimation at larger return periods (i.e., generates longer records), however, the quality of the region (i.e., homogeneity in flood characteristics) generally decreases as more sites are added to the region. For RFFA, the 5T rule for region size (i.e., total station-year of record of the region) states that regions should optimally have five times greater record length than the return period of interest (T) and has been widely accepted as a guideline for optimal trade-off [4,12]. The 5T rule was adopted in this study.

2.5. Generalized Extreme Value (GEV) Distribution and L-Moment Estimation Method

The GEV distribution is used to estimate flood quantiles. The GEV distribution has been determined to be more robust for fitting annual maximum flow at RHBN stations than other commonly used three parameter distributions [39]. The index flood L-Moment parameter estimation method is recommended by many studies for its simplicity, robustness, nearly unbiased estimation, and convenient integration with the GEV and the L-Moment homogeneity test [16,60,61].

2.6. L-Moment Homogeneity Test

The homogeneity test aims to verify if sites in the flood region exhibit similar flood characteristics at an acceptable level of statistical significance. Since L-Moments are considered unbiased statistics of flood data, the L-Moment homogeneity test has received much attention in RFFA applications [4,12,19,30,49,62]. Based on Hosking and Wallis [16], the first step of the homogeneity test is to determine the regional L-Moment ratios t^R, t_3^R, and t_4^R, denoted as the regional L-CV, L-skewness, and L-kurtosis, respectively. For a region comprises N sites, the regional L-Moment t^R (similarly apply for t_3^R and t_4^R) is calculated as:

$$t^R = \sum_{i=1}^{N} n_i t^{(i)} / \sum_{i=1}^{N} n_i \tag{10}$$

where $t^{(i)}$ is the at-site L-Moment ratio for site i, and n_i is the record length for site i.

Dispersion can then be expressed as:

$$V = \left[\sum_{i=1}^{N} n_i \left(t^{(i)} - t^R\right)^2 / \sum_{i=1}^{N} n_i\right]^{0.5}, \tag{11}$$

To assess if the dispersion, V, is within the limit of region homogeneity, two variables are required: μ_V, the expected mean of V; and σ_V, the expected standard deviation of V. μ_V and σ_V are estimated through many reproductions of the original region. To do this, a Kappa distribution fit by L-Moment ratios of 1, t^R, t_3^R, and t_4^R is used to reproduce the N_{sim}, or number of original regions (N_{sim} = 1000 used in this study). Each reproduced region has the same region size (N sites in a region) and the same record length, n_i for site i, with respect to the original region.

For each reproduced region, the dispersion, V, is calculated using Equations (10) and (11). Based on the N_{sim} number of V values, the expected mean σ_V and the expected standard deviation μ_V can be obtained.

Lastly, the homogeneity statistic is defined as:

$$H = \frac{V - \mu_V}{\sigma_V} \tag{12}$$

where H is the homogeneity statistic. A region is regarded as acceptably homogeneous if $H < 1$, possibly homogeneous if $1 \leq H < 2$, and heterogeneous if $H \geq 2$. In this study, $H < 1$ was used to determine if the region was homogeneous.

2.7. Automatic Region Revision Algorithm (ARRA)

For a given target site and ROI and any attribute, the initial flood region formation often still tests heterogeneous. Many studies have reported this situation, and, subsequently, a region revision process is needed to reduce region heterogeneity by editing the initial group membership [12,28,29,44]. The revision process includes steps such as adding, deleting, and replacing site(s) within the initially formed region, subsequently testing for homogeneity after each progressive change. In past studies, this is largely carried out through a heuristic process, meaning there is no set procedure regarding the order of steps or the methodology of revision [12,17,29,63]. For our large-scale study, however, it is ineffective to proceed via a heuristic process for each region, therefore, an automatic region revision algorithm (ARRA) was /designed with the intent of reducing region heterogeneity through an automatic and non-subjective modification of the region membership.

A heterogeneous region is input into the ARRA, and a revised region with improved homogeneity is output from the first iteration. If the output region does not meet the homogeneous criteria (i.e., $H < 1$), the ARRA can be reapplied to the region to further reduce heterogeneity. Each time the region membership is modified, the homogeneity of the membership increases, but the attribute similarity decreases because the newly added site(s) have larger attribute distance(s) compared to the replaced site. As a region should be formed primarily based on attribute similarity, the number of ARRA iterations needs to be constrained to ensure an appropriate trade-off between region homogeneity and attribute similarity. We perform a sensitivity analysis on the number of ARRA iterations used to revise 186 randomly formed initial pooling regions by counting the number of homogeneous regions produced after each ARRA iteration. From this sensitivity analysis, we determine that a maximum of five iterations of the ARRA should be applied (see Section 3.1. ARRA performance). If, after five iterations of the ARRA, a region still tests heterogeneous, this region is regarded as unable to form a homogeneous region.

Figure 4 illustrates methodological procedure followed by the ARRA. The L-Moment homogeneity test is embedded in the ARRA and used to identify sites that should be removed and new sites that should be added to achieve the greatest improvement in region homogeneity. The order of searching for a newly added site depends on attribute similarity, such that shorter attribute similarities are tested first. The process terminates once an improved region is formed and the 5T region size rule is satisfied.

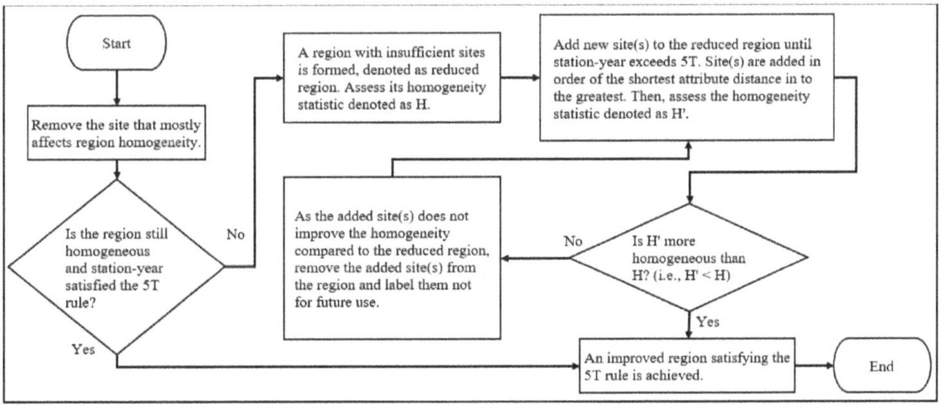

Figure 4. Schematic of the automatic region revision algorithm (ARRA) process.

2.8. Flood Region Identification Process

For each of the five considered attributes, the process of identifying flood regions is demonstrated below. First is the identification of the initial flood region for a study site, which uses the ROI approach to group regions based on attribute similarity alone. To be specific, catchment sites having the shortest attribute distance to the target site are pooled into the initial region. The region size is set to 500 station years of record, which allows for accurate estimation up to the 100-year flood according to the 5T rule. Next, the homogeneity of the initial region is assessed using the L-Moment homogeneity test. If the initial region is heterogeneous, the ARRA is applied to revise region membership, up to a maximum of five iterations. The homogeneity of the revised region is re-evaluated using the homogeneity test. This process was repeated for all 186 study sites, and the total number of homogeneous regions identified for each attribute was determined.

Annual maximum flows for all region members are typically used for the homogeneity test and the subsequent flood quantile estimation. In this study, however, we purposely exclude annual maximum flows at the target site to afford more robust and rigorous evaluations of homogeneity and flood quantiles (i.e., a leave-one-out analysis), and therefore, our methodology can later be applied for ungauged regional analyses.

2.9. Assessing the Accuracy of Regional Flood Quantiles

Estimated regional flood quantiles are compared to "true" flood quantile determined by at-site samples. It is common, in practice, to determine "true" quantiles from at-site FFA when the return period of interest is below half the at-site record length (i.e., a 2T rule) [4]. Comparison of regional and at-site quantiles provides a means to assess the accuracy of regional estimates relative to standard practice.

There were only 11 sites with record lengths greater than 90 years included in this study, therefore, for the purpose of reliable at-site estimation, the return periods selected for comparison could not be extreme quantiles; we selected a range of 20 to 45 years. For each return period, T, the selected sites were those that were able to identify 5T homogeneous regions across all attributes and those having record lengths greater than 2T for reliable at-site estimation. A homogeneous region is easier to form for smaller region sizes, therefore, the number of sites available for analysis for each return period differed, with more sites meeting our criteria at smaller return periods.

Table 1 lists the return periods considered for comparison, the number of sites considered at each return period, and the required record lengths for adequate at-site and regional quantile estimates. It is noteworthy that flood estimation for both at-site and regional methods was subject to sampling uncertainty, with the uncertainty bound decreasing with decreasing return period. Thus, the smaller return periods provided improved reliability for assessing results.

Table 1. Required record length for at-site and regional estimate at different return periods used in analysis.

Return Period for Comparison	Required Record Length for at-Site Estimate	Number of Sites Available	Station-Years of Record for Regional Estimate
20	40	88	100
25	50	47	125
30	60	29	150
35	70	15	175
40	80	14	200
45	90	11	225

$$Relative\ bias = \frac{1}{n}\sum_{i=1}^{n}\frac{Q_i - q_i}{q_i} \times 100\% \quad (13)$$

$$Relative\ RMSE = \left[\frac{1}{n}\sum_{i=1}^{n}\left(\frac{Q_i - q_i}{q_i}\right)^2\right]^{0.5} \times 100\% \quad (14)$$

where Q_i is the quantile of regional estimate for site i, q_i is the quantile of at-site estimate for site i, and n is the number of available sites that analyses for each return period.

3. Results and Discussion

3.1. ARRA Performance

Table 2 shows the resulting number of homogeneous regions produced by the attribute and the number of ARRA iterations. When the ARRA is not applied and the regions are formed based on shortest attribute distance alone, it results in only five to ten sites (of 186) sites that form homogeneous regions across all attributes. Forming homogeneous regions based on attribute similarity alone is, therefore, found to be unproductive, and the use of the region revision process (i.e., the ARRA) to revise initial regions is deemed necessary.

Table 2. The number of homogeneous regions identified for each attribute with target region size 500 station years of record. For each ARRA iteration, bold italicized number(s) indicate the best outcome across the five considered attributes; if two attributes tested equally, they were both best outcomes.

Number of ARRA Iterations	Considered Flood-Related Attributes					Alternative Series (Initial Regions Randomly Formed)
	Geographical Proximity	Flood Seasonality	Physiographic Variables	Monthly Precipitation Pattern	Monthly Temperature Pattern	
0	***10***	6	5	6	***10***	0
1	***26***	22	17	23	21	1
2	49	43	35	50	***54***	9
3	70	50	52	69	***80***	22
4	83	66	69	82	***88***	43
5	89	78	83	***99***	94	63
6	97	98	97	***110***	97	74
7	106	110	104	***118***	105	98
8	106	116	106	***120***	109	112

Once implemented, the number of homogeneous regions the ARRA identifies non-linearly increases with the number of ARRA iterations for all attributes. In general, the number of homogeneous regions increases significantly for one to three iterations of the ARRA and increases less from four to eight iterations. Two to four iterations of the ARRA results in identification of relatively more homogeneous regions when considering geographical proximity, precipitation, and temperature patterning than for flood seasonality and physiographic attributes. For five or more iterations, monthly precipitation pattern produces the most homogeneous regions.

To determine a suitable threshold for the number of ARRA iterations, an alternative series composed of 186 regions, for which membership was randomly formed (i.e., without the use of attribute similarity), is used and one to eight iterations of the ARRA applied (last column, Table 2). Comparing results between the five attributes and the alternative series, we find that attribute similarity is largely irrelevant to the identification of homogenous regions after eight iterations. At five iterations, approximately half of the sites form homogenous regions across all attributes, and the number of regions associated with each attribute remains greater than the alternative series. This suggests reasonable preservation of attribute similarity as a selection criterion. We therefore find a maximum of five iterations of the ARRA to be a suitable balance between maintaining attribute similarity for a region and leveraging the revision power of the ARRA.

With appropriate use of the ARRA (i.e., five iterations), approximately 79 to 99 sites of 186 sites identify homogeneous regions across all attributes. This is significantly higher than the five to ten sites identified prior to the use of the ARRA.

3.2. Identification of Homogeneous Regions

When the ARRA is applied for five times (or less), monthly precipitation pattern identifies the largest number of homogeneous regions among all other attributes, followed by temperature pattern, geographical proximity, and flood seasonality. Physiographic variables produce the fewest number of homogeneous flood regions. Differences among the attribute results are relatively small, where the total difference between the two most extreme results (flood seasonality and monthly precipitation pattern) is 21 sites, which is ~11% of the 186 study sites.

Figure 5 shows homogeneous region identification across Canada for each attribute. Note that sites that could not identify a homogeneous region but may be members of another site's homogeneous region are also indicated in blue. Catchment sites are non-uniformly distributed across Canada, with clusters in southern Canada aligned with urban development and large populations, while remote and sparsely gauged regions are often found in the continental interior and mid to high latitudes of the continental landmass.

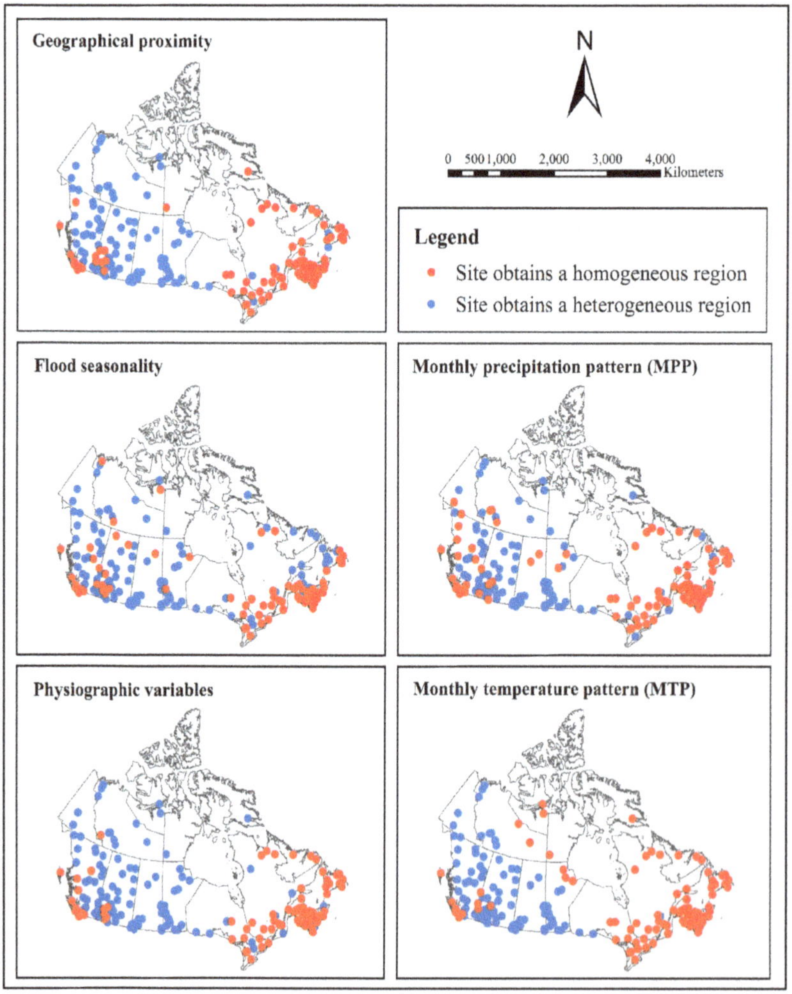

Figure 5. Sites achieving homogeneous regions (red) relative to those that did not (blue), shown by geographic location for each attribute. Note that ARRA was applied up to a maximum five iterations.

Results are generally similar across all attributes at the national scale, with regionalized discrepancies identified. In general, all attributes readily identify homogeneous regions in eastern Canada, while, in western Canada (particularly the interior and the northern regions), the identification of homogeneous regions is more problematic. Catchment sites in eastern Canada are generally clustered in small geographical areas, therefore, they experience more similar flooding behavior. Site clusters are also found in Vancouver Island and southeast British Columbia, where considerable homogeneous regions were also identified across all attributes.

As catchment sites in eastern Canada are more tightly clustered, less variability in flood attributes is expected. Figure 6 presents three boxplots comparing catchment physiographic variables between eastern and western sites with respect to catchment area, water body area in the catchment, and standard deviation of elevation across catchment. The variability in attribute physiography for the eastern sites is noticeably less than that for the western sites, particularly for the standard deviation of elevation across catchment.

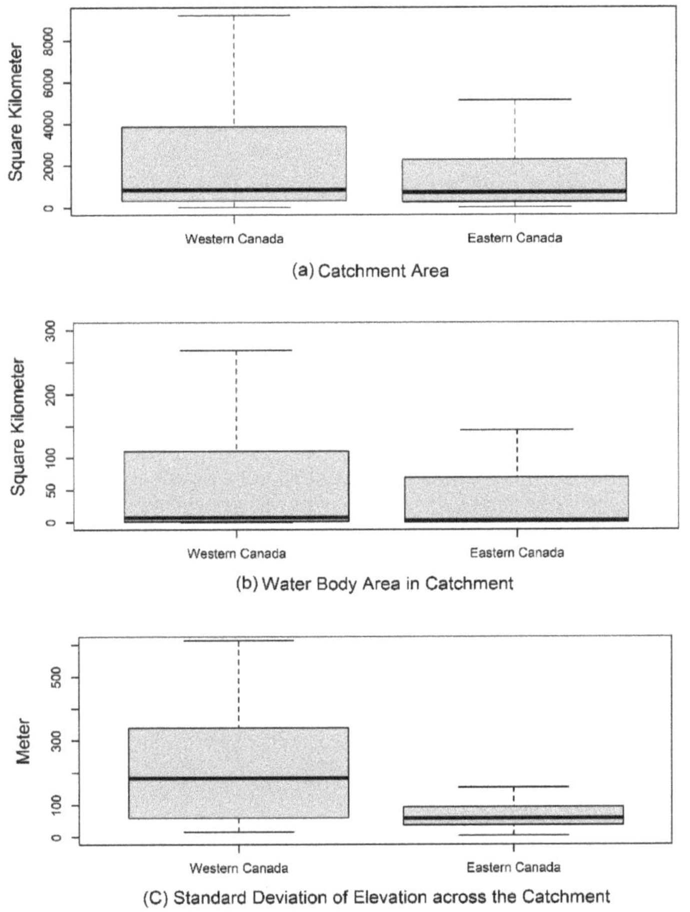

Figure 6. Boxplots comparing physiographic variability of eastern and western sites across Canada. Eastern (Western) computed based on 76 (110) sites; Ontario-Manitoba border is considered the east–west divide, respectively. Boxes represent 25th and 75th percentiles and the median (black line); whiskers extend to extreme values without outliers, where outliers are defined as 1.5 the interquartile range (outliers are removed for scaling purposes).

Some site clusters are found across the southern Canadian prairies in Alberta, Saskatchewan, and Manitoba, where annual peak floods are predominated during the spring snowmelt period. The geographical proximity attribute is typically effective when sites are clustered. Important nival regime influences, such as snowpack accumulation and timing and rate of snowmelt, are reflected in attributes such as monthly precipitation pattern, monthly temperature pattern, and flood seasonality. The regional pooling results, however, show that not many catchment sites within the cluster groups identified homogeneous regions across all attributes. Though site clusters are found in both eastern Canada and the southern prairie region, homogeneity often occurs across large regions, indicating geographical contiguity cannot always warrant effective homogeneous region identification.

Literature indicates that the Canadian Prairies are known for their hydrological complexity, mainly attributed to the presence of potholes and hummocks, which results in highly variable fill and spill runoff process and dynamic effective drainage area, leading to highly non-linear flooding generating mechanisms [26,27,64,65]. Zhang et al. [39] provided statistical evidence that annual maximum flows from prairie RHBN stations are difficult to adequately fit robust distributions as well, including GEV, log Pearson type III, and generalized logistic distributions. This is a strong indication of multiple flood responses occurring at a single catchment site. Ehsanzadeh et al. [66] studied prairie flood response based on nine prairie sites and revealed noteworthy non-linear flood frequency curves.

In addition, flood record length across the Prairies is generally limited (Figure 2). The average record length over 28 prairie sites is 25 years, which is substantially lower than the rest of the 158 sites examined across Canada (having an average record length of 52 years). In order to develop a flood region with 500 station years, more catchment sites must be pooled into these flood regions. This adds an additional challenge for developing homogeneous regions, since more sites leads to more variable flood responses within the flood region.

In Burn [28,29], wherein successfully identified homogeneous regions were formed for southern Manitoba, region identification did not simply rely on attribute similarity measures. A heuristic membership revision process was applied with subjective trial and error to improve region homogeneity. Such region revision approaches are more statistically rigorous than our proposed ARRA, however, they require practitioners to have sound knowledge of local hydrology and are more statistically sophisticated [67]. Our method, on the other hand, is designed to be accessible to all water resource practitioners seeking to perform food frequency analysis.

For the mountainous western Canada region, annual peak floods are predominately snowmelt and rain-on-snow regimes, though rare heavy rainstorms can also trigger annual peak floods in smaller basins [54]. Homogeneous region identification maps are noisy along the cordillera mountain chain for all attributes, namely, it is difficult to interpret a distinctive spatial pattern. In southern British Columbia and Alberta, some sites identify homogeneous regions, however, locations of these sites differ amongst attributes. In central British Columbia and Alberta, and south of Northwest Territories, only flood seasonality and monthly precipitation patterns identify homogeneous regions, and just for a few sites. The western mountain chain is subject to highly variable climate and basin characteristics. Flood generation mechanisms are influenced by combined basin features including catchment size, drainage topography (e.g., channel slope, floodplains, alluvial fans, canyons), localized snow accumulation and distribution, as well as glaciation and avalanches [8]. These basin features, as well as temperature and precipitation, are highly variable spatially and temporally in mountainous environments [54]. Attributes selected in this study capture flood behavior from a limited set of physiographic characteristics and are likely not rigorous enough for catchment regionalization in the mountains.

In northern Manitoba, Northwest Territories, and Nunavut, catchment sites are characterized by cold subarctic climate, barren and tundra rolling landscape, as well as long-lasting (five or six months of the year) snow and ice cover underlain by permafrost [54]. Annual peak floods are primarily snowmelt driven; therefore, the duration and the rate of snowmelt are key characteristics for grouping catchment sites. Homogeneous region identification shows that monthly temperature pattern is more effective than other attributes because it captures timing, rate, and duration of snowmelt driven flood behavior. Some sites also identify homogeneous regions using flood seasonality, possibly because duration and rate of snowmelt are inherently correlated with the average and the variation of peak flood dates.

We find two general and probable causes that account for the inability to identify homogeneous flood regions in Canada. First, the clustering or the proximity of gauge sites has considerable impact on the outcome of identifying homogeneous regions, regardless of the attribute considered. The tendency for attributes to be more similar within smaller geographical regions is significant, despite the fact that regions (for attributes other than geographical proximity) can also include sites that are non-proximal. Second, attributes selected in this study measure a distinctive hydrological feature, however, across large spatial domains (e.g., Canadian landmass), there exist significant local-scale hydrological complexities that influence flood generation mechanisms. For sites that are influenced by multiple hydrological characteristics, our attribute selection is not rigorous enough to capture the particulars of flood behavior and is thus unable to group catchment sites with similar flood frequency regimes. Related to this, Table 2 indicates that most sites identify homogeneous regions as an outcome of ARRA interactions; ARRA revises region membership based on a specified attribute. If the specified attribute does not capture primary flood characteristics, the subsequent ARRA enhancement becomes ineffective.

3.3. Analyzing Membership Characteristics

To gain insight into membership characteristics, two catchment sites along with their region memberships were selected for more detailed case studies. Flood regions for both sites are identified based on flood seasonality and five ARRA iterations, with only one of the two regions being homogeneous.

Target catchment site Water Survey of Canada (WSC) gauge 03MB002–Whale River at 40.2 km from the Mouth in northern Québec, cannot identify a homogeneous flood region with flood seasonality attribute and ARRA iterations. This site and its 12 members are plotted in geographically (Figure 7a) and in flood seasonality space (Figure 7b) and are summarized in Table 3 based on physiography. Figure 8 provides a group of boxplots showing the spread of physiographical variables of this flood region. Flood seasonality space indicates that member sites share similar annual average occurrence dates for flooding, resulting in the formation of this flood region.

Based on geographical proximity (Figure 7a), site membership is supported from a climatic perspective. The map shows that target and member sites broadly span across Canada, from Pacific to Atlantic and from southern British Columbia to the northern edge of the Northwest Territories. All member sites are, however, situated near an ocean or a coastal region and receive substantial annual precipitation (see Table 3). Since member sites span a broad range of latitude, there is variation in annual temperature range that alters the amount and the temporal distribution of rain and snowfall, thus affecting the dominant runoff mechanism during the annual peak flood season. Expected differences in flood behavior are also reflected in the varying physiography among member sites (Table 3).

Four member sites in southeastern British Columbia have high mean annual precipitation and noticeably higher mean annual temperatures compared to other members further north. These four member sites are exposed to more pluvial or mixed rain-on-snow floods. Basin area also substantially varies among the membership; five member sites are small basins (i.e., <500 km^2), whereas the seven others and the target site have basin areas ranging between 3500 km^2 and 49, 000 km^2. Four of the five small basins are located in southeastern British Columbia. The basin compactness ratio (BasinArea/Perimeter^2) is a surrogate measure for time to peak flow and is significantly greater (as expected) for the smaller basins, indicating much shorter routing times than what is seen for the larger basins. A wide spectrum of mean basin slope also exists, ranging from 3.3% to 45.4%, across member sites. Smaller basins in British Columbia mountains are remarkably steeper than member sites from other areas of Canada. Mean basin slope affects time to peak flow, as well as runoff ratios. Member sites that are highly variable in such physiographic characteristics are less likely to exhibit similar flood behavior.

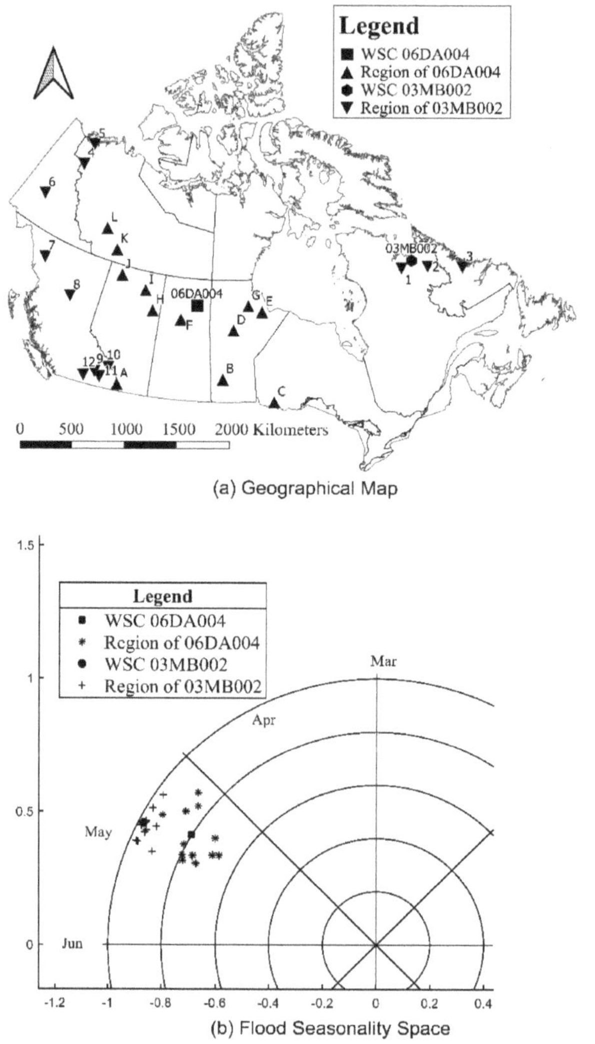

Figure 7. Region membership for study sites 03MB002 and 06DA004, presented by (**a**) geographical extent and in (**b**) flood seasonality space. Members for the 03MB002 (06DA004) region are labeled in (**a**) with numbers (alphabets) as they are referenced in Table 3.

Table 3. Physiographic variables for WSC 03MB002 and WSC 06DA004 and their regions formed based on flood seasonality.

WSC ID	Flood Region of	Province	Map ID	Catchment Area (km^2)	Catchment Perimeter (km)	Compactness Ratio (Area/Perimeter^2) (%)	Mean Basin Slope (%)	Mean Annual Precipitation (mm)	Mean Annual Temperature (°C)
03MB002	Target Site	QC	Target Site	29,124	1417	1.5	2.6	732.3	−4.6
03KC004	03MB002	QC	1	39,371	1901	1.1	3.3	654.9	−5.4
03MD001	03MB002	QC	2	22,440	1627	0.8	3.5	815.2	−4.4
03NF001	03MB002	NL	3	7322	780.6	1.2	8.2	881.0	−3.8
10LA002	03MB002	NT	4	18,746	1173	1.4	18.1	385.9	−6.6
10ND002	03MB002	NT	5	65	44.4	3.3	3.5	220.0	−8.7
09BC001	03MB002	YT	6	48,867	1752	1.6	15.9	456.5	−3.9
08CD001	03MB002	BC	7	3555	488.9	1.5	7.4	562.1	−1.8
07EC002	03MB002	BC	8	5559	597.8	1.6	23.2	648.7	0.2
08NE006	03MB002	BC	9	330	103.3	3.1	45.4	1326	1.4
08NF001	03MB002	BC	10	416	105.1	3.8	31.3	796.1	0.0
08NH005	03MB002	BC	11	442	130.5	2.6	44.5	1218	1.2
08NN015	03MB002	BC	12	233	100.3	2.3	12.1	941.7	2.1
06DA004	Target Site	SK	Target Site	7729	684.0	1.7	2.2	506.7	−2.5
05AA008	06DA004	AB	A	403	105.2	3.6	25.4	753.2	1.9
05LJ005	06DA004	MB	B	348	115.5	2.6	2.5	522.7	1.7
05PB014	06DA004	ON	C	4768	585.6	1.4	2.4	718.7	2.6
05TG002	06DA004	MB	D	886	157.7	3.6	0.8	449.6	−1.4
05UH002	06DA004	MB	E	2191	369.4	1.6	0.4	466.1	−4.4
06BD001	06DA004	SK	F	3670	395.8	2.3	2.6	483.4	−1.5
06FB002	06DA004	MB	G	4274	355.4	3.4	0.4	478.5	−4.7
07CD001	06DA004	AB	H	30,792	1548	1.3	1.5	469.4	0.1
07KE001	06DA004	AB	I	9856	614.0	2.6	0.7	443.1	−0.3
07OB003	06DA004	AB	J	36,901	1278	2.3	0.9	450.8	−0.9
10FA002	06DA004	NT	K	9213	553.4	3.0	0.7	474.9	−3.2
10GB006	06DA004	NT	L	20,696	1146	1.6	0.9	351.3	−4.6

In contrast, target catchment site WSC gauge 06DA004 (Geikie River below Wheeler River) identifies as a homogeneous region consisting of 12 catchment sites, excluding the target site (i.e., leave-one-out analysis). The target site is in northern Saskatchewan with very few other sites nearby. The climatology is described as sub-arctic, cold temperature, with physiography consisting of flat to rolling topography with numerous surface water bodies present in the catchment. The sub-arctic, cold climate causes annual peak flooding that is predominately snowmelt driven; the amount of accumulated winter snowpack, as well as timing and rate of snowmelt are influential to flood generation. The geographical extent of the 12 site membership is shown (Figure 7a), along with flood seasonality (Figure 7b), physiographic values (Table 3), and boxplots of physiographical values (Figure 8). Flood seasonality space indicates the membership has good consistency in the regularity of date of occurrence, suggesting these 12 member sites likely have similar flood type and characteristics. Geographically, member sites are situated in the interior of Canada with most located in mid-to-northern Alberta, Saskatchewan, and Manitoba. This area is subject to prolonged, colder, sub-arctic climate; hence the annual peak flood is a nival flood regime. Although catchment area and perimeter span a large range (Table 3 and Figure 8), catchment compactness ratio, mean basin slope, mean annual precipitation, and mean annual temperature are within the same order of magnitude. The spread of 06DA004 box plots is noticeably smaller than the spread observed for the 03MB002 region among all physiographic variables, which reflects their contrasting results in terms of homogeneity.

The above case studies provide examples of the application of catchment physiographic variables to further investigate membership characteristics, which can potentially diagnose causes for region (homo)heterogeneity. We conduct a similar member physiographic analysis for other considered attributes, and for prairie and mountainous sites that cannot identify homogeneous regions. Member sites in heterogeneous regions often displayed large physiographic variability. Therefore, it is generally found that our selected attributes and ARRA regionalization approach are not rigorous enough to identify homogeneous flood regions for catchments with significant hydrological complexity, which, in Canada, are those primarily located in prairie and mountainous regions.

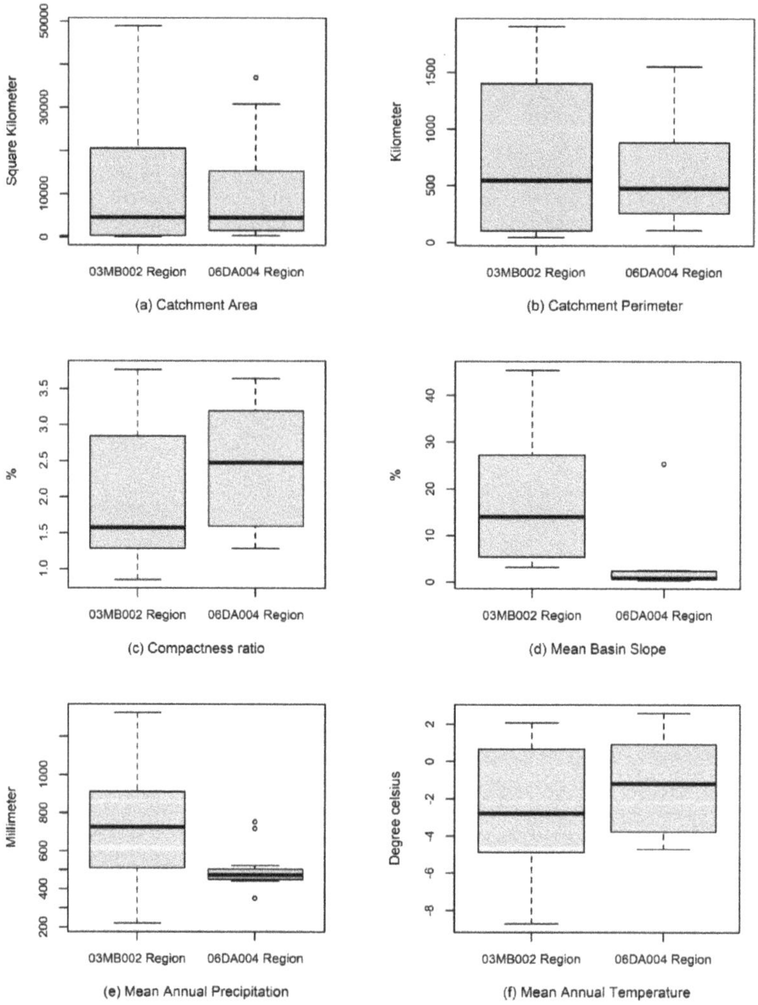

Figure 8. Boxplots for physiographic variability of 03MB002 and 06DA004 flood regions. Boxes represent 25th/75th percentiles with the median (black line); whiskers extend to the extreme values without outliers; outliers (circles in plot) are defined as 1.5 the interquartile range.

3.4. Predictive Measures for Regional Quantile Estimation

Predictive measures for regional quantile estimation are presented as relative bias and relative RMSE for return periods ranging from 20 to 45 years (Table 4 and Figure 9). In general, relative bias across all considered attributes is small for all return periods (ranging from −0.6% to 3.7%). As biases are within ±5% deviation, regional estimation accuracy is considered satisfactory. Bias is generally positive, suggesting that regional estimates tend to overestimate "true" flood quantiles, but are uncorrelated with the magnitude of the flood quantile. Comparing bias among attributes, flood seasonality and physiographic variables exhibit larger bias than geographical proximity, monthly precipitation pattern, and monthly temperature pattern, in general.

Table 4. Relative bias and relative RMSE performance measures (in percentages) for quantiles produced from regionalized estimates. Bold italicized numbers indicate the best outcome for each return period.

Statistic	Return Period	Attribute				
		Geographic Proximity	Flood Seasonality	Physiographic Variables	Precipitation Pattern	Temperature Pattern
Relative Bias	20	***0.3%***	1.0%	0.9%	0.6%	0.6%
	25	0.8%	2.0%	0.5%	***0.4%***	1.0%
	30	***0.5%***	2.5%	3.3%	1.1%	1.9%
	35	1.1%	3.1%	2.8%	−0.6%	***−0.1%***
	40	0.9%	1.8%	2.3%	***0.2%***	0.8%
	45	0.1%	2.1%	3.7%	***0.004%***	−0.2%
Relative RMSE	20	6.5%	6.7%	6.6%	***6.4%***	6.7%
	25	7.9%	7.7%	***7.6%***	7.7%	***7.6%***
	30	***7.5%***	7.9%	8.7%	8.5%	8.0%
	35	9.5%	8.9%	9.6%	8.4%	***8.1%***
	40	9.4%	9.7%	10.0%	8.8%	***8.7%***
	45	***8.6%***	11.1%	13.0%	9.9%	10.1%

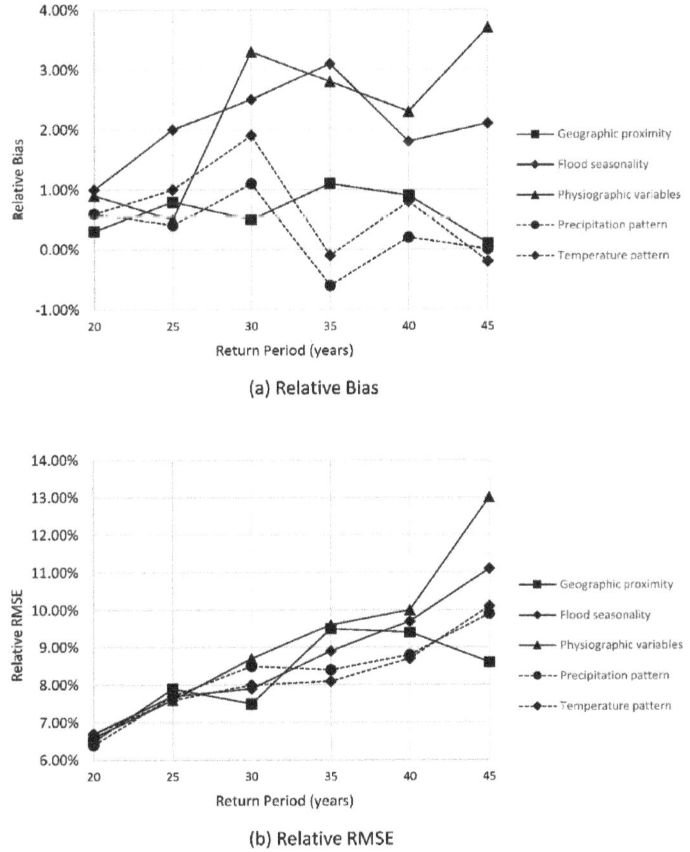

Figure 9. Relative bias and relative RMSE results by return period.

RMSE generally increases with increasing return period across all attributes. Similar RMSE among attributes is found within each return period equal to and less than 35 years. At larger return periods (i.e., 40 and 45 years), flood seasonality and physiographic variables show noticeably larger RMSE than geographical proximity, monthly precipitation pattern, and monthly temperature pattern attributes. Though the "true" quantile is modeled by at-site estimates using accepted methods, estimation uncertainties caused by statistical extrapolation increase with increasing quantiles for both at-site and regional estimates. Therefore, higher relative RMSE at larger return periods is anticipated.

Geographical proximity, monthly precipitation pattern, and monthly temperature pattern perform better across both metrics than flood seasonality and physiographic variables, possibly because regions identified based on the first three attributes often have a higher degree of geographic proximity. Flood seasonality and physiographic measures end up grouping sites across a wider geographical extent, therefore, the degree of hydrological similarity between sites may be lower, resulting in slightly poorer (but acceptable) regional flood estimation results.

Overall, all considered attributes produced satisfactory regional flood quantile estimates for Canada based on acceptable range of bias and a reasonable range of estimation uncertainty. The success in regional quantile estimation demonstrates the applicability of proposed regionalization process based on ROI and ARRA.

4. Conclusions

This study provides insight into five distinctive flood-related attributes for their behavior in identifying homogeneous flood regions across Canada. All considered attributes show similar results regarding the number of homogeneous regions identified and locations where homogeneous regions could be identified. In general, the success of homogeneous region identification is relevant to local hydrological complexities and whether the considered attribute reflects primary flood generation mechanisms and geographic clustering of the sites.

Through combinations of these factors, results of homogeneous region identification are highly distinctive when mapped for Canada. Catchment sites in eastern Canada are generally clustered in small geographic regions and are more likely to exist within similar hydrological environments. Annual peak floods in northern Canada are predominately snowmelt driven, which is sensitive to temperature variation, making monthly temperature pattern important for homogeneous region identification. The Prairie region and the western mountains are subject to highly variable physiographic characteristics, resulting in difficultly in identifying homogeneous regions, regardless of the attribute considered.

Use of a regionalization revision process to revise initial group membership was found to be important. We proposed an automated process, the ARRA, to efficiently revise group membership across large domains and showed it successfully increased the number of homogeneous regions. Flood quantiles obtained from the identified homogeneous regions were reasonably close to estimated at-site "true" quantiles, further demonstrating success of the regionalization process. The ARRA can be readily adopted for other types of regionalization frameworks (e.g., clustering) when subsequent region revision is required.

Findings of this study, on the basis of 186 catchment sites across Canada, provide valuable input on the identification of homogeneous flood regions as well as their attribute behaviors and spatial characteristics. The success of identifying homogeneous flood regions is essential for RFFA and thus for reliable flood quantile estimation. Within the FloodNet project, work on refining RFFA techniques will aid in appropriate sizing of flood resilient infrastructures, which is crucial to proactive protection of lives and properties against flood risk.

Author Contributions: Conceptualization, Z.Z. and T.A.S.; Methodology, Z.Z. and T.A.S.; Software, Z.Z.; Validation, Z.Z. and T.A.S.; Formal Analysis, Z.Z.; Investigation, Z.Z. and T.A.S.; Resources, T.A.S.; Data Curation, Z.Z.; Writing—Original Draft Preparation, Z.Z.; Writing—Review and Editing, Z.Z. and T.A.S.; Visualization, Z.Z. and T.A.S.; Supervision, T.A.S.; Project Administration, Z.Z.; Funding Acquisition, T.A.S. All authors have read and agreed to the published version of the manuscript.

Funding: This research was funded by (i) the University of Manitoba through Graduate Enhancement of Tri-Council Stipends, and (ii) The Natural Sciences and Engineering Research Council (NSERC) of Canada through Canadian FloodNet Project (Grant number: NETGP 451456).

Acknowledgments: The authors gratefully acknowledge Donald H. Burn for providing valuable guidance for this study. The authors extend gratitude to Erika Klyszejko with data support from Environment and Climate Change Canada. The authors acknowledge the helpful comments from five anonymous reviewers that contributed to improving this manuscript. The authors would like to dedicate this study to the memory of Peter F. Rasmussen, who was a key contributor to the journey of this research.

Conflicts of Interest: The authors declare no conflict of interest.

References

1. Salinas, J.L.; Castellarin, A.; Viglione, A.; Kohnová, S.; Kjeldsen, T.R. Regional parent flood frequency distributions in Europe—Part 1: Is the GEV model suitable as a pan-European parent? *Hydrol. Earth Syst. Sci.* **2014**, *18*, 4381–4389. [CrossRef]
2. Salinas, J.L.; Castellarin, A.; Kohnová, S.; Kjeldsen, T.R. Regional parent flood frequency distributions in Europe—Part 2: Climate and scale controls. *Hydrol. Earth Syst. Sci.* **2014**, *18*, 4391–4401. [CrossRef]
3. Ball, J.; Babister, M.; Nathan, R.; Weeks, W.; Weinmann, E.; Retallick, M.; Testoni, I. *Australian Rainfall and Runoff: A Guide to Flood Estimation*; Commonwealth of Australia (Geoscience Australia): Barton, Australia, 2016.
4. Robson, A.; Reed, D. *Statistical Procedures for Flood Frequency Estimation, Flood Estimation Handbook, Vol. 3*; Centre for Ecology & Hydrology: Wallingford, UK, 1999.
5. England, J.F., Jr.; Cohn, T.A.; Faber, B.A.; Stedinger, J.R.; Thomas, W.O., Jr.; Veilleux, A.G.; Kiang, J.E.; Mason, R.R., Jr. Guidelines for determining flood flow frequency–Bulletin 17C. In *U.S. Geological Survey Techniques and Methods, Book 4, Chap. B*; U.S. Geological Survey: Reston, VA, USA, 2018.
6. Public Safety Canada Canadian Disaster Database. Available online: https://www.publicsafety.gc.ca/cnt/rsrcs/cndn-dsstr-dtbs/index-en.aspx (accessed on 13 November 2019).
7. Blöschl, G.; Bierkens, M.F.P.; Chambel, A.; Cudennec, C.; Destouni, G.; Fiori, A.; Kirchner, J.W.; McDonnell, J.J.; Savenije, H.H.G.; Sivapalan, M.; et al. Twenty-three unsolved problems in hydrology (UPH)—A community perspective. *Hydrol. Sci. J.* **2019**, *64*, 1141–1158. [CrossRef]
8. Watt, W.E. *Hydrology of Floods in Canada—A Guide to Planning and Design-NRC Publications Archive-National Research Council Canada*; National Research Council Canada, Associate Committee on Hydrology: Ottawa, ON, Canada, 1989; ISBN 06660128764.
9. Moudrak, N.; Feltmate, B. *Preventing Disaster before it Strikes: Developing a Canadian Standard for New Flood-Resilient Residential Communities*; Intact Centre on Climate Adaptation, University of Waterloo: Waterloo, ON, Canada, 2017.
10. Water Survey of Canada Environment Canada Data Explorer-HYDAT Database 2020. Available online: https://www.canada.ca/en/environment-climate-change/services/water-overview/quantity/monitoring/survey/data-products-services/explorer.html (accessed on 16 December 2018).
11. GREHYS. Presentation and review of some methods for regional flood frequency analysis. *J. Hydrol.* **1996**, *186*, 63–84. [CrossRef]
12. Burn, D.H.; Goel, N.K. The formation of groups for regional flood frequency analysis. *Hydrol. Sci. J.* **2000**, *45*, 97–112. [CrossRef]
13. Wallis, J.R.; Wood, E.F. Relative accuracy of log Pearson III procedures. *J. Hydraul. Eng.* **1985**, *111*, 1043–1056. [CrossRef]
14. Potter, K.W.; Lettenmaier, D.P. A comparison of regional flood frequency estimation methods using a resampling method. *Water Resour. Res.* **1990**, *26*, 415–424. [CrossRef]
15. Stedinger, J.R.; Lu, L.H. Appraisal of regional and index flood quantile estimators. *Stoch. Hydrol. Hydraul.* **1995**, *9*, 49–75. [CrossRef]
16. Hosking, J.R.M.; Wallis, J.R. *Regional Frequency Analysis—An Approach Based on L-Moments*; Cambridge University Press: Cambridge, UK, 1997; ISBN 0521019400.
17. Yang, T.; Xu, C.-Y.; Shao, Q.-X.; Chen, X. Regional flood frequency and spatial patterns analysis in the Pearl River Delta region using L-Moments approach. *Stoch. Environ. Res. Risk Assess.* **2010**, *24*, 165–182. [CrossRef]
18. Jingyi, Z.; Hall, M.J. Regional flood frequency analysis for the Gan-Ming River basin in China. *J. Hydrol.* **2004**, *296*, 98–117. [CrossRef]
19. Merz, R.; Blöschl, G. Flood frequency regionalisation - spatial proximity vs. catchment attributes. *J. Hydrol.* **2005**, *302*, 283–306. [CrossRef]

20. Mediero, L.; Kjeldsen, T.R.; Macdonald, N.; Kohnova, S.; Merz, B.; Vorogushyn, S.; Wilson, D.; Alburquerque, T.; Blöschl, G.; Bogdanowicz, E.; et al. Identification of coherent flood regions across Europe by using the longest streamflow records. *J. Hydrol.* **2015**, *528*, 341–360. [CrossRef]
21. Wazneh, H.; Chebana, F.; Ouarda, T.B.M.J. Identification of hydrological neighborhoods for regional flood frequency analysis using statistical depth function. *Adv. Water Resour.* **2016**, *94*, 251–263. [CrossRef]
22. Zadeh, S.M.; Burn, D.H. A super region approach to improve pooled flood frequency analysis. *Can. Water Resour. J.* **2019**, *44*, 146–159. [CrossRef]
23. Durocher, M.; Burn, D.H.; Mostofi Zadeh, S. A nationwide regional flood frequency analysis at ungauged sites using ROI/GLS with copulas and super regions. *J. Hydrol.* **2018**, *567*, 191–202. [CrossRef]
24. Durocher, M.; Burn, D.H.; Mostofi Zadeh, S.; Ashkar, F. Estimating flood quantiles at ungauged sites using nonparametric regression methods with spatial components. *Hydrol. Sci. J.* **2019**, *64*, 1056–1070. [CrossRef]
25. U.S. Water Resources Council. *Guidelines for Determining Flood Flow Frequency, Bulletin 17B*; Hydrology Committee: Reston, VA, USA, 1982.
26. Muhammad, A.; Evenson, G.R.; Stadnyk, T.A.; Boluwade, A.; Jha, S.K.; Coulibaly, P. Assessing the importance of potholes in the Canadian Prairie Region under future climate change scenarios. *Water* **2018**, *10*, 1657. [CrossRef]
27. Whitfield, P.H.; Shook, K.R.; Pomeroy, J.W. Spatial patterns of temporal changes in Canadian Prairie streamflow using an alternative trend assessment approach. *J. Hydrol.* **2020**, *582*, 124541. [CrossRef]
28. Burn, D.H. Delineation of groups for regional flood frequency analysis. *J. Hydrol.* **1988**, *104*, 345–364. [CrossRef]
29. Burn, D.H. Cluster analysis as applied to regional flood frequency. *J. Water Resouces Plan. Manag.* **1989**, *115*, 567–582. [CrossRef]
30. Zrinji, Z.; Burn, D.H. Regional flood frequency with hierarchical region of influence. *J. Water Resour. Plan. Manag.* **1996**, *122*, 245. [CrossRef]
31. Sandrock, G.; Viraraghavan, T.; Fuller, G.A. Estimation of peak flows for natural ungauged watersheds in southern saskatchewan. *Can. Water Resour. J.* **1992**, *17*, 21–31. [CrossRef]
32. Ouarda, T.; Girard, C.; Cavadias, G.S.; Bobee, B. Regional flood frequency estimation with canonical correlation analysis. *J. Hydrol.* **2001**, *254*, 157–173. [CrossRef]
33. El-Jabi, N.; Caissie, D.; Turkkan, N. Flood analysis and flood projections under climate change in New Brunswick. *Can. Water Resour. J.* **2016**, *41*, 319–330. [CrossRef]
34. Faulkner, D.; Warren, S.; Burn, D. Design floods for all of Canada. *Can. Water Resour. J.* **2016**, *41*, 398–411. [CrossRef]
35. Sandink, D.; Kovacs, P.; Oulahen, G.; McGillivray, G. *Making Flood Insurable for Canadian Homeowners: A Discussion Paper*; Institute for Catastrophic Loss Reduction & Swiss Reinsurance Company Ltd.: Toronto, ON, Canada, 2010.
36. Zahmatkesh, Z.; Jha, S.K.; Coulibaly, P.; Stadnyk, T. An overview of river flood forecasting procedures in Canadian watersheds. *Can. Water Resour. J.* **2019**, *44*, 213–229. [CrossRef]
37. Aucoin, F.; Caissie, D.; El-Jabi, N.; Turkkan, N. *Flood Frequency Analyses for New Brunswick Rivers*; Fisheries and Oceans Canada: Moncton, NB, Canada, 2011.
38. FloodNet Floodnet–NSERC Network–Enhanced Flood Forecasting and Management Capacity in Canada. Available online: https://www.nsercfloodnet.ca/ (accessed on 11 March 2020).
39. Zhang, Z.; Stadnyk, T.A.; Burn, D.H. Identification of a preferred statistical distribution for at-site flood frequency analysis in Canada. *Can. Water Resour. J./Rev. Can. Ressour. Hydr.* **2020**, *45*, 43–58. [CrossRef]
40. Ashkar, F.; El Adlouni, S.E. Adjusting for small-sample non-normality of design event estimators under a generalized Pareto distribution. *J. Hydrol.* **2015**, *530*, 384–391. [CrossRef]
41. Ashkar, F.; Ba, I. Selection between the generalized Pareto and kappa distributions in peaks-over-threshold hydrological frequency modelling. *Hydrol. Sci. J.* **2017**, *62*, 1167–1180. [CrossRef]
42. Durocher, M.; Zadeh, S.M.; Burn, D.H.; Ashkar, F. Comparison of automatic procedures for selecting flood peaks over threshold based on goodness-of-fit tests. *Hydrol. Process.* **2018**, *32*, 2874–2887. [CrossRef]
43. Durocher, M.; Burn, D.H.; Ashkar, F. Comparison of Estimation Methods for a Nonstationary Index-Flood Model in Flood Frequency Analysis Using Peaks Over Threshold. *Water Resour. Res.* **2019**, *55*, 9398–9416. [CrossRef]

44. Mostofi Zadeh, S.; Durocher, M.; Burn, D.H.; Ashkar, F. Pooled flood frequency analysis: A comparison based on peaks-over-threshold and annual maximum series. *Hydrol. Sci. J.* **2019**, *64*, 121–136. [CrossRef]
45. Burn, D.H. An appraisal of the "region of influence" approach to flood frequency analysis. *Hydrol. Sci. J.* **1990**, *35*, 149–165. [CrossRef]
46. Burn, D.H. Evaluation of regional flood frequency analysis with a region of influence approach. *Water Resour. Res.* **1990**, *26*, 2257–2265. [CrossRef]
47. Burn, D.H.; Whitfield, P.H. Changes in floods and flood regimes in Canada. *Can. Water Resour. J.* **2016**, *41*, 139–150. [CrossRef]
48. Burn, D.H.; Whitfield, P.H.; Sharif, M. Identification of changes in floods and flood regimes in Canada using a peaks over threshold approach. *Hydrol. Process.* **2016**, *30*, 3303–3314. [CrossRef]
49. Burn, D.H. Catchment similarity for regional flood frequency analysis using seasonality measures. *J. Hydrol.* **1997**, *202*, 212–230. [CrossRef]
50. Burn, D.H.; Zrinji, Z.; Kowalchuk, M. Regionalization of catchments for regional flood frequency analysis. *J. Hydrol. Eng.* **1997**, *2*, 76–82. [CrossRef]
51. Klyszejko, E.; Environment and Climate Change Canada, Ottawa, Canada. Personal communication, 2016.
52. McKenney, D.W.; Pedlar, J.H.; Papadopol, P.; Hutchinson, M.F. The development of 1901-2000 historical monthly climate models for Canada and the United States. *Agric. For. Meteorol.* **2006**, *138*, 69–81. [CrossRef]
53. Historical Monthly Climate Grids for North America. Natural Resources Canada. Available online: https://cfs.nrcan.gc.ca/projects/3/3 (accessed on 22 April 2019).
54. Buttle, J.M.; Allen, D.M.; Caissie, D.; Davison, B.; Hayashi, M.; Peters, D.L.; Pomeroy, J.W.; Simonovic, S.; St-Hilaire, A.; Whitfield, P.H. Flood processes in Canada: Regional and special aspects. *Can. Water Resour. J.* **2016**, *41*, 7–30. [CrossRef]
55. Brimley, B.; Cantin, J.F.; Harvey, D.; Kowalchuk, M.; Marsh, P.; Ouarda, T.M.B.J.; Phinney, B.; Pilon, P.; Renouf, M.; Tassone, B.; et al. *Establishment of the Reference Hydrometric Basin Network (RHBN) for Canada*; Environment Canada: Ottawa, ON, Canada, 1999.
56. Whitfield, P.H.; Burn, D.H.; Hannaford, J.; Higgins, H.; Hodgkins, G.A.; Marsh, T.; Looser, U. Reference hydrologic networks I. The status and potential future directions of national reference hydrologic networks for detecting trends. *Hydrol. Sci. J.* **2012**, *57*, 1562–1579. [CrossRef]
57. Nathan, R.J.; McMahon, T.A. Identification of homogeneous regions for the purposes of regionalisation. *J. Hydrol.* **1990**, *121*, 217–238. [CrossRef]
58. Noto, L.V.; Loggia, G. La Use of L-moments approach for regional flood frequency analysis in Sicily, Italy. *Water Resour. Manag.* **2009**, *23*, 2207–2229. [CrossRef]
59. Eslamian, S.S.; Hosseinipour, E.Z. A modified region of influence approach for flood regionalization. In Proceedings of the World Environmental and Water Resources Congress 2010: Challenges of Change, Providence, RI, USA, 16–20 May 2010; pp. 2388–2414.
60. Vogel, R.M.; Fennessey, N.M. *L* moment diagrams should replace product moment diagrams. *Water Resour. Res.* **1993**, *29*, 1745–1752. [CrossRef]
61. Bobée, B.; Rasmussen, P.F. Recent advances in flood frequency analysis. *Rev. Geophys.* **1995**, *33*, 1111–1116. [CrossRef]
62. Zrinji, Z.; Burn, D.H. Flood frequency analysis for ungauged sites using a region of influence approach. *J. Hydrol.* **1994**, *153*, 1–21. [CrossRef]
63. Atiem, I.A.; Harmancioğlu, N.B. Assessment of regional floods using L-moments approach: The case of the River Nile. *Water Resour. Manag.* **2006**, *20*, 723–747. [CrossRef]
64. Shook, K.R.; Pomeroy, J.W. Memory effects of depressional storage in Northern Prairie hydrology. *Hydrol. Process.* **2011**, *25*, 3890–3898. [CrossRef]
65. Muhammad, A.; Evenson, G.R.; Stadnyk, T.A.; Boluwade, A.; Jha, S.K.; Coulibaly, P. Impact of model structure on the accuracy of hydrological modeling of a Canadian Prairie watershed. *J. Hydrol. Reg. Stud.* **2019**, *21*, 40–56. [CrossRef]

66. Ehsanzadeh, E.; Spence, C.; van der Kamp, G.; McConkey, B. On the behaviour of dynamic contributing areas and flood frequency curves in North American Prairie watersheds. *J. Hydrol.* **2012**, *414–415*, 364–373. [CrossRef]
67. Prieto, C.; Le Vine, N.; Kavetski, D.; García, E.; Medina, R. Flow Prediction in Ungauged Catchments Using Probabilistic Random Forests Regionalization and New Statistical Adequacy Tests. *Water Resour. Res.* **2019**, *55*, 4364–4392. [CrossRef]

MDPI
St. Alban-Anlage 66
4052 Basel
Switzerland
Tel. +41 61 683 77 34
Fax +41 61 302 89 18
www.mdpi.com

Water Editorial Office
E-mail: water@mdpi.com
www.mdpi.com/journal/water

www.ingramcontent.com/pod-product-compliance
Lightning Source LLC
Chambersburg PA
CBHW040311270926
51850CB00007BC/1043
* 9 7 8 3 0 3 6 5 4 6 4 4 5 *